AF683874

FIRST AID TREATMENT OF SNAKE BITES

FIRST AID "DO NOTs"

Unfortunately, there are many traditional first aid treatments of snake bite, published in books and taught in first aid courses **which do more harm than good.**

DO NOT PANIC

DO NOT make any cut, scratch or incision with a razor, knife or any other sharp object at the site of the bite or anywhere else (this will introduce infection, could damage important anatomical structures such as nerves, tendons and blood vessels, and could lead to continued very severe bleeding if the snake venom has made your blood non-clotting).

DO NOT suck at the wound with your mouth or with any suction apparatus.

DO NOT apply icepacks to the bitten area.

DO NOT apply electric shocks to the bite or anywhere else!

DO NOT use a conventional very tight constricting band, belt or tourniquet around the bitten limb. This will cut off the blood supply and, if left for more than a couple of hours, may cause irreversible damage to the bitten limb leading to gangrene which may require amputation.

DO NOT drink alcohol, take herbal medicines or asprin.

WHAT YOU SHOULD DO

Reassure the patient. Most people survive snakebites.

Remove all rings and bracelets from the bitten part of the body. Some snake poisons cause swelling and they may become painful.

Apply a pressure bandage (ideally with a crepe bandage, but thin strips of material or clothing can be used) and immobilise the limb with a splint. A diagram showing how to apply a pressure bandage can be found over the page. Do not tie the bandage too tight; you should be able to push a little finger under the bandage. If the fingers or toes go cold or blue the bandage is too tight.

Get to medical help as quickly as possible, but **do not run.** Walk or be carried and get transportation as soon as possible.

If the snake has been killed take it with you in a box or thick bag, the doctors will want to examine it to decide what snakebite treatment you require. Do not handle the snake, even dead snakes have bitten people. If you do not have the dead snake but someone else saw it, take them with you. If you are on your own try to remember what the snake looked like - What colour was it? Did it have any pattern? How long was it? Where was it?

If the patient is unconsious lay them on their side so they do not inhale vomit if they are sick.

Remember that there are over 80 kinds of land snake in PNG, most of them like pythons and treesnakes have no poison although they will bite to defend themselves. **Even if you are bitten by a venomous snake, there is only about a 50% chance that a dangerous amount of venom will have been injected by the snake.** Only 5 kinds of land snakes in PNG can kill you, but often when they bite they do not inject any poison or they inject only a small amount.

WEI BILONG LUKAUTIM MANMERI SAPOS SINEK IBIN KAIKAIM EM

ITAMBU TRU!

Wanpela bikpela hevi em olsem - planti taim tumas sampela lain ol i tok giaman tru long wei bilong lukautim manmeri sapos sinek ibin kaikaim em. Dispela kain krangi tok em i wok long bagarapim manmeri moa yet.

ITAMBU long kirap nogut

ITAMBU long katim sikin bilong manmeri husat sinek i kaikaim em. Dispela wei bilong katim sikin wantaim reza, naip, oa narapela samting, em i nogut tru. Bihain bai traipela soa ikamap, na sampela masel bilong bodi bilong em bai bagarap. Em tu, sampela poison bilong sinek em inap mekim blut bai i ron hariap tru. Sapos yu katim sikin, manmeri bai lusim planti blut tumas.

ITAMBU long kaikaim dispela soa sinek i wokim, bilong kamautim poison bilong sinek long maus bilong yu. Itambu tu long usim ol kain liklik masin long kamautim poison.

ITAMBU long karamapim dispela soa long ais.

ITAMBU long pairapim soa long sampela paua oa ilectrik masin.

ITAMBU long banisim soa tait tru wantaim gumi oa belt oa narapela roap.

ITAMBU long dringim bia na spirit, oa kaikaim aspirin, oa dringim wara bilong ol lip bilong bus.

BAI YU WOKIM WONEM?

Sapos sinek i kaikaim wantok bilong yu, yu mas wokim olsem:

Tok isi long em. Planti manmeri tru inap winim birua bilong sinek, na kamap sitrongpela gen.

Sapos wantok igat sampela bilas long han na bodi, olsem ring bilong pinga, or sen long bun bilong han bilong em - yumas rausim. Poison bilong sampela sinek bai inap swelapim sikin, na sikin bai pait na pen nogut tru.

Usim wanpela kain plasta olikolim "pressure" bendis. Sapos yu lukluk gut long narapela pes namel, bai yu lukim piksa inap soim yu long wokim dispela samting.

Sapos yu laik wokim "pressure" bendis, bai yu inap usim creip bendis, oa yu ken usim sampela hap laplap bilong klos bilong yu. Maski taitim bendis tumas. Yu mas larim inap spes ananit long em, bilong putim liklik pinga bilong yu. Long ples bilong bodi sinek ibin kaikai, wokim wanpela liklik stik bai inap pasim bendis, na ino inap wokabaut long dispela soa.

Hariap tru na kisim wantok igo long haus sik, tasol **wokabaut**, yu noken larim em long ron. Sapos yu inap karim em long beksait bilong yu, oa sapos car istap namel, em bai gutpela.

Sapos yu bin kilim sinek, putim em long wanpela bokis oa raisbeg, na karim em igo long haus sik. Dokta bai lukim, na kisim save long wonem kain marasin em i mas givim long sikman. Tasol maski long holim sinek long han bilong yu - planti taim sinek husat ibin dai pinis isave sutim poison bilong em yet, na bagarapim narapela man. Sapos sinek i ronowei pinis, tingting bek long taim yu bin lukim em. Wonem kala tru bilong em? Em igat makmak long sikin bilong em oa nogat? Em longpela oa sotpela tasol? Na sinek em ibin stap wea, olsem long giraun, or water, or ananit long lip long giraun?

Sapos man idai, silipim em long sait bilong em. Nogut bihain man i troaut, na em ino inap pulim win.

Igat olsem antap long eitpela ten kain kain sinek bilong giraun long PNG. Planti tru, olsem "python" na ol sinek bilong diwai, ino gat poison. Tasol sampela taim ol i trai long kaikaim yu long tit bilong ol. **Na em tu, sapos wanpela poison sinek tru ibin kaikaim yu, yu mas tingting olsem - 50% long olgeta man sinek ibin kaikaim em, em ino dai, bilong wonem, poison bilong sinek ino go insait long blut bilong man.** Igat faivepela kain "giraun sinek" tasol insait long PNG inap kilim man, na planti taim ol i kaikaim man nating, na poison bilong em ino go insait long blut.

HOW TO APPLY AN "AUSSIE BANDAGE"

SINEK BAIT FIRST AID

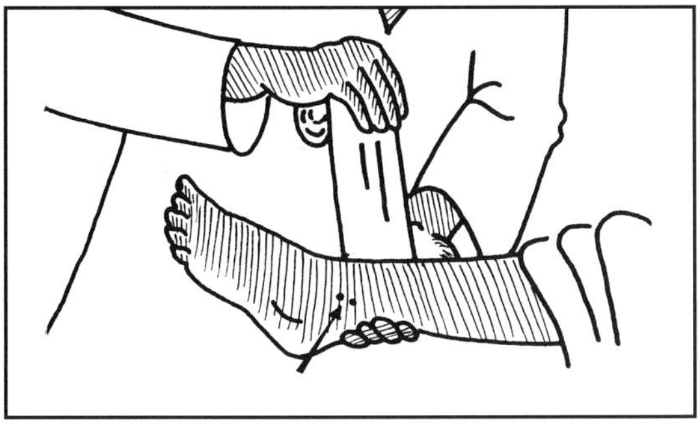

1 Putim wanpela bikpela bendis oa laplap oa sitrongpela bikpela lip antap we ol mak bilong tit istap. Sapos man igat longpela trausis, OK, yu noken rausim trausis bilong man bikos em bai mekim poison igo kwiktaim long blut. Man ino ken wokabaut.

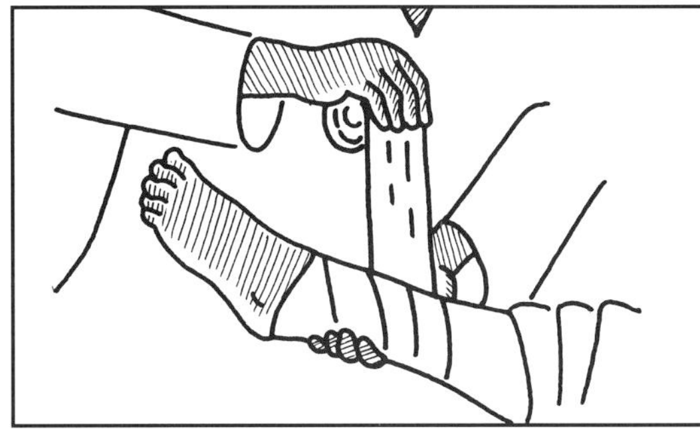

2 Bendis imas istap pas olsem yu bai putim long skru sapos man i pundaun na skru i lus.

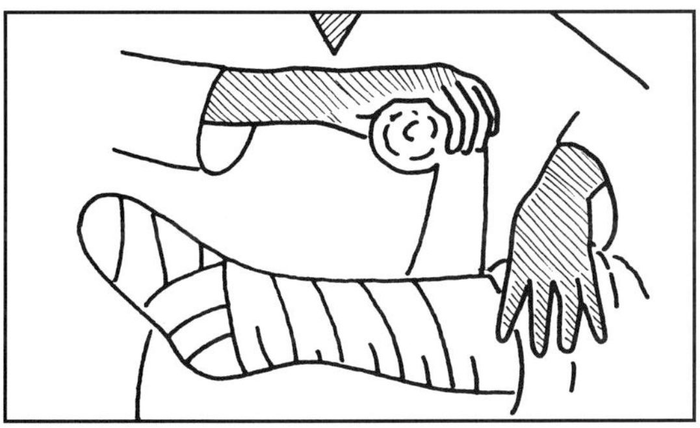

3 Pasim longpela bendis raunim lek inap igo long 24 inses antap long mak bilong tit bilong sinek.

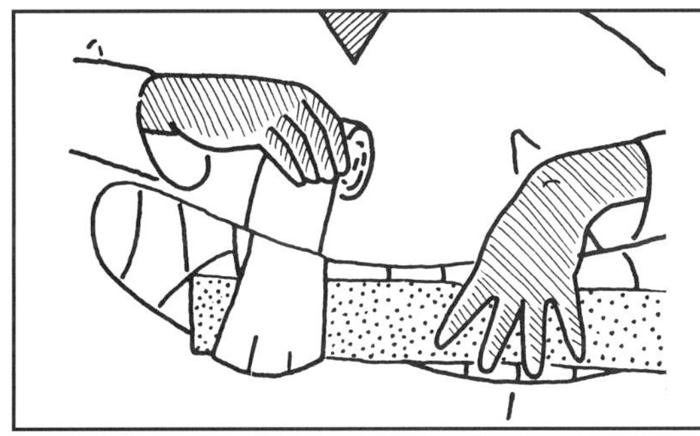

4 Putim sampela stik, oa pangal, oa plenk samting ananit long lek bai man inoken wokabaut or brukim skru.

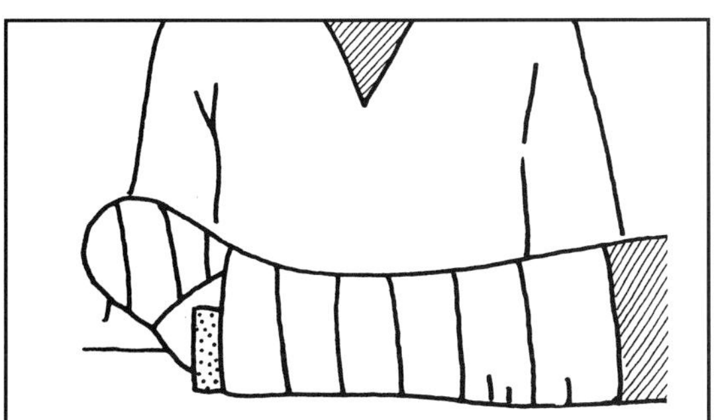

5 Usim wanpela laplap oa bendis bilong pasim sitrong plenk, diwai, oa pangal.

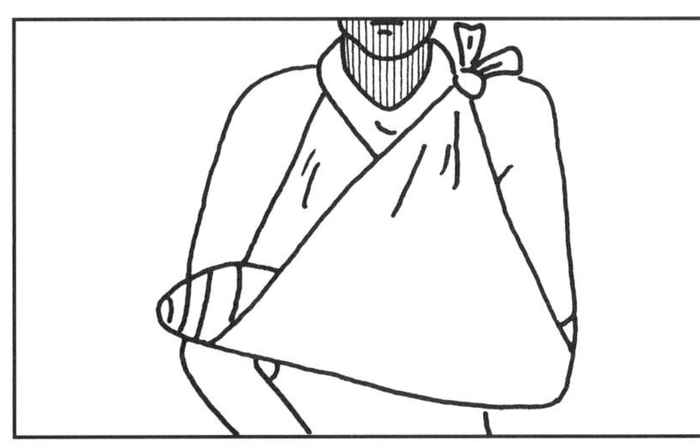

6 Sapos sinek ibin kaikaim man long han bilong em, olrait, putim bendis antap long han igo inap long skru. Putim wanpela diwai long skru, na silipim han long sling.

Bihain long wokim dispela, man imas igo long aidpost oa long helt senta, oa bikpela haus sik. Yupela mas karim man sinek ibin kaikaim long lek. Sapos sinek ibin kaikaim han, man iken wokim olsem long piksa namba sikis, na bihain, wokabaut.

Sapos yu tasol istap taim sinek ibin kaikaim yu, trai na wokim bendis olsem long dispela toksave. Na taim yu laik wokabaut, yu mas usim diwai stik olsem man igat lek i bagarap. Noken putim dispela kek we sinek i kaikaim bai igo daun long giraun. Sapos sinek ibin kaikaim han bilong yu, yu ken wokim bendis na sling olsem dispela toksave na bihain, wokabaut stret.

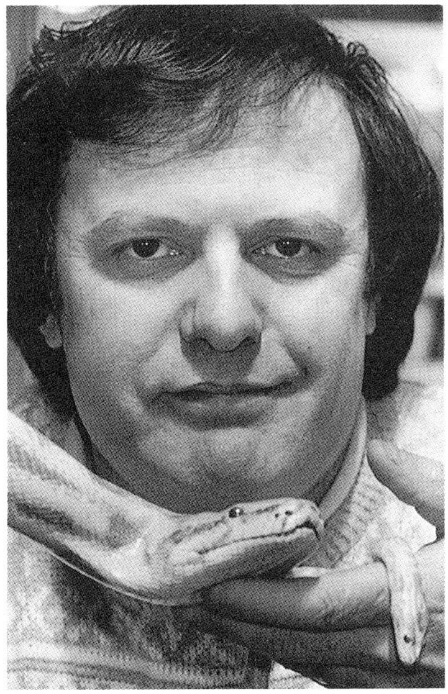

*Dave Lester with the first UK captive breeding of albino Burmese or Golden pythons (*Python molurus bivittatus*) in 1991.*
Photo courtesy of the Wolverhampton Express and Star.

This book is dedicated to the lasting memory of
Dave Lester who died tragically in May 1994, aged 41.
Dave was internationally regarded as a leading pioneer in the captive care, husbandry
and breeding of rare and difficult reptile species, especially boids.
He was also a great personal friend and he will be sadly missed.

Also in tribute to my father, Mo, who died suddenly
on 17th June 1995 and never saw this book published.
He would have been so proud.

ACKNOWLEDGEMENTS

There are many people I should acknowledge here but I am especially grateful to Prof. David A. Warrell, Phil and Brenda Willmott-Sharp, Ilaiah Bigilale and Roger and Rosa Middleton. Without these six people this book would never have materialised.

Firstly David Warrell who sent me to PNG in 1990 and 1992 to capture snakes for the venomous snakebite project and supported me in times of difficulty.

Phil and Brenda are not only my publishers, they are also my friends and they provided accomodation, meals and a vehicle in Port Moresby. Phil also became press-ganged as a fellow snake catcher and the Willmott-Sharp residence became a temporary home for not only myself but also bags and boxes of serpents awaiting identification, photography and venom extraction. It was Phil and Brenda who initially suggested the idea for this book since PNG lacked a publication dealing with its entire snake fauna.

Ilaiah Bigilale, Acting Curator of Natural History at the National Museum of PNG, became a firm and enthusiastic friend, always ready to offer advice and assistance. The co-operation of the National Museum and its snake holding facilites became an integral part of the Venomous Snakebite Project and access to the museum's preserved collection proved invaluable.

Roger and Rosa Middleton (Kulili Estates) generously provided not only accomodation and meals during my several stays on Kar Kar Island but also plantation workers which enabled me to investigate large areas of Kaviak and neighbouring plantations in my search for specimens. Without their kindness and support, and the assistance of their son Ian and the plantation workers, none of the recent venom work on the small-eyed snake *Micropechis ikaheka,* carried out at Liverpool School of Tropical Medicine, would have been possible.

Financial support in 1990 was provided by a Fellowship from the Christensen Fund (Palo Alto, California, USA) and I am also grateful for the use of facilities and assistance provided by Matthew Jebb (former Director) and Larry Orsak (current Director) and their staff at the Christensen Research Institute, Madang, PNG.

I am grateful to Prof. David Warrell and Dr David Lalloo of Oxford University Centre for Tropical Medicine, for providing the up-to-date sections on snakebite first-aid treatment and clinical management. In a country where venomous snakebite is an ever present possibility it is most important that correct procedures be publicised and encouraged, and equally important that dangerous practises are discouraged. Dr Lalloo and his wife, together with fellow Port Moresby-based Oxford clinicans Andrew Trevett and Arnold Drury provided accomodation, advice and, in the case of Arnold, welcome company on night drives.

Also in PNG thanks should go to the staff of the National Museum and Art Gallery, especially Matthew who proved to be an excellent field assistant; Iamo Ila, Guy Kula, Mick Raga, Vagi Genorupa, Lawrence Oro (Dept. of Environment and Conservation); Institute of PNG Studies; Dr James Menzies (Dept., of Biology, UPNG) and Prof. Sirus Naraqi (Dept of Medicine, UPNG); PNG Defence Force Goldie River Barracks - Capt. Richard Bosi (Australian Army), Sgt Cory Frye (US Army) and Sgt David and the NCO's of the 'Warrior' Training Camp; the Sisters and staff of St Gerrard's Hospital, Vaifa'a, Mekeo; John Cheung of Island Airways; Graham and Marie Davis at the Bluff Inn, Paul O'Dowd at Malolo and now Jais Aben, Bishop David Hand, and Celsus Levo, formerly of the forestry department. Also thanks to Operation Raleigh and all the other staff members, venturers and sponsors of the expedition 6B, Papua New Guinea for the logistical support which made it possible for me to undertake initial investigations of the herpetofauna of Western Province in 1986. My thanks also to Dr James Menzies (UPNG) who read the entire manuscript and made numerous valuable comments. The Times of PNG, Wantok, Post Courier and National newspapers, Roger Hau'ofa of PNG Radio Kalang and also EMTV for helping to 'get my message across', and the valuable support, interest and assistance of villagers, townsfolk and foreign residents around the country, especially in Madang, Central, NCD and Western Provinces.

In the UK thanks go to Dr David Theakston of the Alistair Reid Venom Research Unit at the Liverpool School of Tropical Medicine, Colin McCarthy at the British Museum (Natural History), Mrs Ramdoyal at the PNG High Commission in London, Brian Groombridge at the IUCN Monitoring Centre, Cambridge; Kodak UK Ltd., Jessops and Colab (both Wolverhampton). Thanks also to Fred Parker, Roy Mackay and Dr Bernie Hudson in Australia and

Dr Samuel McDowell (Rutgers University, New Jersey, USA) for data and advice. Dr McDowell and the Society for the Study of Reptiles and Amphibians are thanked for granting permission for the use of original figures from the Journal of Herpetology. I have found Dr McDowell's extensive published data invaluable in the compilation of this text.

This manuscript was tirelessly read, proofed and commented upon by Barbara O'Shea (my mother) and in part by Tricha Murphy (University College, Cork). The French translations were done by Denise Gorton and Denis & Betty Lovering and the Russian translation by Peter Rooney (University of Wolverhampton).

CITES information was provided by Brian Groombridge of the IUCN World Conservation Monitoring Centre in Cambridge, UK.

Information on type specimens was obtained from Dr Rainer Günther (Museum für Naturkunde der Humboldt Universität zu Berlin, Germany), Dr C. Smeenk (National Museum Natuurhistorisch, Leiden, Holland), Dr Franz Tiedemann (Naturhistorisches Museum Wien, Austria) and Dr M. Geisthardt (Museum Wiesbaden, Germany). Standardisations of English common names for certain obscure elapids were arrived at in collaboration with Patrick David (Saint-Brice, France), Ivan Ineich (Paris Museum) and Van Wallach (MCZ), who also provided unpublished data on new typhlopids.

Photographs were generously provided by Dr Allen Allison (Bishop Museum, Honolulu, Hawaii); Ken Aplin (Western Australian Museum); Dave Barker (Vida Preciosa, Boerne, Texas); Dr Bill Branch (Port Elizabeth Museum, South Africa); Steve Donnellen (South Australian Museum); Dr Harold Heatwole (North Carolina State University, USA); Hank Jenkins (Australian National Parks Dept.); Dr Micheal McCoy (Honiara, Guadalcanal, Solomon Islands); Dr James Menzies (University of PNG); Dr Harold Voris (Chicago Field Museum, Illinois, USA), Prof David Warrell (Oxford University), John Weigel (Australian Reptile Park) and Daniel Heuclin (France).

Finally I should like to thank my employers, the management of the West Midland Safari and Leisure Park, Worcestershire, England, for allowing me to take substantial periods of unpaid leave in order to conduct my field work overseas.

Design, layout and typesetting by Independent Design, Port Moresby.
Printed by Craft Print Pte Ltd., Singapore.

This is contribution number 147 from the Christensen Research Institute,
PO Box 305, Madang, Papua New Guinea.

© "Snakebite and its treatment in Papua New Guinea" D.A. Warrell & D.G. Lalloo
© photographs Mark O'Shea (except as otherwise stated)

Published in Papua New Guinea by Independent Publishing, Independent Group Pty Ltd., PO Box 168, Port Moresby.

ISBN 9980-916-96-6
National Library of Papua New Guinea

Published for Independent Group Pty Ltd in Singapore by Beaumont Publishing Pte Ltd, 9 Joo Koon Circle, Singapore 629041

ISBN 981-00-7836-6

© 1996 All rights reserved. No part of this work may be reproduced or used in any form or by any means - graphic, electronic or mechanical, including photocopying, recording, taping or information storage and retrieval systems - without written permission of the publishers.

Juvenile Green tree python – *Morelia viridis*. Photo: Mike McCoy

A GUIDE TO
THE SNAKES OF PAPUA NEW GUINEA

by Mark O'Shea

Adult Green tree python – *Morelia viridis*. Photo: Mark O'Shea

A GUIDE TO THE SNAKES OF PAPUA NEW GUINEA

CONTENTS

Foreword

Papua New Guinea is often referred to as a land of diversity. It boasts an amazing variety of species of flora and fauna including many superlatives: the largest pigeon, the longest lizard, the smallest parrot, the tallest banana species, to name but a few. However, while much has been written about PNG's beautiful birds of paradise, its moths and butterflies, its orchids and rainforests, little has been written about its snakes; despite it being home to some of the world's most deadly and most beautiful species. Sadly this ignorance about the snakes of Papua New Guinea (and indeed snakes in general) has led to the unnecessary killing not only of entirely harmless snakes, but also of those which are potentially beneficial to humans; the reflex action upon sighting a snake, no matter what it looks like, is to kill it in 'self-defence'.

At a time when so much damage is being done to our natural environment, it is important to find ways to conserve our country's unique flora and fauna. "A Guide to The Snakes of Papua New Guinea" is therefore an invaluable book. Its style and format are easy to understand, yet it does not trivialise the subject, and it is the first book of its kind to bring together the information currently available concerning the identification, distribution, natural history and snake-bite risk posed by PNG snakes. The photographs and colour-coding make it user-friendly, both for the lay-man and the specialist alike, and I feel that such a book will be of great importance in the endeavour to protect our endangered species by promoting education and awareness.

Hon Parry Zeipi M.P.
Government of Papua New Guinea

Hon. Parry Zeipi held the ministerial portfolio for Environment and Conservation from 1987 - 1988 and again from 1992 - 1995.

AIMS OF THE GUIDE

A Guide to the Snakes of Papua New Guinea is intended to fill a gap in the zoological literature of Papua New Guinea. Various publications have been written dealing with aspects of the snake fauna of PNG such as the venomous species (Worrell 1966; Slater 1968) or those of a particular province (Parker 1982) but to date there has not been a single book dealing with all the snakes known to occur within the political borders of Papua New Guinea and the reader is often forced to refer to publications pertaining to nearby countries such as Australia (Cogger 1969 and others) or the Solomon Islands (McCoy 1980). It is hoped that this book will bring together most of the information currently available concerning the identification, distribution, natural history and snakebite risk posed by PNG snakes within a single publication and makes that information available in as easy to understand a manner as possible without actually trivialising the subject.

Almost 100 species of snakes are known to occur on land in PNG, in its creeks, rivers, lakes and estuaries and in its territorial waters.

This book is aimed at a wide audience, from high school graduates to mining and forestry company employees, from local government officers, police and defence force personnel to bushwalkers and backpackers and from health centre workers to field biologists and herpetologists. Much has been written about PNG's beautiful Birds of Paradise, its huge birdwing butterflies, and the popular mammals have also received their share of publications. Books on the colourful and extremely diverse fish have been published but the snakes of this fascinating country have long been either feared and killed out-of-hand, avoided or simply overlooked. Yet not only does PNG have some of the world's most deadly snakes, it is also home to some of the most beautiful and I felt it was time that all the known land snakes and those seasnakes which enter Papua New Guinean waters were documented together in a single publication.

What does this book set out to accomplish?
Basically the aims of the publication are threefold which should benefit, respectively, the human population, the snakes themselves and the scientific community.

The primary aim is to lower the level of snakebite morbidity and mortality in PNG through education and awareness. In the 3.5 years from mid-1989 until the end of 1992 some 250 serious snakebites, mostly from Central Province or NCD, were treated at Port Moresby General Hospital. Relatively few terminated fatally due to the skill and care of the medical and nursing staff and the facilities available to them. In rural health centres around PNG nursing staff have fewer facilities and often insufficient antivenom to treat snakebites and possibly more deaths occur. It was estimated that 13 people died of snakebite in Central Province during 1992 but certainly there were deaths which were never included in the official statistics since they occurred in the bush, in remote villages or on the road. It is true that Central Province and NCD probably have the worst snakebite problem but Western Province is also a high risk area and almost no data exists for many of the other southern provinces within the range of the more dangerous species. When this is compared with an annual death rate in Australia, which is home to many of the same, or closely related, dangerous snakes, of 1 or 2 people nationally it can be seen that much could be done to improve the snakebite situation in PNG.

Foremost on the list must be education; to enable the population to differentiate between dangerous and harmless snakes and education regarding the correct first aid procedures. Inevitably when someone is bitten they claim the snake to be a "Pap black". This snake is real enough but although it is now very rare throughout much of its original range it is still the snake which instills the most fear amongst nationals and foreign residents alike. Research on human blood serum samples from snakebite victims, carried out at the Liverpool School of Tropical Medicine in England, has demonstrated that the species responsible for the majority of serious snakebites in PNG is the Papuan taipan. The venom of the taipan is much more dangerous than that of the Papuan blacksnake. In addition the venom of the taipan causes continual bleeding from the site of the bite and anywhere else where the skin has been broken. The practise of 'razor-cutting' the bitten limb is highly dangerous since it may lead to massive blood loss and possibly death. (*See* snakebite first aid section).

An ability to correctly identify dangerous snakes and then undertake the correct first aid treatment on the way to hospital should go some way towards reducing serious complications and death.

The second aim is one of conservation. The old adage that "the only good snake is a dead snake" has certainly worn thin in the more enlightened 1990's. To kill snakes just because they are snakes is no longer acceptable, especially as most species encountered and killed are harmless, and some are actually beneficial. Some species, notably pythons, are endangered and are therefore protected by national and international law. An ability to differentiate between snake species in the field removes the need to kill every snake "to be on the safe side".

A further aim concerns a desire for more information. It is certain that we do not know everything there is to know about the snakes of PNG. "... The Snakes of Papua New Guinea" can only contain information which is already known and the snake faunas of whole areas, provinces and regions of PNG have hardly been studied. Parker's (1982) exhaustive work in Western Province provides a checklist for the province of 44 terrestrial species but only eight species are recorded for neighbouring Southern Highlands Province and one from Enga. Whilst it might be expected that the snake fauna of this more montane province will be less diverse than that of Western Province, there are certainly more than eight species within the Southern Highland and one within the Engan provincial boundaries. More fieldwork is badly needed to build up a complete picture of the snake fauna of PNG.

Therefore, the third aim of this book is to entice, cajole and encourage people living or working around Papua New Guinea to help add to our knowledge concerning the snake fauna of their areas. "... The Snakes of Papua New Guinea" is providing the foundations, the base level, so that anyone finding a snake which does not fit the descriptions or the distribution maps can make a contribution by informing us at the National Museum. Providing such data, photographs, live or dead specimens will make the second edition more accurate and increase everybody's knowledge in the long term. This plea for information is dealt with in more detail later in the book. However, in asking for information regarding the snake faunas of remote regions the second aim, that of conservation, should always be borne in mind and snakes should not be killed indiscriminately 'for science'.

It is hoped that this book will find its way onto many bookcases since, to use a biological term, it fills a 'niche' in our knowledge.

THE SNAKE DIVERSITY OF PAPUA NEW GUINEA

Papua New Guinea possesses 93 species of terrestrial, freshwater and marine snakes belonging to six of the world's 13 snake families. A further dozen species have been recorded from Irian Jaya, but not yet from PNG, and a secretive elapid is also known from islands in the western Solomon Islands close to Bougainville, increasing the potential snake fauna to 106 species. In addition, two snake-lizards are also included in this publication since they are often confused with snakes.

The known reptile fauna of PNG is large and diverse, numbering approximately 300 species, almost 0.05% of the reptile fauna of the world (Table 1). Of this number approximately one third are snakes, including seasnakes.

Table I: A synopsis of the class Reptilia.

	Approximate number of species	
Class REPTILIA	Worldwide	PNG
Order Testudines		
Suborder Cryptodira (hidden-necked turtles)	182	8
Suborder Pleurodira (side-necked turtles)	62	6
Order Crocodylia (crocodiles, alligators & gharial)	22	2
Order Rhynchocephalia (New Zealand tuataras)	2	0
Order Squamata		
Suborder Sauria (lizards)	c.3000	c.190
Suborder Amphisbaenia (worm lizards)	130	0
Suborder Serpentes (snakes)	c.2700	c.93
Total	c.6100	c.300

Snakes, lizards and worm lizards form the largest order of reptiles, the Squamata. Worm lizards are largely tropical in distribution but absent from the Australasian region. In New Guinea it is easy to differentiate between snakes and lizards (*see* Table II). The only area of possible confusion are the almost limbless snake-lizards of the family Pygopodidae (and the S.E. Asian/Irian Jayan dibamid lizards, family Dibamidae) which have been included in the species accounts of this guide since they superficially resemble snakes and are often encountered in gardens in the coastal provinces.

Table II: Characteristics distinguishing snakes from lizards in New Guinea

LIZARDS	SNAKES
1. Usually with two pairs of well developed limbs.[a]	1. Limbs absent although vestigal pelvic spurs may be present in boas and pythons.
2. External ear openings present.[b]	2. External ear openings always absent.
3. Eyes protected by moveable eyelids.[c]	3. Eyelids absent, eyes protected by immoveable transparent 'brille'.
4. Tongue usually notched or weakly forked.[d]	4. Tongue always strongly forked.
5. Jaws rigid, mandible fused at chin.	5. Jaws flexible and articulatory, mandible not fused at chin.
6. Ventral scales in several longitudinal rows.	6. Ventral scales in a single longitudinal row.[e]
7. Capable of voluntary tail loss (autotomy) and regeneration.[f]	7. Incapable of shedding or regenerating tail.[g]
8. No venomous species.[h]	8. Many venomous and non-venomous species.

[a] Several species of diminutive skink lizard have short, reduced limbs and the pygopodid lizards (2 species, genus *Lialis*) lack forelimbs and possess hindlimbs which are reduced to scaly flaps (*see* Snake-lizard accounts).
[b] Absent in two species of skinks (genus *Sphenomorphus*).
[c] Skinks of genus *Cryptoblepharus* possess snake-like transparent brilles over their eyes.
[d] Monitor lizards (genus *Varanus*) possess long forked tongues.
[e] Except in primitive blindsnakes (family Typhlopidae), wartsnakes (genus *Acrochordus*) and some seasnakes (family Hydrophiidae).
[f] Not in monitor lizards (genus *Varanus*).
[g] A few Mexican and African snakes also practise 'caudal autotomy'.
[h] The only venomous lizards are confined to Mexico, Guatemala & southwest USA.

The snake-lizard family **PYGOPODIDAE** is confined to the Australasian region with 32 of the 34 species endemic to Australia. The two other species, both belonging to the genus *Lialis*, are found in New Guinea where they are known locally as ‘friendly snakes’ or ‘pencil snakes’, due to their inoffensive nature and their rigid pencil-like bodies. Although these reptiles appear at first glance to be legless they are true lizards with vestigal hindlimbs reduced to scaly flaps. Their forelimbs are completely absent. The pygopodids are important since leglessness is considered an advanced trait in lizards and snake-lizards are believed to have evolved from the geckoes. Apart from the lack of limbs, lizards of the genus *Lialis* possess a number of distinctly lizard-like characteristics, most notably the presence of eyelids and external ear openings; a notched rather than forked tongue; ventral scales similar in shape and size to dorsal scales and the ability to voluntarily break off and regenerate the tail, a defensive trait known as ‘caudal autotomy’. Pygopodids of the genus *Lialis* are primarily predators of small skinks and they are capable of killing and devouring relatively large prey animals in relation to their own body size.

The primitive family **TYPHLOPIDAE** comprises a number of small, fossorial (burrowing) snakes which are occasionally encountered on the surface and are often not recognised as snakes by the layman. Although generally called blindsnakes they are not actually blind since they do possess eyes and are capable of determining light levels and forming basic visual images. Typhlopids are specialised inhabitants of leaf-litter and loose soil, possessing smooth cycloid scales, which are uniform in size and shape both dorsally and ventrally; large smooth scales on the head including a large transparent scale which completely covers the small eye and a small mouth with short recurved teeth confined to the upper jaw. External clues to their fossorial existence include a sharp spine on the end of the short tail, which serves as an anchorage point when the snake is forcing its rigid body through the soil or along the inner galleries of a termite mound, and in some species, a shovel-like rostral scale on the snout tip for burrowing. Prey consists of soft-bodied invertebrates. Blindsnakes are usually discovered during gardening activities, digging, moving vegetation debris or rocks, concrete slabs or oil drums, but specimens may also be encountered on the surface at night after especially heavy rain when flooding of their burrows forces them to move upwards. Reproductively typhlopids are oviparous but one species, *Ramphotyphlops braminus*, is believed to be parthenogenetic since only females have been found. In parthenogenetic species a mature female is capable of laying fertile eggs without the need for a male. The small size of this species makes it possible for specimens to secrete themselves inside the root systems of pot plants or agricultural crops to be transported overseas and this is probably the means by which this species colonised New Guinea since its distribution is centred around areas of prolonged human interaction.

The typhlopids in New Guinea are represented by the primarily Australasian genus *Ramphotyphlops* and the more world-wide tropical genus *Typhlops*.

The **BOIDAE** of Papua New Guinea comprise ten species in two subfamilies, the Boinae (boas) and the Pythoninae (pythons).

The three boas of the genus *Candoia* (two occurring in PNG and a third species further east in the Solomons) are especially interesting because they are believed to be most closely related to the boas of the Caribbean. Current theory suggests that their ancestors reached the Pacific before the Central American isthmus was formed, and were then subsequently transported on rafts of vegetation, across the Pacific, on the equatorial currents from east to west, before being carried south to the Samoan, Fijian and Solomon Islands. From there they are thought to have 'island-hopped' to New Guinea and eastern Indonesia but they did not reach Australia. These Pacific boas are an extremely diverse group containing both stout-bodied, short-tailed terrestrial specimens and slender-bodied, long-tailed arboreal specimens, even within the same species. The boas are viviparous and their prey ranges from frogs and lizards to small mammals.

The pythons are much better represented in the Australasian region than the boas with eight species occurring in New Guinea. The taxonomy of the group is also a source of constant revision and controversy (*see* individual accounts). The Australasian pythons have adapted to most habitats including deserts in Australia. New Guinea species occur from sea-level to altitudes of over 2000m, in a variety of habitats including dry savannas, swamps, lowland rainforests, montane rainforests, plantations and gardens, on isolated islands and even into built up areas. The Pythoninae also contains the largest snakes in PNG with several species achieving lengths in excess of 2.0m and at least two growing to over 3.0m. As with the boas there are both arboreal and terrestrial species. Pythons possess heat-sensitive pits on their lip scales which are sensitive to the body temperature of their mammalian prey. Using these pits a python is capable of stalking and capturing its warm-blooded prey, even in total darkness. Such prey is then killed by constriction before being swallowed whole. The prey of pythons consists primarily of mammals ranging from mice to wallabies, although some lizards and even other pythons are recorded as being taken. In contrast to the boas, the pythons are oviparous, laying clutches of leathery-shelled eggs which the female will incubate within her coils.

It is also the subfamily Pythoninae which contains the most vividly coloured and attractive snakes in PNG. However, being large and visible these species are also often the most persecuted by man. It is important to realise that pythons and boas are non-venomous and therefore a bite, whilst painful and bloody when received from a large specimen, is not life-threatening. It is also important to remember that members of this family are considered 'threatened' worldwide and therefore afforded international protection under the Convention for the International Trade in Endangered Species (CITES).

The **ACROCHORDIDAE** is a small family with only three species which are confined to the Indo-Australian region, two of which occur in PNG. Generally known as file snakes, wart snakes or elephant trunk snakes due to their loose, rough tuberculate texture and appearance the acrochordids are fully aquatic inhabitants of freshwater, estuarine or inshore marine habitats. They rarely venture onto land where their unusually 'baggy' skin and the lack of the large ventral plates (found in terrestrial snakes) reduces locomotion to an awkward crawl. For all their apparent clumsiness on land file snakes are extremely agile in the water using their flattened baggy bodies for propulsion in much the same way that seasnakes use a paddle-like tail. In common with other fully aquatic snakes the file snakes possess small eyes and valvular nostrils positioned dorsally. The granular or tuberculate scales on the head and body are believed to have a sensory purpose which may be connected with prey location and recognition. The prey of acrochordids is comprised entirely of fish which are grasped and constricted. Acrochordids are viviparous.

The largest and most diverse snake family is the **COLUBRIDAE**. World-wide it contains approximately 2000 species, roughly two thirds of the total snake fauna, yet it is very poorly represented in the Australasian region. Although five subfamilies are present, four of these are represented by a single genus. The fifth subfamily contains six monotypic (single species) genera. Despite its relatively limited representation the members of the Colubridae are fairly diverse in their ecology. The non-venomous keelbacks (13 species), so called because of their rough, ridged scales, are inhabitants of freshwater watercourses while the mildly venomous, yet harmless, mud and mangrove snakes (5-6 species) generally prefer estuarine conditions. The non-venomous treesnakes (6 species) possess eyes with round pupils and are diurnally active while the mildly venomous brown treesnake possesses vertically elliptical 'cat-like' pupils and is nocturnal in its habits, as are the semi-arboreal and terrestrial non-venomous ground snakes (7 species). Of these only the brown treesnake may be considered even a potential threat to man. Prey taken by colubrids ranges from crabs, fish, frogs and small lizards to other snakes, mammals and birds. However, none of the New Guinea colubrids achieve any great size, even a 2.0m brown treesnake remaining fairly slender. Reproductively the treesnakes, ground snakes and keelbacks are oviparous and the mud and swamp or mangrove snakes are viviparous.

The most diverse snake family in the Australasian region is the **ELAPIDAE**. Elsewhere in the world this family primarily contains highly venomous species such as cobras, mambas, kraits and coral snakes. However, this family achieves its greatest diversity in the Australasian region where its members have occupied many of the niches which elsewhere would have been occupied by colubrids, (poorly represented) or vipers (entirely absent). Many of the New Guinea elapids occupying 'colubrid niches' are small, virtually harmless leaf-litter inhabitants. Even so, this family also contains the most dangerous snakes in New Guinea and foremost in this respect are the taipan, the Papuan blacksnake, the death adder, the brownsnake and the small-eyed snake. There are several other elapids (Müller's snake and the whip snakes) which are considered large enough to constitute a potential threat to human health. Almost all the terrestrial members of the Elapidae in New Guinea are placed in one subfamily (Oxyuraninae). However, the two amphibious sea kraits, *Laticauda*, are considered distinct enough to warrant a separate subfamily, the Laticaudinae. One small inoffensive elapid, *Parapistocalamus* from Bougainville, is believed to be more closely related to the terrestrial S.E. Asian and American coral snakes (not to be confused with seasnakes) is placed with them in the subfamily, Elapinae.

The elapids and their relatives the hydrophiid seasnakes (*see* below) possess short immovable fangs in the front of their upper jaws. The fangs are hollowed modified teeth which are connected, via a duct, to the venom gland under the eye (*see* Snake Skulls and Snakebite). When a venomous snake strikes the fang enters the skin and venom is transported from the gland, along the duct and down the hollow fang to enter the wound via an orifice in the anterior edge of the fang just above the tip. The venoms of New Guinea front-fanged snakes are neurotoxic (either pre- or post-synaptic or both), myotoxic or haemotoxic or a combination of several in effect (*see* Individual species accounts, Snakebite treatment section and Glossary).

The seasnakes belong to the family **HYDROPHIIDAE** which is believed to be closely related to the Elapidae. Some authors consider the links between terrestrial Australasian elapids and the seasnakes stronger than any existing between the Australasian elapids and Oriental elapids to the extent that they place the terrestrial venomous snakes of Australia and New Guinea (excluding *Parapistocalamus*) in the Hydrophiidae as distinct from the Elapidae. Such arrangements are irrelevant here and the more traditional classification will be adhered too.

Over twenty species of true seasnakes have been recorded around the New Guinea coastline and while many are rarely encountered, a few have been implicated in human snakebite morbidity and mortality both in marine and riverine situations. The high toxicity of seasnake venoms means that all species, whether aggressive or inoffensive, should be considered highly dangerous.

Seasnakes are specialised marine inhabitants possessing a laterally compressed body and 'paddle-shaped' tail for propulsion. Locomotion is by a form of lateral undulation and some species can travel in reverse.

They also exhibit adaptations which enable them to dive rapidly to great depths (up to 100m) and either resurface equally as quickly or remain submerged for up to two hours. Whilst submerged valvular nostrils, the valve at the entrance to the glottis (airway) and projections of the rostral scale, which seal the lingual fossa (the notch-like opening at the front of the mouth through which the tongue passes), prevent entry of water to the lungs. Posterior positioning of the heart, the structure of the respiratory system which includes an air storage or saccular lung and the ability to draw in oxygen and pass out carbon dioxide and nitrogen through the skin, all assist in the seasnake's ability to dive and surface rapidly or remain submerged for prolonged periods. Other adaptations to marine life include the possession of sub-lingual (under the tongue) salt-excretory glands.

The scales of many marine species are juxtaposed (non-overlapping) to reduce infestation with parasites such as sea lice to a minimum and frequent skin sloughing further assists in reducing any parasite burden. In many species the broad ventral scales, familiar on terrestrial snakes, are much reduced in size. The result of these two marine adaptations is reduced mobility on land and truly marine species such as the ocean-going yellow-bellied seasnake, *Pelamis*, are virtually helpless when cast up on a beach after a storm.

Prey consists mainly of fish, especially eels, although a few species have become adapted to 'sucking-up' fish eggs from the sandy sea bed and it is possible that in time these inoffensive species will eventually dispense with their venom apparatus.

Seasnakes are viviparous giving birth out at sea.

FACT OR FICTION ?

The snakes are a much mis-understood group of animals and without wishing to venture into a lengthy lecture on their natural history and biology, subjects expertly dealt with by other authors (Shine 1991 and Glasby, Ross and Beesley 1993 - *see* Further Reading), it is perhaps worth laying a few mis-conceptions to rest.

Snakes are slimy ! No. Snakes are all dry to the touch. Their scales shine but they are composed of a similar material to finger-nails and they are either silky to the touch in the case of smooth-scaled species or rough in the case of keelbacks or wart snakes. The only time snake scales are wet is when they have been swimming. Snakes do not move along on layers of slime, they employ a variety of modes of locomotion from the familiar undulating serpentine motion to side-winding over loose sand. Heavy bodied snakes such as large pythons move in straight lines but again an undulating body motion, brought about by the synchronised movement of ribs, muscles and large ventral scales, is evident. Some aquatic snakes with their ventral scales reduced to the size of the smaller dorsal scales are virtually helpless on land whilst specialised arboreal species possess ridges or 'corners' on their broad ventral scales which enable them to scale vertical coconut palm trunks.

The snake's tongue is a sting ! No. The forked or bifurcated tongue is a highly sensitive sense organ which picks up particles of the atmosphere from around the snake's tongue or touches and 'tastes' objects in its path before returning to the Jacobson's organ situated in the roof of the snake's mouth for analysis. Tongue flicking in a snake is a form of information gathering and not a threat. The forked tongue is probably the primary sense organ of a snake.

The snake's tail is a sting ! Absolutely untrue ! Some snakes, such as blindsnakes, and also snake-lizards have sharp tips, even spines on the end of their tails. These serve as aids for locomotion when the snake is forcing itself through the earth and do not contain any poison.

Snakes hypnotise their prey ! No. Snakes always stare, they cannot blink. Snakes lack eyelids but possess instead a transparent covering over the eye known as a 'brille'. This covering, which is replaced and sloughed off when the snake periodically sheds the outer layer of its skin, enables the snake to see but prevents damage to the eyeball when the snake is crawling through vegetation.

Snakes are deaf ! Technically no. They can detect vibrations through the chin and body but it is true that snakes do not possess external ears.

Some snakes are blind ! No. The blindsnakes have extremely small eyes covered by a large transparent scale but they can determine light intensities and whilst other snakes do not possess vision on a par with mammals or birds, or possibly even some lizards, and they may not be able to see colour, they are aware of movements.

Snakes guard their young ! The only snakes in New Guinea where this might appear to apply are the pythons which will coil around and incubate their eggs. After the eggs have hatched the female will have no further interaction or interest in the young.

Snakes will attack on sight or chase you ! Not true, snakes would much rather escape and generally, given a wide berth, this is what they will do. However, a few species, including the taipan and the black and brownsnakes, have a reputation for aggression and may rush towards an enemy, biting before fleeing themselves.

All snakes are poisonous (venomous) ! Totally untrue although the Australasian region does possess a high proportion of venomous snakes. However, many of the technically venomous species are either too small or secretive to represent any threat. **Pythons, boas, treesnakes and many others are totally non-venomous.**

SNAKE SKULLS AND SNAKEBITE

The presence of teeth does not necessarily indicate that a snake is venomous. Almost all snakes, with the exception of African egg-eating snakes, possess teeth although the typhlophid blindsnakes of Australasia only possess maxillary (upper jaw) teeth, the mandible (lower jaw) being toothless (fig. 1).

All remaining Australo-Papuan snakes possess six rows of teeth as follows;

Two rows on the upper outer jaw (maxillary teeth);
Two rows on the upper inner jaw (palatine teeth);
Two rows on the lower jaw (mandibular teeth).
Fangs, if they are present, are located on the maxilla.

Snake teeth are generally fairly simple, recurved 'hooks', the purpose of which is to maintain a generous grip of the prey whilst it is constricted, swallowed or envenomed. Lacking limbs the mouth and the coils of the body are the only means by which snakes can control and manipulate prey.

To allow the degree of movement required to enable snakes to manoeuvre and manipulate prey, prior to swallowing, the jaws of most snakes exhibit considerable flexibility. The lower jaw, in particular, is not fused together as a single unit at the chin, as is usual in most other vertebrates including lizards. The mandibles, comprising dentary and compound bones, are separated from each other and free to move forwards and backwards or pivot outwards on the slender quadrate and supratemporal bones which attach the entire lower jaw to the skull (fig. 6).

This ability to move the two halves of the lower jaw independently allows snakes to consume large prey animals whole, in the case of pythons several times the width of their own heads, and also draw prey down the throat by alternating advancements of the separate halves of the lower jaw along the prey as it passes, usually head-first, into the mouth. This latter motion is also assisted by the flexible upper jaws and in the case of venomous species such as elapids or viperids, the fangs, which almost 'stab-walk' the prey down the throat.

A similar degree of kinesis or movement to that found in non-venomous snakes is also exhibited by the pygopodid lizards, genus *Lialis*, enabling them to consume fairly large prey, in relation to their own body size, such as skinks (Patchell & Shine 1986).

The teeth of the lower jaw are attached to the anterior bone of the mandibular series, known as the dentary. The outer teeth of the upper jaw are located on the maxilla and in some instances a single tooth from each side may be positioned on the premaxilla. Fangs (modified teeth for the injection of venom, *see* later) are also positioned on the maxilla. In the front-fanged snakes (elapids, hydrophiids and viperids) the fangs are located on the anterior end of a much shortened maxilla. Snakes lacking front fangs may possess fairly elongate maxilla but those snakes with front fangs require short maxilla which can be repositioned easily, brought upwards and forwards to bring the fangs into line for the strike. In the vipers (not represented in the Australo-Papuan region) the large fangs are hinged allowing them to be swung forward through almost 180°, to a horizontal position, when the snake strikes. When the fangs are at rest, the mouth closed, viper fangs take up much of the available space posterior to themselves with the result that the bones posterior to the maxilla are toothless.

In the elapids the somewhat shorter fangs are not hinged but fixed in position. However, the maxilla is still capable of kinesis bringing the fangs into a forward strike position. There is still a toothless gap immediately posterior to the fang known as a diastema. Posterior to the diastema, on the rear half of the maxilla, there may again be a few teeth but owing to the shortened nature of the maxilla most of the remaining upper, outer teeth may be found on the ectopterygoid which follows the maxilla in elapids. The fangs of rear-fanged snakes are located on the posterior tip of the maxilla (fig. 5) and are preceeded by simple solid teeth. The upper, inner rows of teeth are located on the palatine (anterior) and pterygoid (posterior) bones in the roof of the mouth.

Snake teeth are not located in sockets as in mammals or crocodiles. They are fixed along the edge of the jawbones the result being that they are frequently shed or lost when a snake strikes or feeds.

Snakes have been largely classified on the nature of their teeth, the following super-groups or taxa having been used:

Aglyphous teeth (Aglypha, figs. 2-4): simple, solid teeth with no venom carrying capacity; (pythons, boas, filesnakes and non-venomous colubrids i.e. keelbacks and treesnakes in the Australo-Papuan region).

Opisthoglygous fangs (Opisthoglypha, fig. 5): grooved fangs with an open, posteriorly or anteriorly positioned, channel for the conduction of venom, these fangs are positioned on the rear of the maxilla; (rear-fanged colubrids, brown treesnake, mud and mangrove snakes in the Australo-Papuan region).

Proteroglyphous fangs (Proteroglypha, figs. 6 & 7): grooved fangs with the groove almost closed and barely visible, fixed in position on the anterior of the maxilla; (elapids and seasnakes).

Solenoglyphous fangs (Solenoglypha): entirely enclosed, hollow, hinged fangs which function like a hypodermic syringe, positioned on the anterior edge of the maxilla; (true and pit vipers, absent from the Australo-Papuan region).

Of course the opistoglyphs, proteroglyphs and solenoglyphs also possess numerous simple aglyphous teeth. In the case of the front-fanged snakes, although only one fang on either side is functional, reserve fangs are often present, in readiness to replace fangs shed naturally, swallowed with prey or lost when striking. The result, when a snake bites, may be three or even four fang punctures but generally only one fang on either side will conduct venom.

Venom is probably a modified form of saliva designed to subdue or kill the prey and then begin or speed up the processes of digestion. The possession of venom means that a snake need not, necessarily, be stoutly built, like a python, to deal with fairly large prey animals and, in the case of venoms which cause an acceleration of the digestive processes, the period during which the snake is itself vulnerable to attack, due to its immobility whilst digesting, is shortened. This latter is probably more true of large bodied vipers such as the puff adder of Africa with their cytotoxic venoms, than the neurotoxic elapids of the Australasian region.

If the above are the primary and secondary purposes of venom the third use must be defence. However, except for the Afro-Asian spitting cobras which have developed a means by which they deliberately utilise venom for defence, defence by biting and the injection of venom is, probably for most venomous snakes, a useful, though not always utilised, additional bonus. If molested or cornered, with no place to escape, venomous snakes, and non-venomous snakes, will often resort to biting their perceived aggressor. Most snakes seem capable of controlling the quantity of venom they inject and defensive bites may only result in a percentage of the available venom being injected. The amount may vary from 0% (often termed 'dry bites' since no venom is injected) upwards to that which would be expended during a predatory bite intended to kill prey.

In front fanged venomous snakes the venom glands are situated under and behind the eye on the outside and above the maxilla and the ectopterygoid. Venom glands are connected to the venom canal of the fang by a venom duct (*see* photograph). In rear-fanged snakes, which are more primitive, the 'venom' is derived from the Duvernoy's gland situated between the eye and the angle of the jaw. Even in non-venomous snakes such as the colubrids of the subfamily Natricinae (keelbacks genus *Tropidonophis* spp.) the saliva may be slightly toxic. These species feed on frogs with permeable skins and their saliva, absorbed through the skin of the prey, does appear to subdue the prey prior to ingestion. A defensive bite received from one of these snakes with saliva seeping into the wounds made by the enlarged aglyphous teeth, may result in localised tenderness, discolouration and swelling. This indirect form of envenoming by a technically non-venomous snake may be quite alarming and some species of rear-fanged Asian keelbacks in the Natricinae (genus *Rhabdophis*) are now considered dangerous to man.

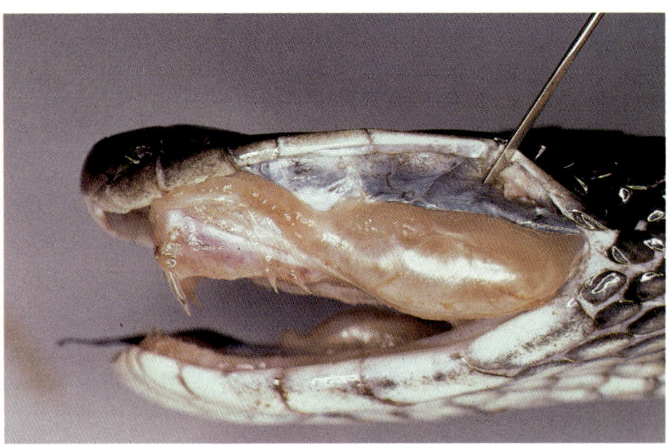

Dissection of a Papuan taipan showing enlarged venom gland, venom duct and fang

Snake venoms are extremely complex compounds, the biochemistry of which is beyond the scope or aim of this book. Suffice it to say that the world of venomous snakes contains neurotoxins which affect neurological and neuromuscular processes; haemotoxins which affect the blood; cytotoxins which cause the breakdown and lysis of tissues; myotoxins which affect the function of, or cause the destruction of, muscles, nephrotoxins which damage the kidneys and cardiotoxins which attack the heart. Within each category are

NEW GUINEA SNAKE SKULLS & DENTITION (not to scale)

(Simple teeth - light shading Fangs - dark shading)

Dentition in modern snakes
In most snakes more advanced than the blindsnakes (typhlopids etc.) there are six rows of teeth; four in the upper jaw, on the outer maxilla (and ectopterygoid) and inner palatine and pterygoid, and two in the lower jaw on the dentary bones.

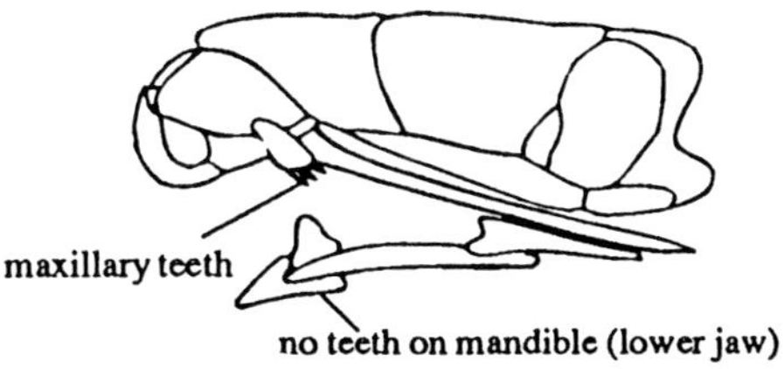

Fig. 1. A typical typhlopid skull.
Typhlops diardii - Diard's blindsnake
(after Smith 1943)

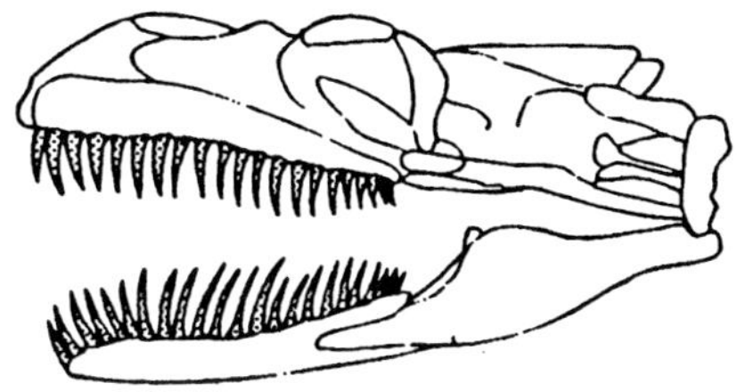

Fig. 2. A typical boid skull.
Morelia boeleni - Boelen's python
(after McDowell 1975)

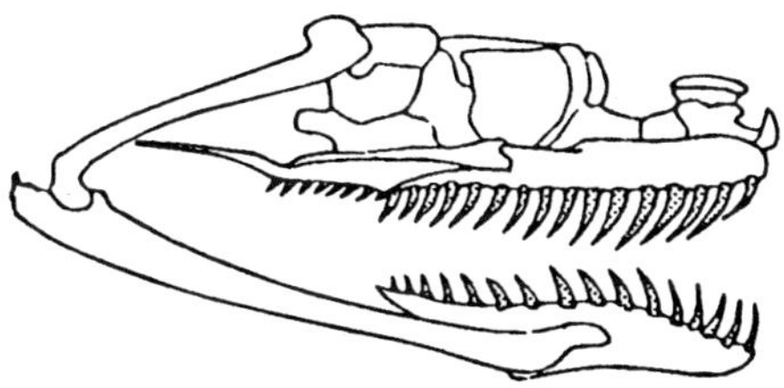

Fig. 3. A typical acrochordid skull.
Acrochordus arafurae - Arafura filesnake
(after McDowell 1979)

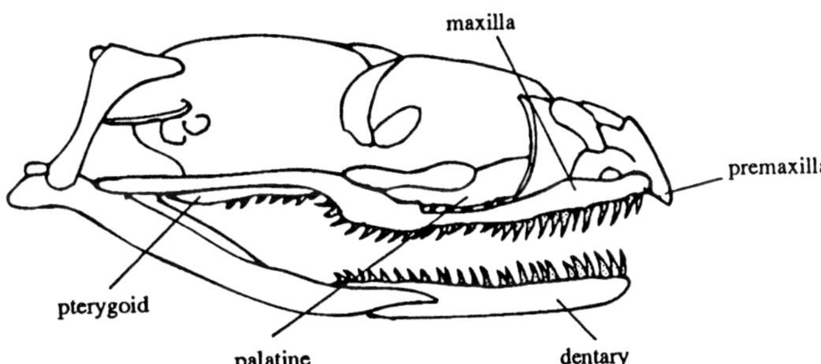

Fig. 4. A typical non-venomous colubrid skull.
Dendrelaphis punctulatus - Common treesnake
(after Plant *in* Ehmann 1993c)

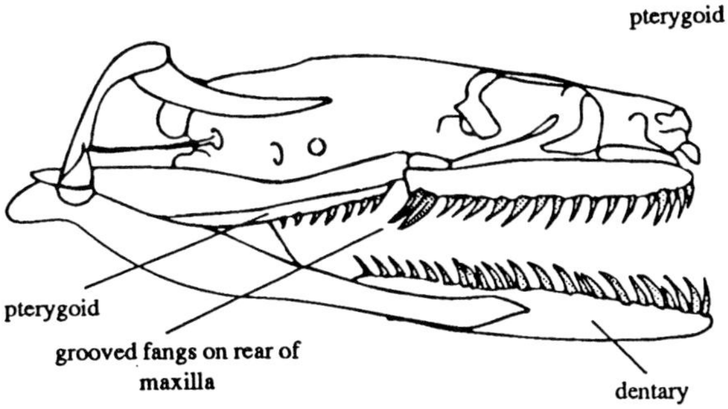

Fig. 5. A typical mildly venomous rear-fanged colubrid skull.
Cerberus rynchops - Bockadam
(after Plant *in* Ehmann 1993c)

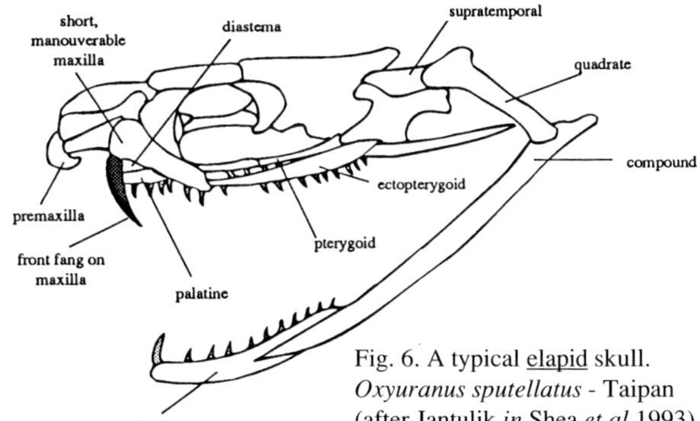

Fig. 6. A typical elapid skull.
Oxyuranus sputellatus - Taipan
(after Jantulik *in* Shea *et al* 1993)

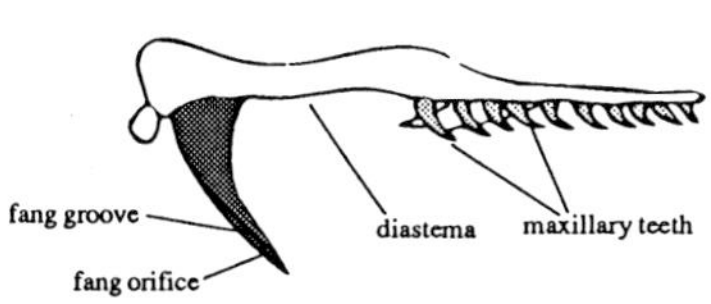

Fig. 7. A typical elapid maxilla.
Aspidomorphus lineaticollis - Crowned snake
(after Parker & Grandison 1977)

further classifications i.e. pre-synaptic neurotoxins which affect the transmission of neuromuscular impulses before the synapse or post-synaptic neurotoxins which cause paralysis beyond the synaptic gap. No snake venom contains all of the above but usually a snake venom will contain a combination of several different toxins i.e. the venom of the taipan (*Oxyuranus scutellatus*) contains a powerfully pre-synaptic neurotoxic (causing paralysis) and a procoagulant (a haemotoxin causing prolonged bleeding). It also contains a post-synaptic neurotoxin, is weakly haemolytic and may also be myotoxic or cardiotoxic.

SORCERY, SNAKES AND SNAKEBITE

Around the world snakes are associated with many contrasting beliefs: death and longevity; disease and fertility or virility; good and evil; demons and gods. They appear suddenly, often silently and in the dark, and disappear almost as quickly, leaving behind confusion, awe, fear and sometimes death. It is not surprising that the snake in all its forms exerts a powerful hold over the minds of people of many cultures, races, religions and differing degrees of technological advancement. I am well aware, by listening to the words of generally well educated Europeans, just how little is actually known by the 'general public' regarding the natural history of snakes and how many unlikely stories are attributed to them. To a villager in the tropics the threat of snakebite is often a real danger and the snake is a much feared and/or respected animal. Snakes have a powerful hold over the minds of those people who dwell alongside them. Certain villagers have learned the ways of the serpent and in so doing have earned the respect, fear and homage of their fellow villagers. The skills of these 'wise men' are jealously guarded and people turn to these men (and women) to treat snakebites, and also, on occasion, to cause them. These people may be called witchdoctors, magic men or sorcerers and rural Papua New Guinea certainly has its fair share. I have encountered two or three such people during my field work in the southern provinces and on one occasion was left in the position of having my field assistant 'hexed' into a trance in front of me.

Early anthropologists studying the culture and social structure of PNG encountered many such magic men and myths. These early observations were documented in a huge array of anthropological books and papers, a full apraisal of which is beyond the aims of this book. However, Campbell (1969) and Faliu (1987) did make a study of the relationships which existed between rural people in PNG, the sorcerers and the snakes. This short section is drawn almost entirely from the papers of these two authors and is intended to provide a 'potted version' of the position of the snake in rural PNG society and explain in part the problem that exists for modern medicinal practices in village society.

RURAL PNG UNDERSTANDING OF ILLNESS

In rural Papua New Guinean culture there are roughly two types of sickness; those occurring naturally - minor illnesses such as *endeni vivi* or 'body sickness' of the coastal Trans-Fly people or 'sick nothing' of the highland Arapesh, which may be attributed to changes in the weather, eating the wrong food or common maladies relating to childbirth or infancy, and those caused supernaturally - more serious illnesses and accidents, occurring in societies accepting natural causation for less serious maladies, or all illnesses, including minor ailments, in those societies which do not.

Supernatural illnesses may be caused by:

a) the activity of spirits or the soul of dead ancestors i.e. the *sovai* of the Orokaiva people or the *sua* of Motu culture;
b) the activity of unrelated or independent malevolent spirits, bush spirits or mythical beings i.e. the *paipai* of the Roro people and the *ove-hahu* of the Gulf Province Elema).

These alien spirits may manifest themselves in the bodies of animals, including snakes, in order to injure or 'take' a human.

Spirit interest in the living world may be caused by:

a) breaking a cultural taboo (spirits of the dead);
b) breaking a place taboo, entering a forbidden or sacred area (spirits of the bush).

Spirits aggrieved in either manner are apparently open to restitution with gifts or prayers and illnesses resulting from their anger are usually non-fatal.

When the patient does not recover the cause is considered to be stronger and sorcery is blamed. All deaths, excepting deaths amongst the very young or old, are believed to have resulted from sorcery.

Treatment of minor ailments is carried out either by relatives or specially gifted people such as the *dapu* of the Purari delta or the *heduru tauna* of the Central Province Motu. These practitioners use herbal remedies, prayer, incantations and trances to drive out or settle the anger of the spirits involved in all cases not involving direct sorcery. In cases of sorcery they try to advise the relatives which sorcerer may be involved and how he may be cajoled to reverse his spell.

RURAL PNG UNDERSTANDING OF SNAKES

Other than large pythons (the *morans*) and 'blacksnakes', most people do not appear to differentiate particularly between various snake species. It seems to matter little whether a snake is venomous or not but rather, whether it may be used as an instrument of sorcery.

Snakes are common place in hereditary stories in tribal society and in most cases the snakes concerned are non-venomous. Pythons are especially prevalent in such stories which also feature such common beliefs as the ability of men and women to turn into snakes (a theme also encountered by myself in West Africa) and snakes embodied with spirits which may turn into humans, always males, and go on to cohabit with human women and father human/spirit children.

Snakes may also represent the incarnation of mythical beings. To the Arapesh and Kapauku from the PNG Highlands, the Kiwai of Western Province and the Gulf Province Orokolo and Busama, snakes may be the dwelling places favoured by ghosts, spirits of departed ancestors, evil demons or bush spirits. Other examples cited by Campbell (1969) include a demon in the form of a huge python which takes men from the men's house and is much feared by the Mae Enga, and a creature called the *pathogu* which resembles a strong old woman dwelling beneath the ground who surfaces to steal children. According to the Marind-anim the *pathogu* can take on the guise of a snake to go about its work, as can the *uri* trickster of the Waropen who also believe in a man-eating snake in the Woisimi River.

Of less malicious intent some Trobriand Islanders consider the snake to be the reincarnation of an ancestral chief but his visit to a village is considered a bad omen and food and prayer must be offered up to tempt him away before ill befalls the village. The close approach of such a serpent was believed the cause of an epidemic on Kiriwina. Killing a sacred snake containing the spirit of an ancestral chief will bring particularly serious consequences to the village concerned.

Still in the Milne Bay Province inhabitants of some islands in the d'Entrecasteaux group will rapidly leave a house if a snake enters. Rossel Islanders (Louisiade Arch.) believe their primary god Wonajo and his lesser gods dwell as snakes during the day. Snakes were considered sacred and are not killed.

These cases are interesting since the islands of Milne Bay Province are entirely devoid of highly venomous snakes, the only front-fanged species (elapids) which do occur being secretive semi-fossorial members of the genera *Toxicocalamus* and *Aspidomorphus*. Rossel Islanders seem to accept that snakes pose no threat but the islanders from the Trobriand and d'Entrecastaux Archipelagoes still fear snakes. This may in part be due to the eastern isolation of Rossel Island and the comparative close proximity of the Trobriands etc. to the mainland with its populations of large, dangerously venomous species.

However, in most cases snakes are greatly feared but not because of any venomous capabilities. The fear is based on their supernatural relationships and for this reason non-venomous snakes may be feared as much as, possibly even more than, venomous species since merely the sighting of a snake may be a bad omen without any requirement for the snake to attack or bite the observer.

However, in many cultures large pythons are viewed more as benevolent rather than malevolent. Even on islands like Guadalcanal in the Solomons there are reportedly female python spirits preying on crop-damaging rodents. This is an interesting situation since pythons are not distributed through the Solomon Islands.

The Western Province coastal Kiwai recognise several snake spirits with beneficial purposes. *Soido*, the promoter of agriculture, dwells under the ground and may surface as a snake. *Maigidubu* is a snake-like creature which leaves tracks in the gardens following the yam planting festival which are seen as good omens for the crop. The *etengena* spirits may occur as birds or snakes and guard the trees of the plantation by biting trespassers. As in many other cultures snakes are associated with fertility and the Kiwai bury pieces of snake along with the sugar cane shoots when planting the crop.

In Madang the Tangu recognise the spirits of their ancestors as *puoker* which dwell in pythons and guard ancient sites from trespassers. The inhabitants of the volcanic island of Manam were also said to consider snakes to be reincarnations of the dead. In the Owen Stanley Mountains of Central Province similar guardians of the Fuyughe people protect sacred sites. The Fuyughe differentiate between the sacred guardian pythons and those which are hunted for meat. I have heard similar stories from people now settled around Port Moresby, but originally from the Woitape region, who recognise pythons, including Boelen's python *Morelia boeleni*, as sacred in a particular area but legitimate game for the pot slightly further away.

RURAL PNG UNDERSTANDING OF VENOMOUS SNAKES

Snakebites and snakebite fatalities are generally considered to be the work of a sorcerer. The species most commonly implicated in sorcery is the blacksnake or 'Pap black' but there appear to be several types of blacksnake in the cultures of the southern Central and Western Province coastal people. The true Papuan blacksnake, *Pseudechis papuanus*, is a large and highly venomous elapid but it may be confused with dark grey or black specimens of the taipan, *Oxyuranus scutellatus*, which are much commoner than 'true' blacksnakes today on the coastal savannas of Central Province, or the extremely active, diurnal whip snakes, *Demansia* spp., non-venomous keelbacks, *Tropidonophis mairii*, and the black specimens of the common treesnake, *Dendrelaphis punctulatus*, which occur in the southern Trans-Fly of Western Province.

In Mekeo and Roro culture (northern coastal Central Province) the sorcerer employs several methods to cause a blacksnake to bite someone. He may boil the snake in a pot with an item of the victims personal belongings, a piece of clothing or bodily secretions of some kind, or he may use a snake stone which kills anyone it touches, with the exception of the sorcerer himself.

To the Kiwai of Western Province the sorcerer is the 'blacksnake man', the *ove-devenar*. He makes a model of a snake, speaks the intended victim's name and places that person's excreta in the mouth of the snake model. The model is placed in the path of the victim and the suitably outraged serpent will bite and kill that person.

All snakebites can draw blood and may be considered, by the victim and his relatives, to be the result of sorcery by the victim and his relatives. If the magic man or sorcerer can save nineteen out of twenty victims he is considered a powerful man but it should be realised that many non-venomous snakes (treesnakes, keelbacks and pythons) will bite defensively and the mildly venomous cat snake is a common cause of non-life-threatening snakebite in villages. If nineteen bites are caused by non-venomous species and one by a taipan the sorcerer's ability to save 19 out of 20 patients is not a sign of his skill since the result would have been the same without his intervention.

SNAKES AND SICKNESS

Apart from being the tools of sorcerers and cause of snakebite, snakes may be implicated in other illnesses. The Mafulu (Owen Stanley Mts., Central Province) consider stomach ache to be caused by the presence of a snake in the gut. The Mafulu also used to undergo a nose-piercing ceremony during which a living 300-450mm specimen of local 'grass snake' was drawn backwards through the hole in the septum. It seems probable that the snake species concerned may have been a slender green treesnake, most likely *Dendrelaphis calligastra*, since a grass snake or keelback, *Tropidonophis* spp., is stouter and has strongly keeled scales which could have made the process quite painful.

SNAKES AND DANCE

The Waropen perform a dance wherein all the dancers form a long writhing serpent to illustrate the story. These were the people who believed the Woisini River to contain a huge man-eating serpent and the writhing line of people could equally represent a strong river. Several fast and dangerous rivers in PNG have been dubbed 'Eaters of Men'.

In New Britain the Baining, who fear snakes generally, spend months collecting live snakes to be handled freely on the night of a single ceremony. Parallels may be drawn with similar annual ceremonies in India, where cobras are worshipped and sprinkled with coloured dyes; Italy, when the statue of the Virgin Mary is carried shoulder-high, draped with snakes; the Hopi Indians of southwestern USA and possibly even the Appalachian evangelist snake-handling cults. Again, however, the people of New Britain are putting themselves in little real danger since the most venomous snake present is the Müller's snake (*Aspidomorphus muelleri*) which is secretive and rarely responsible for bites. It is unlikely that similar cults or dances would survive long in the Mekeo region!

SNAKES AND SNAKEBITE

Campbell conducted a number of interviews with village elders from Central and Gulf Provinces in an effort to learn more about the power of the snake in their cultures. He considered he had compiled sufficient data on only three groups and his results are summarised here.

Motu -Koitabu (Port Moresby)

The snakes:
Duba - dark snakes - possibly either Papuan blacksnake, whip snakes or dark taipan (Kila Kila);
Larana Karo - long blacksnake - probably Papuan blacksnake or dark taipan (Barune);
Kabagi - red snake - the taipan, a probable reference to the orange vertebral stripe (Barune);
Asenamo Api or *Api* - death adder;
Navara - python (Koitabu);
Lavara - python (Motu).

Pythons are considered to have magical powers and if they are found near the house or in the gardens they are believed to be protective and are not killed, but encountered in the bush they may be killed for the pot.

Snakebites, like other accidents, are caused by a problem within the social group, the breaking of a taboo or some other 'crime' such as adultery. The usually protective spirits, *birava*, allow the person to be affected but the illness is generally non-fatal and intended as a warning. More serious snakebites are caused by the *vada* or sorcerer who may be a Koitabu in another village or a Koiari further inland. Deaths are always believed to have been caused by sorcery.

Treatment involves taking the victim to the 'helper practitioner', the *hedura tauna* or *babalau tauna*, who would prick the area with barbed leaves (*atoro or mahita*), a pointed bamboo or a special sharp red stone used for cutting wounds (*vasiga*). Questioning of the victim and relatives determines how the *birava* were upset and rituals performed to placate them. Serious snakebites required the attendance of a sorcerer from another village who would rub leaves on the bite and administer potions.

Keakalo (Marshall Lagoon)
The snakes:
Gelema rupa - black coloured snake - the Papuan blacksnake;
Relena gamara - brown coloured snake - the taipan;
Vanaame - a short snake which springs - the death adder;
Mumu - a snake with legs, possibly either a monitor lizard or the blue-tongued skink which is an elongate bodied, short limbed lizard which will gape widely, protrude a large blue tongue and rush forward to deliver a forceful bite. Such bites may go septic and have serious consequences if ignored. [This may be the reason why blue-tongued skinks are believed to be poisonous by some people in Madang (*pers. obs.*).]
Maatha - a harmless snake in houses, coconut and banana trees - possibly a treesnake *Dendrelaphis*;
Mulapala - seen in the jungle - any number of possibilities;
Kapari - the python which is again often considered sacred and found in many legends.

Snakebites are believed to be caused by accident; because of some infringement of a social taboo or village law such as theft, cheating or ill-treatment of someone, or by sorcery which usually proves fatal.

Treatment is by traditional methods or *gauguri* and carried out by one of the few practitioners who know the skills. It may involve splitting a stick into four and ritually burning the right thumbnail, right big toenail, left thumbnail, left big toenail with the burning end of the wood. If the patient shouts and stretches the limb in response to the pain this is a good sign (the patient is not yet paralysed). Special leaves are then rubbed until black and added to warm water which the patient drinks. If the patient's body goes cold or he vomits blood the situation is very serious and due to sorcery. Mucus (awatha) in the back of the throat may be removed with a cockatoo feather but this is a very serious sign of sorcery induced snakebite and will probably result fatally.

Maopa villagers cut around the snakebite with a sharp stick, possibly a fore-runner of the harmful razor-cutting of today.

Serious snakebites require the attendance of the *mega mega auri* "the men who send snakes". These, usually elderly, men are consulted over the bite. They chew *paia* (a treebark) and *wamela* (a vine) and blow them into the nostrils, ears, mouth and over the body of the victim while adminstering a massage with coconut flesh, coconut milk and *kiki* leaves. The *mega mega auri* can also cause snakes to seek out and bite victims.

Snakebite may be avoided by following a ritual before venturing into the bush which consists of stamping the feet, chanting and annointing the body with the chewed leaves of *paia ivoa* (ginger) or *wamala*, all of which have an odour thought to repel snakes. Ginger is commonly considered a good snake deterrent.

Mekeo (Kairuku)
The snakes:
Afi - meaning "sharp-eyed" - the death adder of the swamps and the wet season;
Auguma - meaning "to bite again" - a sure reference to a large elapid such as the taipan or blacksnake. Two types are known, one living above ground and the other below. Vaifa'a (Beipa) people also recognise a third type of blacksnake.

Snakebite is always believed to be due to sorcery. Repeated bites, i.e. the bite of the *auguma* (taipan, blacksnake) and the presence of blood are serious signs.

Treatment initially involves trying to find who caused the snakebite to happen. Relatives consult with the *imaonge* or fortune-teller who tries to ascertain the name of the sorcerer by pulling the victim's fingers one at a time whilst naming sorcerers. If the finger cracks the named one is not the sorcerer, if the finger does not crack the sorcerer is the one named. If this fails a similar process is carried out by balancing cowrie shells. If this fails another *imaonge* is consulted.

It can be seen that simply determining the name of the sorcerer responsible can take a considerable time and in cases of serious snakebite this will weigh heavily against the chances of survival for the victim.

If the sorcerer is identified he must be presented with a gift and he will also balance cowrie shells. If this fails another sorcerer is required. Treatment involves coconut potions and lotions and, again, ginger plant leaves.

Sorcerers are generally widowers who live alone but in some societies they may be young men whose power wains with age and competition from up and coming magic men. Power over snakes may be achieved by mixing human body parts with snake blood and leaves in a clay pot. Burying the contents in a fire would send the spirits to harm the victim although the sorcerer can do as much by merely touching the person. Snake stones are also extremely powerful, causing touched persons to be bitten by snakes whilst protecting the sorcerer from the same fate. The night is the time for sorcery and in many societies people will not leave their houses after dark for fear of sorcery. In Western Province sorcerers and their acolytes frequent burial grounds by night, eating the food and drink left for the dead during the three months of mourning. To see the sorcerer at night is to invite serious consequences.

Campbell ended his appraisal of sorcery and snakebite by recognising that whilst valuable time may be lost while relatives and victim consult fortune-tellers and sorcerers these beliefs are fairly ingrained in village practices and cannot be ignored. He suggested that the newer, modern medicine men must try to work alongside the magic men rather than against them, encouraging good first aid practises and discouraging harmful ones like razor-cutting, and also promoting a rapid retreat to the *haus sik* (hospital) if the signs suggest that the snakebite is serious and beyond the powers of village practitioners.

CONSERVATION OF SNAKES IN PNG OR 'SAPOS YU PAINIM SINEK, YU NO KEN KILIM' !

Snakes are probably killed for four main reasons - out of fear, for food, for fun or for financial gain - the 4 'F's if you like. Only the first two of these, in my opinion, can be justifiable and even then some of the reasoning behind the 'fear' may be questionable.

From fear

The immediate reaction of some people coming across a snake is to reach for the nearest stick and kill it, even if it means that the holder of the stick is putting himself in an unnecessarily dangerous position in order to kill the snake. It is almost as if people believe it is their 'duty' to kill every snake they see. Drivers will even swerve across a road to hit a snake in the opposite lane. This reaction stems from the belief that all, or most, snakes are dangerous, even evil. To be a snake is almost a 'crime'. What isn't often realised is that many species such as the pythons and boas are not venomous and cannot kill by biting even though they will often defend themselves by biting and drawing blood if picked up or molested. The majority of snakes encountered are non-venomous and even many of the smaller venomous species are virtually harmless to man. In fact only 8.5% of all land snake species in New Guinea can be considered highly dangerous to man - seven highly dangerous species from a total of 81 species (excluding marine seasnakes and sea kraits).

Not only are most snakes harmless to man, many are actually beneficial. The most obvious group in New Guinea are the pythons. They feed primarily on small mammals and particularly on rodents such as rats and mice. Being 'nature's mouse-trap', a python should be welcomed during its nocturnal hunts around houses and out-buildings but instead it is usually killed. Another example might be the freshwater keelback, a small rough-scaled snake found in road-side creeks and damp areas which is often killed as a suspected 'Pap black'. The keelbacks as a group are of Oriental origin having spread to New Guinea from S.E. Asia and speciated into over a dozen different species. In common with certain snakes and other animals of Australasian origin they eat frogs. For many frog-eating species in Australia and New Guinea the introduction of the cane toad from S. America in the 1930's as a crop-pest control for sugar cane was a disaster since native amphibian predators could not cope with the powerful 'bufotoxin' which exuded from the enlarged parotoid glands on the rear of the toad's head. Snakes, goannas and marsupials used to preying on native amphibians would die if they attempted to consume cane toads but the keelbacks appear to be able to prey on small cane toads without being poisoned. The keelbacks are of Oriental origin and their Asian relatives prey on species of toads which also produce 'bufotoxin' as a defence. Although the cane toad would appear to produce larger quantities of this poison, of higher toxicity, the keelbacks of Australasia may have inherited an apparent partial immunity from their Asian ancestors. However, keelbacks are small snakes, capable of predating only small toads. It is also unlikely that a few snakes could make any serious dent in the huge cane toad populations which exist on Daru Island, around Port Moresby or along the Madang coast, to name just three invasion points, although keelbacks remain a valuable ally in the control of these unwelcome colonisers.

The message is **do not kill harmless snakes**. Of course the killing of a dangerous species in the garden or around the house, especially where children might be playing, is understandable but the unnecessary killing of non-venomous pythons, boas, watersnakes, treesnakes etc. is undesirable.

At this juncture it would be a good idea to analyse why people dislike or fear snakes and put these fears into context.

In Western European society, where snakebites are rare and deaths though snakebite almost unheard of, the fear or loathing of snakes seems to centre around the misapprehension that snakes may be slimy and unpleasant to the touch.

In tropical countries where snakebite mortality is either a fairly common, or at least not an infrequent event, people avoid snakes for an entirely different reason: some snakes kill people!

Fear and loathing of snakes is further reinforced by the Bible which portrays the serpent as the villian, but not all religions despise the snake and a few actually revere and project it as sacred.

The human fear of snakes is again not helped by the fact that in certain areas, at certain times of the year, snakes may be very common. Nor does it help that snakes are virtually silent and tend to appear when and where they are least expected: something which tends to cause sudden panic.

Even so a fear of dangerous snakes is a rational fear and one I respect, unlike the irrational fear of the European. Such a fear is a very sensible survival factor and one to be encouraged amongst people who live in close proximity to dangerous species. However, I try to channel this fear into what I term a "healthy respect" for snakes which basically entails leaving them alone.

It is usually argued that people do not know how to tell dangerous snakes from harmless snakes. This is the very reason for this book and I hope anyone reading it will, in future, be able to differentiate between which snakes are dangerous and which are potentially beneficial, or at least of no threat to humans.

For food

Occasional snakes, even pythons, killed in the bush to feed a hunter or his family are 'fair game' and the impact of their loss on the wild population is negligible. However consideration of a few guidelines could ensure that the populations are not adversely affected i.e.

– try not to take gravid (pregnant) females or collect eggs;
– do not hunt threatened species like pythons intensively or for commerical purposes such as the sale of 'bush meat';
– do not take protected species such as the Boelen's python when other species are available.

For fun

Sport hunting does not generally extend to snakes as they are probably not a very demanding quarry, but in the USA (Texas and neighbouring states) a particularly unpleasant form of 'sport' takes place each year - the Rattlesnake Round-up. It results in the slaughter of thousands of snakes and untold ecological damage.

This sort of activity couldn't happen in PNG could it - or could it ?

On numerous occasions as I drive the roads I have come across the aftermath of what appears to be some sort of game. The unwilling participant is a

The snake game: a 5m Papuan olive python dragged on to the Ramu Highway at Nadzab, Morobe Province.

snake, usually a large python although any moderately sized snake seems to suffice. The 'game' consists of noosing a python in the bush or in a garden, with a piece of string and after clubbing it, dragging it out and stretching it across a busy road. The other participants in the game watch as the snake is repeatedly run over by passing vehicles. The unnecessary and untimely termination of a valuable rat-eaters life!

Although a driver cannot be expected to swerve and endanger his passengers to avoid a snake in his path some people seem to get pleasure out of driving over snakes and PMV drivers have been seen, with their vehicles full of people, going backwards and forwards over some small carpet python.

For financial gain

Whether it be for skins, 'medicinal products', meat or the pet trade, few wild snake populations can stand commercial harvesting for long. Large pythons are especially at risk and the populations of pythons of some Asian and African countries have been severely depleted by skin hunters in modern times. Fortunately Papua New Guinea has an enlightened view regarding commercial exploitation of its wildlife, especially by outside commercial interests. The only use of snakeskin within PNG which springs immediately to mind is the use of Arafura file snake (*Acrochordus arafurae*) skin on Western Province kundu drums. Drums from other provinces usually use the skins of goannas (monitor lizards, *Varanus* spp., usually *Varanus indicus,* the spotted mangrove monitor) or green tree pythons (*Morelia viridis*). It should be realised that these two species, or their derivatives such as skins, are internationally protected and subject to export and import restrictions even as tourist souvenirs, (*see* CITES below) so whilst you can own one and keep it on your wall in PNG you might get into trouble if you try to import it into your home country. In PNG such a limited use of skins for village drums is unlikely to affect local populations provided they are not produced in excessive commercial numbers for export.

Conservation of snakes

People generally think of conservation as referring to tigers, pandas, elephants or birds of paradise - large, emotive, awe-inspiring or colourful creatures and generally either covered in fur or feathers. It is a great deal easier to raise public awareness to the plight of these 'popular' animals than it is to the conservation of scaly creatures like snakes, lizards or crocodiles. A sceptical public will ask, why conserve snakes ?

Snakes have a part to play in the natural history, not least as biological controls on the populations of smaller creatures such as rats or bandicoots or wallabies. If the ecology of a region is an effective, well-oiled machine then snakes comprise several components in that machine, as predators of small animals and prey for some larger creatures. Take away the snakes and you risk breaking the machine or at least causing it to function less effectively in the future.

With this book you should now have the means to determine whether a snake is dangerous or not. If you see a snake in the garden or in the bush don't reach for the nearest heavy object to kill it, reach for this Guide and find out whether it is indeed a dangerous species. In many Australian towns and cities there are groups of experienced snake experts who are quite prepared to be called out to capture and relocate not only harmless species but also dangerous snakes. It may be many years before such a network as this could be established in Port Moresby, Lae or Madang but there are people in PNG who are interested in snakes and both my publisher and I, and several PNG based friends have been called out on numerous occasions to capture, remove and relocate snakes from a variety of domestic situations. Many village communities have a 'snake man' or magic man who may also be able to take snakes away.

However, if the reader can learn to accept snakes it is quite likely that he/she may also learn to appreciate and live alongside them. Give snakes a chance!

CITES and conservation

Many potentially threatened species of animals and plants are protected by an international body called CITES - the Convention for the International Trade in Endangered Species of Wild Fauna and Flora. CITES was formed in the early 1970's and PNG ratified the treaty and joined in 1976. The aim of CITES is to monitor, control and, in some cases, restrict the international trade in certain threatened or endangered life forms. Those PNG species of reptiles protected by CITES are listed below. Note that the CITES Appendix designations listed for species here are for PNG populations - *Crocodylus porosus* elsewhere in its range falls into Appendix I and certain Asian countries apply a third designation, Appendix III, to their populations of snakes which are undesignated in PNG i.e. Indian populations of *Cerberus rynchops* (Colubridae: Homalopsinae). [Source: Groombridge 1988].

Appendix I (Endangered, trade prohibited)
Family Chelonidae (sea turtles)

Caretta caretta	Loggerhead turtle
Chelonia mydas	Green turtle
Eretmochelys imbricata	Hawksbill turtle
Lepidochelys olivacea	Pacific Ridley's turtle
Natator depressa	Flatback turtle

Family Dermochelyidae (leatherback seaturtle)

Dermochelys coriacea	Leatherback turtle

Appendix II (Threatened, limited trade with a license)
Family Crocodylidae (crocodiles)

Crocodylus novaeguineae	New Guinea freshwater crocodile
Crocodylus porosus	Indo-Pacific saltwater or estuarine crocodile

Family Scincidae (skink lizards)

Corucia zebrata	Solomons monkey-tail skink lizard

Family Varanidae (monitor lizards or goannnas) (Robert Sprackland *pers. comm.*)

Varanus beccarii	Aru Black tree monitor
Varanus bogerti	Bogert's black tree monitor
Varanus doreanus	Blue-tailed monitor
Varanus finschi	Bismarck mangrove monitor
Varanus gouldii	Gould's sand or argus monitor
Varanus indicus	Pacific mangrove monitor
Varanus jobiensis	Peach-throated or Sepik monitor (formerly *V. karlschmidti*)
Varanus prasinus	Emerald tree monitor
Varanus salvadori	Tree crocodile monitor
Varanus similis	Spotted monitor (formerly synonym of *V. timorensis*)
Varanus telenesetes	'Rossel Is. tree monitor'

Family Boidae (boas and pythons)

Candoia aspera.	New Guinea ground boa
Candoia carinata	Pacific tree and ground boas
Apodora papuana	Papuan olive python
Bothrochilus boa	Bismarck ringed python

Leiopython albertisi	D'Albertis white-lipped python
Liasis fuscus	Brown water python
Morelia amethistina	Amethystine python
Morelia boeleni	Boelen's python
Morelia spilota	Carpet python
Morelia viridis	Green tree python

Quite often foreign residents, returning to their countries of origin, will seek to take with them their 'pet' python, usually the green tree python (*Morelia viridis*). Be aware that to export such an animal the exporter/owner will require an import licence from their home country. This import permit will usually only be issued following prior sight of the export documentation. Papua New Guinea generally does not allow the export of its wildlife for purposes other than genuine research so obtaining an export license may be a very time-consuming process. Also, some countries do not allow the importation of CITES animals from the Indo-Australian region and Australia generally does not allow the importation of non-Australian fauna for private ownership.

Quarantine laws which are applied in the UK for mammals and birds do not apply to reptiles but laws in other countries may be different.

For the private individual who has a pet snake, and who is leaving PNG, it would be a lot easier for them to release the specimen where they found it or pass their pet on to the National Museum of PNG (which has a live snake collection) or another resident with an interest in, and a knowledge of, reptiles and their captive requirements.

In addition to the international CITES regulations Papua New Guinea enforces its own laws relating to the possession of certain native species. It is well known that it is illegal for a non-National to even possess a Bird of Paradise feather. It is less well known that the beautiful Boelen's python, which occurs in scattered locations along the central New Guinea mountain chain, is subject to the same national regulations which makes possession of the species, or its derivatives such as its skin, by a non-National a criminal offence. This is the most protected reptile in PNG and since specimens are found fairly infrequently (*see* Request for Information at the back of this book,) this is one species which should certainly not be killed.

SNAKE BITE AND ITS TREATMENT IN PAPUA NEW GUINEA

David A Warrell and David G Lalloo

Centre for Tropical Medicine, University of Oxford, UK and
Faculty of Medicine, University of Papua New Guinea

Medically–important venomous snakes in PNG

Death adders (genus *Acanthophis*) are widely distributed in the coastal areas and highlands including the Ramu Valley. They have been blamed for isolated fatal cases of envenoming in many parts of PNG, including remote tribes in the Highlands (reported by anthropologists) and may be responsible for the majority of snake bites in Irian Jaya. On the Papua New Guinea side, taipans (*Oxyuranus scutellatus canni*) are now by far the most important cause of snake bite in Central Province, with death adders responsible for perhaps 10–15% of bites and Papuan blacksnakes (*Pseudechis papuanus*) reliably implicated in only about 5% of cases. There has been a dramatic change since the 1950s and 1960s when C.H. Campbell blamed Papuan blacksnakes for the majority of cases of envenoming admitted to Port Moresby General Hospital. The small–eyed snake (*Micropechis ikaheka*), which is widely distributed in PNG, was proved to be responsible for one fatal case and has been implicated in a number of other cases of severe envenoming, especially around Madang and Kar Kar Island. The eastern brown snake (*Pseudonaja textilis*) has been identified as a cause of a few bites as have the whip snakes (*Demansia* spp.) and the colubrid brown treesnake (*Boiga irregularis*). Sea snakes, especially the banded sea krait *Laticauda colubrina*, are abundant around the shores of Papua New Guinea and a few bites and fatalities have been attributed to sea snake bites in fishermen, including bites by *Enhydrina schistosa*, 100 km up the Ramu River.

Incidence of snake bite

Although Gajdusek (1977) reported that snake bite was the commonest cause of death in some Fore villages in the Highlands and fatal snake bite is a familiar problem in almost all parts of the mainland, the burden of snake bite morbidity and mortality does not now seem to be heavy except in Central Province. In the 1960s C.H. Campbell (1966; 1967; 1969) reported an average of 20 cases of snake bite per year admitted to Port Moresby General Hospital; in 1986–8, Currie reported 86 cases per year and from 1989 to 1993 Lalloo and Trevett recorded 139 per year. Although these total numbers are not large compared with other parts of the world, the proportion of severely envenomed cases is remarkable; 36% of envenomed patients required ventilation.

Clinical features of snake bite

Circumstances of the bites: In Papua New Guinea, patients are usually bitten while walking through long (kunai) grass or on marshy ground. As a result, they very rarely see the snake responsible and are practically never able to kill it and bring it to hospital. In Lalloo *et al's* series of 335 snake bites, only 3.6% brought the dead snakes, in the remainder the diagnosis was confirmed by a specific enzyme immuno-assay. Eighty–six per cent of bites were inflicted on the feet or lower legs.

SYMPTOMATOLOGY

TAIPAN (*Oxyuranus scutellatus canni*). Usually, there is little pain or swelling at the site of the bite (Fig. 1), but there may be persistent bleeding from the fang marks. Painful enlargement of local lymph nodes may develop within the next 1–2 hours. The commonest early symptoms of systemic envenoming are vomiting and syncope. Patients may collapse unconscious as soon as one minute after the bite; patients may recover after a few minutes as with a simple faint or may remain unconscious for 30 minutes or more.

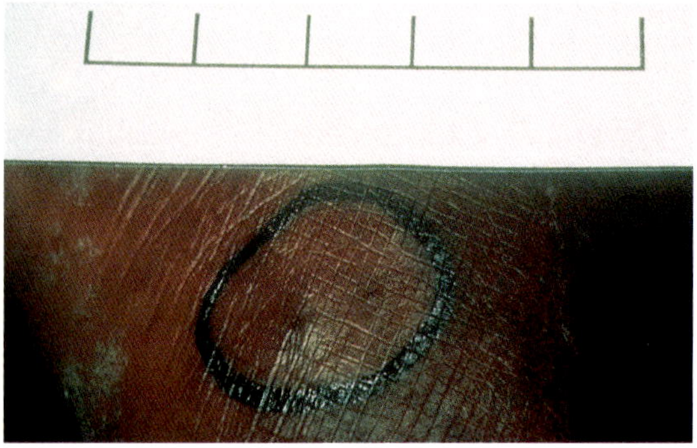

Fig 1 : Papuan taipan bite. At the site of the bite there are two fang marks, in this case 1 cm apart, with no local swelling (scale in cm). © DA Warrell

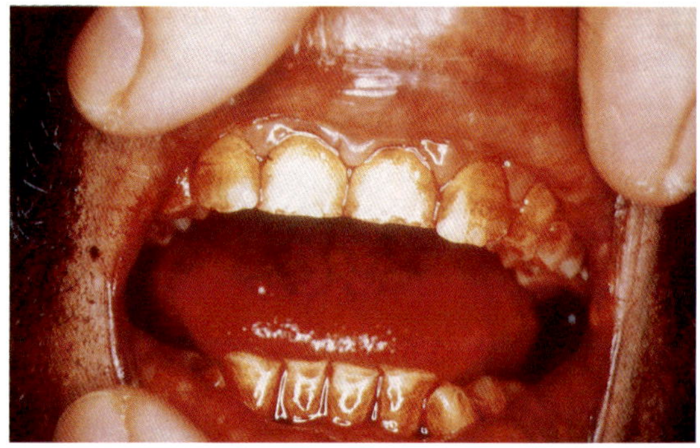

Fig 2 : Spontaneous systemic bleeding from the gums in a patient envenomed by a Papuan taipan. © DA Warrell

Symptoms of incoagulable blood (bleeding from old wounds and sites of new trauma such as venepunctures) and of haemorrhaging activity (bleeding from gums (Fig. 2), nose and blood in vomitus, urine or stool) may develop within a few hours of the bite. Other systemic symptoms include headache, dizziness, weakness, epigastric and muscle pains. Symptoms of neurotoxicity may appear within 30 minutes of the bite. Initially there is a heavy feeling of the eyelids, then blurring of vision, frank drooping of the eyelids (ptosis), paralysis of eye movements (Fig. 3) and difficulty in speaking, protruding the tongue, opening the mouth fully and swallowing saliva. Eventually, there may be paralysis of breathing with respiratory arrest, unless breathing is supported artificially, and generalised flaccid paralysis. Patients with extensive paralysis who may appear unconscious or even 'dead' may retain the ability to flex their fingers and toes on command. Renal failure occurs in a small percentage of patients.

DEATH ADDER (*Acanthophis* spp.). As with taipan bites, there are negligible local symptoms, but unlike these bites there is no persistent bleeding from the fang marks. Paralysis usually develops more rapidly and is more readily and rapidly responsive to antivenom and to anticholinesterase drugs. Symptoms of haemostatic disturbance weigh against the diagnosis of death adder bite.

PAPUAN BLACK SNAKE (*Pseudechis papuanus*). In the few authenticated cases of envenoming by this species, symptoms were milder but otherwise indistinguishable from those of taipan bite. There is slight local swelling. Vomiting, abdominal pain and tender enlargement of lymph nodes draining the bitten limb are followed by neurotoxic symptoms, spontaneous bleeding and, rarely, renal failure. The coagulopathy is mild compared with that in taipan victims. Thrombocytopenia is reported.

SMALL–EYED SNAKE (*Micropechis ikaheka*).
In the very few cases of proven *M. ikaheka* bites, symptoms included paralysis, bleeding and coagulopathy. Studies of the neurotoxins in *M. ikaheka* venom suggest that anticholinesterase drugs may be beneficial. Some physicians suspect that rhabdomyolysis may occur, resulting in myoglobinuria.

EASTERN BROWN SNAKE (*Pseudonaja textilis*). Symptoms are indistinguishable from those of taipan bites. They include neurotoxicity, severe haemostatic disturbance including marked thrombocytopenia and renal failure.

Bites by other species

In PNG, bites by brown treesnakes (*Boiga irregularis*) and Müller's snakes (*Aspidomorphus muelleri*) have caused mild pain and local swelling. In Guam more severe symptoms, including life–threatening

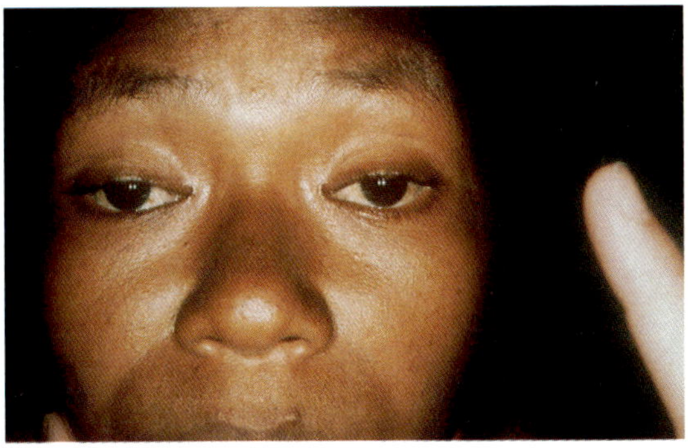

Fig 3: Ptosis and external ophthalmoplegia in a patient envenomed by a Papuan taipan. The eyelids are drooping (ptosis) and the patient cannot look to her left towards the doctor's finger. © DA Warrell

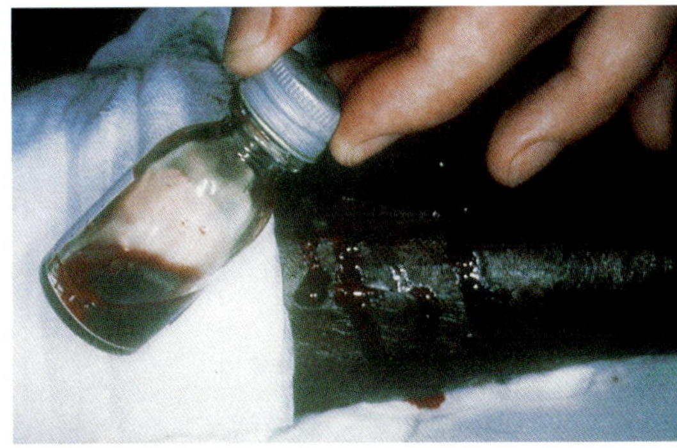

Fig 4: Twenty minute whole blood clotting test (WBCT20). © DA Warrell

neurotoxicity in small children, have resulted from bites by *B. irregularis.*

Sea snake bites

Familiar symptoms of sea snake bite have been described in Papua New Guinea – a painless bite with negligible local symptoms followed by progressive paralysis with muscle pain and tenderness and myoglobinuria.

Evolution and recovery from envenoming

Laboratory investigations

A neutrophil leucocytosis is commonly associated with significant systemic envenoming. In patients bitten by taipans, Papuan black snakes, small-eyed snakes and eastern brown snakes, the blood may become incoagulable within 30–60 minutes of the bite. The simplest way to assess coagulopathy is with the **20 minute whole blood clotting test** (WBCT20) (Fig. 4).

> A few ml of venous blood are placed in a clean, dry glass test tube and left vertical and undisturbed on the bench for 20 minutes. At the end of that time the tube is tipped. If there is defibrinogenation or if an anticoagulant/antiplatelet factor is present the blood will be liquid, clear evidence of systemic envenoming and an absolute indication for antivenom treatment.

Other tests of haemostasis may also be abnormal: prolonged prothrombin, thrombin, partial thromboplastin and bleeding times, decreased concentration of fibrinogen, other clotting factors and platelet count, increased concentration of fibrin(ogen) degradation products and other tests of fibrinolysis. Frank or occult blood may be detected in vomitus, urine or faeces. The haemoglobin may fall if there is bleeding. Intravascular haemolysis is most unusual; pink plasma suggests haemoglobinaemia; dark urine may be explained by haematuria, haemoglobinuria and/or myoglobinuria in the rare cases where there is rhabdomyolysis (as in the case of sea snake bite). Serum/plasma concentrations of muscle enzymes (creatine phosphokinase, aminotransferases) and myoglobin will be elevated. Transient electrocardiographic abnormalities (ST, T segment changes, evidence of septal ischaemia, bradyarrhythmias) are found in about 10% of cases of taipan bites and have been reported with eastern brown snake bites. Biochemical evidence of mild renal dysfunction may be observed in patients bitten by taipans and Papuan black snakes. Sea snake bite victims may develop acute tubular necrosis.

TREATMENT

First aid treatment: In PNG the most popular methods continue to be application of tourniquets, local incisions and the use of herbal remedies. Community education must discourage these practices which are ineffective, time–wasting and potentially dangerous. The principles of first aid are to transport snake bite victims as quickly and passively as possible to the nearest medical facility (dispensary, peripheral health station, district hospital etc); to attempt to delay the evolution of life–threatening envenoming at least until the patient is in the hands of medical personnel and to alleviate severe early symptoms of envenoming.

First aid must be carried out as soon as possible after the bite by the victim or by those who happen to be around at the time.

GENERAL RECOMMENDATIONS:

1. **Reassurance:** in PNG most snake bite victims will fear impending death after any snake bite and may also believe that the bite was inflicted through sorcery. Firm reassurance is essential with emphasis on the efficacy of modern medical treatments. Even with the most dangerous species of snake, not all bites result in the injection of significant amounts of venom. In PNG, at least one–third of patients bitten by venomous snakes, who have fang marks indicating that the fangs pierced the skin, will not develop significant envenoming.

2. **The wound site should not be tampered with, incised** (Fig. 5), **sucked, squeezed, massaged or rubbed.**

3. **Immobilise** the bitten limb with a splint or sling.

STRUAN SUTHERLAND'S PRESSURE IMMOBILISATION METHOD ("AUSSIE BANDAGE")

This first aid method has gained great popularity in Australia and some other countries. In animal studies it was shown to prevent the systemic spread of venoms of Australasian elapids and other venomous snakes. No formal clinical trials have been carried out but clinical experience of the method over more than ten years suggests that it is safe (unless the compressing bandage is applied too tightly or left in place for an unnecessarily long period) and seems to delay the onset of systemic envenoming. **The method is recommended for the first aid of bites by Australasian elapids.** Two practical difficulties limit the usefulness of this method in developing countries. First, the cost of providing long crepe or other stretchy bandages to the poor rural people most at risk and secondly, the difficulty/impossibility of the victim's applying pressure immobilisation to a bitten upper limb if they are alone at the time of the accident.

Technique (see page ii). The end of a long (2 metres) stretchy, crepe, elasticated or Biscard bandage, 2–3" wide is placed over the fang marks and applied round the limb as shown in the figures, incorporating a makeshift splint (e.g. a piece of wood or other rigid material). The whole limb is finally encased. The binding should be tight but not so tight that (i) a finger cannot be inserted easily beneath the bandage or (ii) the peripheral pulse (at the wrist or ankle) is occluded and cannot be felt.

4. **Transport to a medical facility: exercising or movement of the bitten limb should be avoided.** This increases the systemic absorption of venom. Ideally the snake bite victim should be carried on a stretcher and transported by boat or motor vehicle. In countries like PNG which have a limited road

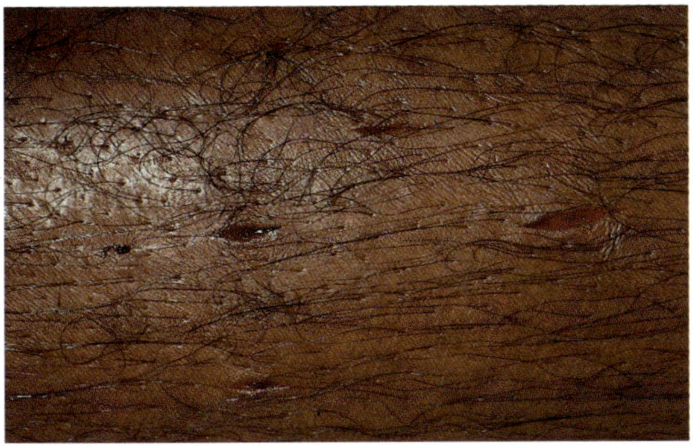

Fig 5: Razor-cut incisions are frequently made by friends and family in the mistaken belief that the venom will be 'bled out'. This is a dangerous procedure, especially in the case of snake venoms causing continual bleeding, i.e. taipan or blacksnake, since it may lead to excessive blood loss and shock.

Fig 6: Death adder victim being transported to medical care on a makeshift stretcher.

Photograph courtesy of James Menzies, University of Papua New Guinea

network and where large areas of the country remain relatively inaccessible, there may be great difficulties in transporting the patients to medical care (Fig. 6).

5. **If the snake has been killed**, take along its dead body which is very valuable evidence for the medical staff. However, pursuing the snake and trying to kill it after a bite is **not recommended !** Dead snakes should be handled with great care and carried on a stick, in a plastic bag or a box. They should not be touched with bare hands as even a dead snake can bite by reflex.

6. **Avoid traditional first aid methods !** Tight tourniquets, cauterization, incision, excision or amputation of the bite site, suction by mouth, vacuum pump or syringe, combined incision and suction, injection or instillation of chemicals such as potassium permanganate, application of electric shocks, ice packs, traditional herbal or folk remedies, are all useless and potentially dangerous.

The use of tourniquets (tight belts or bands round the upper part of the bitten limb) is potentially dangerous as, by depriving the limb of its blood supply for a long period, it may become gangrenous. In practice, such tourniquets, applied by country people, are generally ineffective.

First aid treatment of early systemic envenoming

Patients bitten by Papua New Guinean elapids may develop the following symptoms before they reach hospital.

1. Pain – this may be severe especially if the bite is in a finger or toe. **Treatment** – paracetamol tablets by mouth. Aspirin is not recommended as it may lead to gastrointestinal bleeding.

2. Nausea, retching, vomiting and abdominal pain. These symptoms may be attributable to systemic envenoming or to emetic herbal remedies taken by mouth. There is a risk that the patient may inhale their vomit, especially if drowsy and developing paralysis of the swallowing and breathing muscles. **Treatment** – chlorpromazine or some similar anti–emetic phenothiazine given by mouth, suppository or by injection. [1]

[1] In patients with incoagulable blood, intramuscular injection may lead to development of large haematomas in the muscle. In these patients, injections should be given intravenously (but not in the case of adrenaline).

3. Collapse, fainting/syncope. An early and sometimes sustained fall in blood pressure may cause the patient to collapse unconscious. **Treatment** – lay the patient flat, raise the legs, make sure that the airway is clear. 0.1% (1 in 1000) adrenaline/epinephrine can be given intramuscularly or subcutaneously if available in a dose of 0.5–1.0 ml for adults or 0.01 mg/kg for children.

4. Early bulbar and respiratory paralysis. If swallowing becomes difficult, saliva and other secretions begin to pool in the throat and there is difficulty clearing the throat, coughing and taking deep breaths, the patient must be laid on their side and the airway must be kept clear. The jaw should be elevated to prevent the tongue blocking the throat and, if available, an oral airway should be inserted. If breathing movements become weak and the patient's lips and tongue appear blue (cyanosed), oxygen should be given if available and mouth–to–mouth artificial respiration should be considered.

Medical treatment in hospitals or dispensaries

The three most urgent questions to be asked of a victim of snake bite are

1. Where (in which part of your body) were you bitten?
2. How long ago were you bitten?
3. Have you brought the snake and, if not, can you describe it?

Details of the circumstances of the bite and the symptoms evolving since the bite should be recorded. Physical examination should include the following.

1. *Signs of haemostatic disturbances*: examine the gums, nose, sputum, vomitus, urine and stool for evidence of bleeding. If the blood is incoagulable patients may bleed at the site of chronic ulcers, recent healing wounds and scabs and venepunctures.

2. *Neurotoxicity*: Ask the patient to follow your finger upwards. Ptosis (bilateral paralysis of the muscles which elevate the eyelids) is an early sign of neurotoxicity. The patient will be unable to retract the eyelid when they look upwards; as a result the pupil will become hooded. Later the eyes may be paralysed so that they cannot follow a finger to the sides or up and down. Sequential paralysis of the muscles which open the mouth and cause swallowing, protrusion of the tongue, speaking and finally breathing will follow.

3. *Other signs*: Tender, enlarged lymph nodes draining the bitten limb (for example the groin for bites on the lower limb, in the axilla for bites in the upper limb) indicate systemic spread of venom. Papua New Guinean elapids cause little or no local swelling at the site of the bite. There may be persistent bleeding. Blood pressure should be checked. In Papua New Guinea, systemic envenoming by taipans, Papuan black snakes and eastern brown snakes causes incoagulable blood. The quickest and simplest bedside test for detecting severe coagulopathy or the presence of a potent venom anti–platelet/anticoagulant factor is the **20 minute whole blood clotting test** (see page 25).

Diagnosis of the biting species

1. Identification of the dead snake brought by the patient (see species account sections of this book).

2. Commonwealth Serum Laboratories snake bite diagnosis kits (Fig. 7). This cleverly designed kit can produce a positive result indicating one of the four genera/species: taipan, Papuan black snake, death adder or eastern brown snake. Wound swab (taken from close to the fang sites), wound aspirate (from the fang marks using a 25 gauge needle), urine or serum can be used. In the case of Papuan taipan bites, a positive result may come up within a minute of adding the sample to the kit.

3. Patients' or bystanders' description of the snake may sometimes be helpful. Ask for a description of the snake if it was seen. Do not suggest particular colours or markings by direct questioning. Get the witnesses to volunteer the details. Useful descriptive points are the orange or dorsal stripe of the taipan, the short, squat build with thin tail and large, well demarcated head of the death adder and the whitish appearance of the small–eyed snake. Unfortunately, in PNG, few patients catch more than a fleeting glance of the snake and descriptions tend to be unreliable and misleading. On the southern side of PNG, most biting snakes are described as "Pap blacks" or by the local names (such as "auguma" in Mekeo). Taipans, whip snakes and even harmless keelbacks and treesnakes are commonly described as "Pap blacks".

Antivenom treatment

The most important decision in the management of a snake bite patient is whether or not to give antivenom.

Indications for antivenom treatment

1. Neurotoxic symptoms (ptosis etc).
2. Haemostatic disturbances (spontaneous systemic bleeding for example from gums or nose, non–clotting blood in the WBCT20).
3. Shock (low blood pressure).
4. Impaired consciousness of any cause.
5. Passage of red, brown, black or "Coca–cola" coloured urine.

Choice of antivenom

In PNG, the following antivenoms are sometimes available, all from Commonwealth Serum Laboratories, Parkville, Victoria, Australia.

1. Polyvalent (Fig. 8)
2. Taipan
3. Death adder
4. Black snake

The geographical distribution of the important venomous snakes in PNG may be helpful. Thus, there are no taipans or Papuan black snakes reported from Highland or Momase (Morobe, Madang and Sepik) Provinces. Only death adders and perhaps small–eyed snakes are found in the highland areas. If you are confident (from the geographical distribution, description of the snake and clinical features of envenoming) of the species diagnosis, choose a monospecific antivenom (eg. taipan, death adder). If you are **uncertain** choose polyvalent antivenom.

Initially, one ampoule of antivenom should be administered by slow intravenous injection or diluted in a convenient volume of isotonic fluid, by intravenous infusion over about 30 minutes.

There is always a risk of antivenom reactions which are likely to be especially severe if the patient suffers from an allergic/atopic disease such as asthma, hay fever or eczema, or if they have reacted before to equine antiserum (for example, anti–tetanus serum or antivenom).

Premedicate with an antihistamine such as promethazine ("Phenergan") or chlorpheniramine ("Piriton"). Always have adrenaline (0.1% 1 in 1000) ready drawn up in case of acute anaphylactic reactions to antivenom. Most antivenom reactions begin within 30 minutes of starting intravenous antivenom treatment. Early reactions consist of itching, urticaria, tachycardia, vomiting, fever, bronchospasm and shock. Reactions are readily reversed with intramuscular or subcutaneous adrenaline (see previous page) followed by parenteral antihistamine (for example intravenous chlorpheniramine maleate).

Late antivenom reactions which resemble serum sickness may develop 5–12 days later and consist of recurrent urticaria, swelling of the joints, lymphadenopathy, fever, proteinuria and sometimes neurological symptoms. Treat with antihistamine, or, if symptoms persist or are severe, with corticosteroids.

Monitoring the dose of antivenom

Repeat the dose of antivenom if:

1. the blood remains incoagulable (WBCT20) 6–12 hours after the first dose;
2. if neurological symptoms progress;
3. if shock persists (unless of course this shock was the result of an antivenom reaction).

Ancillary treatments

Neurotoxic envenoming – if possible, all patients with neurotoxic signs should be tested with anticholinesterase drugs, ideally with the "Tensilon test" also used in patients with suspected myasthenia gravis. Atropine sulphate (0.6 mg for adults, 50 mg/kg for children) is given first by intravenous injection followed by edrophonium chloride (10 mg in adults, 0.25 mg/kg in children) by slow intravenous injection. If there is a convincing improvement in muscle power (such as ability to retract the eyelids when looking upward or the strength of grip etc) after about 10 minutes, patients should be maintained on neostigmine. Experience suggests that most death adder victims will show rapid improvement in their neurotoxic signs following anticholinesterase, but that taipan victims do not respond.

Despite antivenom and anticholinesterases, many patients may progress to respiratory paralysis and will require **assisted ventilation**. In hospitals, the indications for endotracheal intubation or tracheostomy are inability to clear secretions from the throat (bulbar paralysis) and respiratory failure. Mechanical or manual ventilation (using an Ambu anaesthetic bag) for 1–3 days usually allows patients to recover from the paralytic effects of snake venoms. Renal failure may develop in patients with severe envenoming. In PNG it is rarely severe enough to require more than conservative treatment.

Fig 7: Commonwealth Serum Laboratory snake venom detection kit - the test will identify the venom of taipan, blacksnake, brownsnake or death adder within a few minutes. One kit is designed to test three snakebite cases and costs approximately A$300.

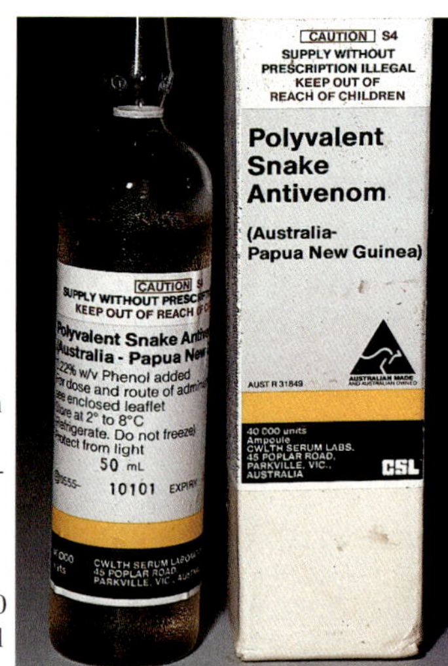

Fig 8: Commonwealth Serum Laboratory Polyvalent (Australia - New Guinea) Antivenom in 50ml ampoules (1991 price per ampoule A$807.30 - Theakston & Warrell 1991).

HOW TO USE THE SPECIES ACCOUNTS

All the species accounts adhere to the following format. Important species such as the pythons and the highly dangerous species are dealt with as individual accounts whilst groups of closely related species; keelbacks, treesnakes, burrowing snakes etc. are dealt with as species groups (i.e. genera or subfamilies) with shortened individual accounts providing details of variation within each group. All scientific terminology included in these accounts is explained in the Glossary at the end of the book.

COMMON ENGLISH NAME

Where these are available common English names are included. In some instances a snake may be known by more than one common name and a decision as to which one to use has been made. For instance the amethystine python is also known in PNG as the giant python (which may be confused with the Papuan olive python) and in Australia at the scrub python. Since the name amethystine python is the more internationally accepted name this is the name used here. Although the PNG name for the carpet python is Moran this name has also been omitted as it is frequently used for any python. In the case of the D'Albertis python the name white-lipped python is also widely used and both are included.

For many snakes occurring in PNG, endemic blindsnakes, keelbacks, ground snakes, secretive elapids and the seasnakes, no recognised common names exist. Since this Guide is designed to be user-friendly for non-herpetologists or non-scientifically trained persons who find Latin names difficult I have sought to provide common names for these snakes from a number of sources as demonstrated below. (Note: standardisations for minor elapid common names were arrived at in collaboration with David & Ineich.)

a) From other literature sources - Ehmann (1992) provides names for all his Australian species and some of the seasnakes are also applicable to PNG i.e. Reef shallows seasnake for *Aipysurus duboisii*;
b) From the general geographical distribution of the species i.e. 'Northern New Guinea blindsnake', *Ramphotyphlops erycinus*, 'Southern New Guinea blindsnake', *R. polygrammicus* or 'Bismarck ground snake', *Stegonotus heterurus*;
c) From the distribution of an apparent local endemic i.e. 'Setekwa River forest snake' for *Toxicocalamus grandis,* 'Mt Rossel forest snake' for *T. holopelturus* or 'Mamberamo River watersnake' for *Heurnia ventromaculata*;
d) From some characteristic of the species (sometimes reflected in the scientific name) i.e. 'Yellow-bellied blindsnake', *Ramphotyphlops flaviventer*;
e) From the name of the herpetologist or collector honoured in the scientific name i.e. 'Buerger's Forest snake', *Toxicocalamus buergersi* or 'Pruess's Sepik snake', *T. preussi preussi*.

It will be noticed that all names coined by myself in b, c, d, and e, have been placed in inverted commas. This will enable the reader to different between recognised common names and those first published in this book.

Scientific name – References to name changes are included in brackets.

VENOMOUS

Each account is assigned to one of the following:
Non venomous - no venom glands or fangs. However, most species possess teeth and will bite if molested. Bites from large pythons can be painful and bloody. The top right hand corner of accounts in this category are colour-coded GREEN (except the pygopodid snake-lizards which are coded YELLOW to distinguish them from snakes).
Mildly venomous - species with fangs positioned in the back of the upper jaw, rear-fanged, and possessing

venom glands known as Duvernoy's glands. Large specimens are capable of biting a human but the venom is not generally considered dangerous; unless the bite is to an infant. (*see* Venom section in Brown treesnake account). Colour-coding for this group is ORANGE.

Venomous - the smaller elapid species possessing fixed fangs in the front of the upper jaw and venom glands but generally considered either inoffensive to man or possessing weak venoms or mouths too small to deliver a serious bite to a human. However, certain of the larger species in this group should be treated with caution eg. *Aspidomorphus muelleri*, *Demansia* spp. and *Toxicocalamus loriae*. The top right corners are colour-coded RED.

Highly venomous - snakes with fixed fangs in the front of the upper jaw and venom glands which are capable of delivering serious or fatal snakebites. The colour-coding for these snakes is RED with a skull & crossbones to indicate HIGHLY DANGEROUS.

Average/Maximum length – Average and known maximum lengths for the species if known.

Description

a) **physique:** The overall shape of the snake i.e. whether the body is short and stumpy, slender and elongate; the shape of the head and whether it is distinct from the neck or the same width; the length of the tail; size of the eye and any other relevant characteristics.

b) **colouration:** Colouration and patterning of the body and head including geographical or other variations.

c) **scalation:** Important scale counts (*see* diagrams) eg. scales around the dorsum of the body, whether keeled or smooth; ventral scale count from throat to anal plate; condition of anal plate and number and condition of scales under the tail; number of supralabial scales, including those contacting the eye, and presence or absence of other relevant scales. DMB = dorsal scales at midbody; V = ventral scales; SC = subcaudal scales.

Habitat – The preferred habitat or habitats of the species concerned.

Habits – Notes on the behaviour, temperament, activity cycle, mode of reproduction and ecology of the species.

Diet – Food taken by adults and juveniles.

Taxonomic note – Discussions on the validity of particular species or any recent name changes. Primarily of interest to herpetologists rather than field naturalists.

Seas (Sea kraits and seasnakes only) – The general marine distributional spread of the species.

Distribution within PNG

The known distribution of the species, or group of related species, within the borders of Papua New Guinea. It should be realised that many provinces have been poorly sampled and the distribution maps may be improved by local contributions (*see* A Plea for Information).

Key to Province abbreviations used: (see page 199 for map of Papua New Guinea)

CENT:	Central Province	GULF:	Gulf Province
E.HIGH:	Eastern Highlands Province	MAD:	Madang Province
ENB:	East New Britain Province	MAN:	Manus Province
ENGA:	Enga Province	M.BAY:	Milne Bay Province
E.SPK:	East Sepik Province	MOR:	Morobe Province

NCD:	National Capital District	SIMBU:	Simbu (or Chimbu) Province
N.IRE:	New Ireland Province	WEST:	Western Province
N.SOL:	North Solomons Province	W.HIGH:	Western Highlands Province
ORO:	Oro (or Northern) Province	WNB:	West New Britain Province
S.HIGH:	Southern Highlands Province	W.SPK:	West Sepik (or Sandaun) Province

Distribution outside PNG – The distribution of PNG species, or closely related species or subspecies, outside the country.

The remaining three subsections are only included in the front-fanged venomous accounts.

Confusing species – The elapid accounts include this section which indicates species which may possibly be confused with the account species. Refer to the descriptions in the accounts mentioned.

Venom – The composition, effects, yield and toxicity of the species' venom is recorded together with notes on the potential danger posed by the species.

Antiserum/initial dose – The relevant Commonwealth Serum Laboratory (Melbourne, Victoria) antivenom and the initial dose required.

PASIN BILONG KISIM SAVE LONG DISPELA BUK

Ol sinek igat tupela nem bilong em. Wanpela nem yumi save kolim olgeta taim. Narapela nem i olsem teknikal nem, bilong ol man husat i wok long kisim save long pasin bilong sinek. Tupela nem istap wantaim antap stret long olgeta pes bilong dispela buk.

Tasol, igat wanpela isi pasin bilong painim sinek long buk. Lukluk tasol long piksa bilong ol sinek istap wantaim tok save bilong pasin bilong em.

Dispela toksave long sinek igat wanwan kala bilong em, olsem:

Non-venomous:	(Yelo kala) Em olsem, bai yu ken save dispela em i no sinek, em i olsem lizad husat inogat leg bilong em. (Grin kala) Sinek hia em inogat poison bilong em.
Mildly venomous:	(Orenj kala) Dispela sinek igat poison, tasol poison bilong em i liklik, na poison inogat sitrong tumas long em.
Venomous:	(Redpela kala) Kain sinek olsem igat poison bilong em.
Highly venomous:	(Redpela kala wantaim hetbun bilong man) Dispela sinek igat olsem traipela poison nogut tru.

Hap bilong ol sinek

Sapos yu laik save klia long as ples bilong ol sinek insait long PNG, yu mas lukluk gut long mep namba istap ananit long olgeta nem bilong sinek, insait long dispela buk. Mep namba bai inap soim yu long olgeta hap sinek bai istap long em. Kisim dispela mep namba, na painim mep insait long buk.

Wokim olsem bai yu ken save sapos wanwan kain sinek istap namel long hap bilong yu.

IDENTIFICATION KEY TO THE SNAKE FAMILIES OF PAPUA NEW GUINEA

(incorporating legless lizards)

The keys in this Guide are intended to aid the reader in establishing the identity of any snake encountered in Papua New Guinea. The keys are known as dichotomous keys since each section is a 'couplet' or 'dichotomy' comprising two statements, one of which should apply to the specimen being examined. An example might be:

1a	Dorsal scales at midbody number more than 19	go to couplet 2
1b	Dorsal scales at midbody number 19 or fewer	go to couplet 3

Depending on the number of dorsal scales present at midbody the reader is referred to either couplet '2' or '3' and so on until the snake has been identified either to genus or species level. The reader should then refer to the relevant species account in the main body of the book. Diagrams illustrating head and body scalation accompany the key to assist the non-herpetological reader with identification.

Warning: Before using this key the possibilities that the snake to be identified might be venomous and therefore potentially dangerous must be considered and all precautions taken to prevent a bite. It must be remembered that apparently 'dead', even decapitated, venomous snakes, can and have delivered fatal bites when incorrectly handled. For persons, health workers etc, who find they are frequently called upon to identify unknown snakes, the provision of several pieces of clean transparent plastic fuel piping or hose of various widths would be a wise option. A 'suspect' snake could be 'posted' headfirst into the pipe and gripped to prevent its escape. In this position it can be safely examined quite closely.

A snake should be considered highly dangerous if it exhibits one or more of the following characters:
a) a flattened 'paddle-like' tail;
b) a pair of fangs in sheaths of skin in the front of the upper jaw;
c) raised 'horn-like' supraocular scales over the eyes (fig. 9);
d) the loreal scale is absent (figs. 7 - 9).

THE LOREAL SCALE

The scales on the heads and bodies of snakes, and lizards, are not arranged indiscriminately. There are recurring arrangements and patterns of scales which are common to particular groups of related species. Each scale, especially the enlarged 'scutes' of the head, has a name (figs. 1 - 4) and the presence or absence, or size and position of an individual scale, or the number of scales in a particular series, serve as important aids in the identification of specimens.

One of the problems with snake identification in Australia and New Guinea is the innocuous appearance of some of the most dangerous species. Whereas, in other parts of the world, snakes such as the puff adder, rattlesnakes and cobras are immediately recognisable, many of the dangerous snakes of Australia and New Guinea resemble harmless species in their general appearance. Probably only the seasnakes and sea kraits, with their flattened 'paddle-shaped' tails, and the short stumpy death adder, with a raised 'horn-like' supraocular scale above each eye, are instantly recognisable as dangerous.

However, there is one scale, the presence or absence of which can be used with almost complete certainty to differentiate between venomous front-fanged **elapids** (the dangerous terrestrial snakes of Australasia) and the harmless species. This scale is known as the **loreal scale** and it is the most important scale for snake identification in Australia and New Guinea. The loreal scale is shaded in figure 2.

Three conditions exist:

A: **The loreal scale is present.** By its presence the loreal scale prevents contact between the nasal scale and the preocular and also between the prefrontal and the supralabials (figs. 2 & 5). This condition is found in all the harmless snakes in PNG with the exception of a single species of mangrove snake, *Fordonia leucobalia* (fig. 6) from southern and southwestern coastal PNG. Most harmless snakes possess a single loreal scale on each side of the head although in some of the pythons and boas the loreal scale is fragmented into numerous small granular scales (fig. 12).

B: **The loreal scale is absent.** The preocular scale is in contact with the nasal scale preventing contact between the prefontal* and the supralabial scales (fig. 7). This is the usual condition found in the elapids, the family of front-fanged venomous snakes which includes some of the most dangerous species in PNG.

C: **The loreal scale is absent.** The prefrontal scale is in contact with at least one supralabial scale preventing contact between the preocular and the nasal scales (fig. 8). This condition is less commonly found in New Guinea elapids being present in *Furina tristis*, *Rhinoplocephalus boschmai* and some specimens of *Toxicocalamus loriae* and *Parapistocalamus hedigeri*.

* It should be noted that some species in the diminutive burrowing elapid genus *Toxicocalamus* have the prefrontal and preocular scales fused into a single scale (*see Toxicocalamus* account). However, these snakes, although technically venomous, are not considered dangerous to man.

BASIC SNAKE HEAD SCALATION

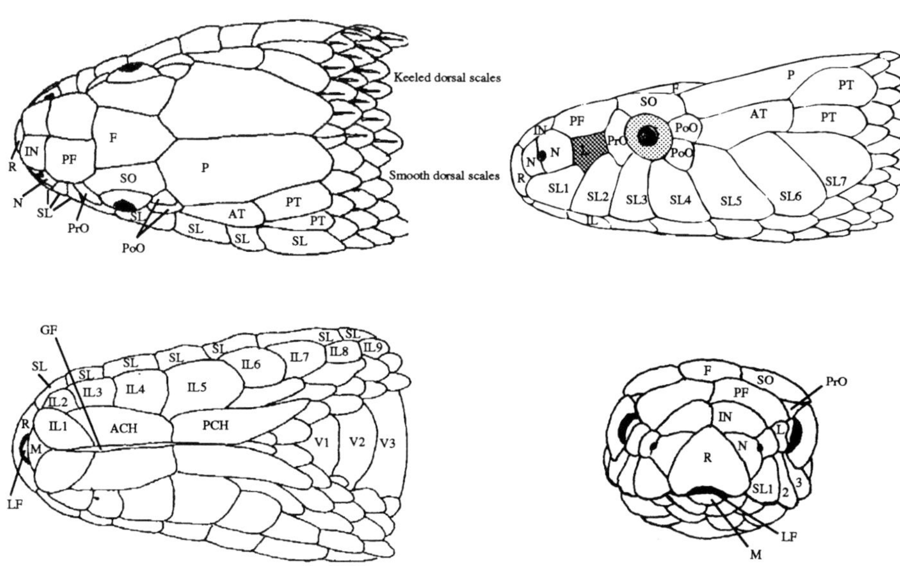

KEY: ACH - anterior chin shield. AT - anterior temporals. F - frontal. GF - gular fold. L - loreal (shaded in Fig 2). IL - infralabials. IN - internasal. LF - lingual fossa. M - mental. N - nasal. P - parietal. PCH - posterior chin shield. PF - prefrontal. PoO - postoculars. PrO - preocular. PF - prefrontal. PT - posterior temporals. R - rostral. SL - supralabials. SO - supraocular. V - ventrals.

COLUBRID SNAKE HEAD SCALATION

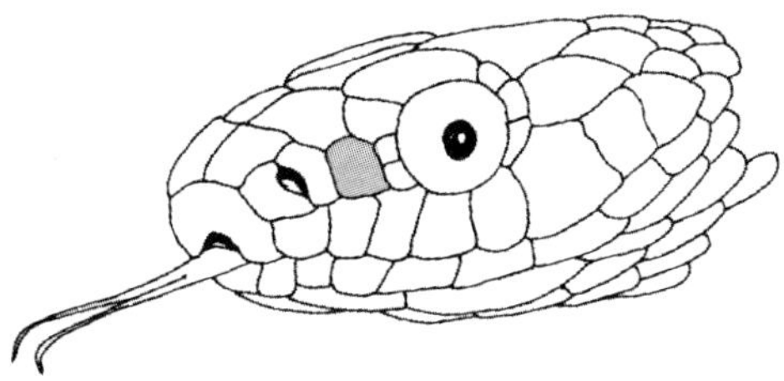

Fig 5: *Tropidonophis statasticus* - a keelback. Typical colubrid head scalation consisting of large regular scales including the loreal scale (shaded). Notice tongue protruding through lingual fossa without the requirement to open the mouth.

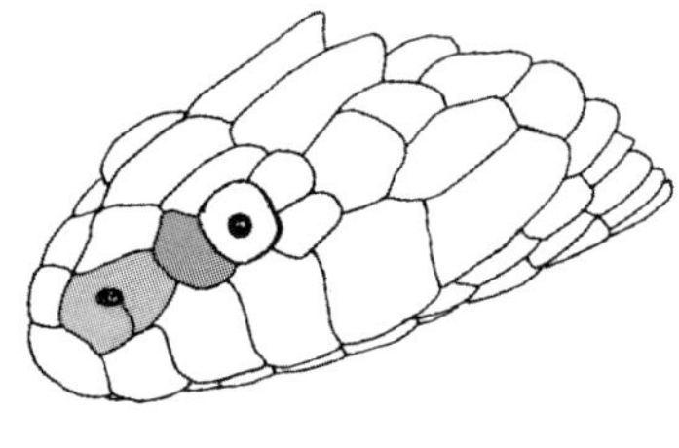

Fig 6: *Fordonia leucobalia* - a mangrove snake. A typical colubrid head scalation resembling an elapid in the absence of the loreal scale and contact between the preocular and nasal scales (shaded).

ELAPID SNAKE HEAD SCALATION

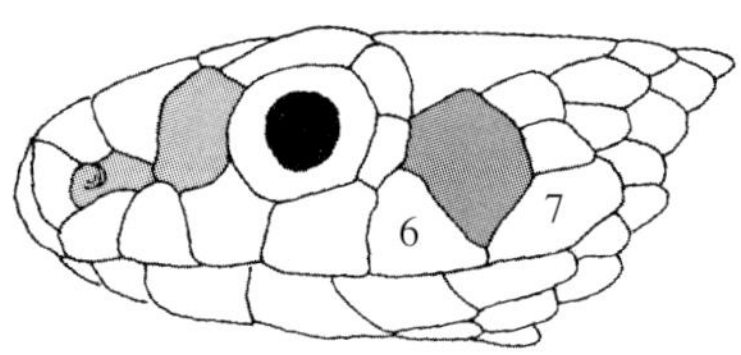

Fig 7: *Demansia atra* - a whipsnake. Typical elapid head scalation consisting of large regular scales, but lacking a loreal scale allowing contact between the preocular and nasal scales (light shaded). Also typical elapid temporalabial scale (dark shaded) between 6th and 7th supralabials but excluded from lip (absent in *Pseudonaja*).

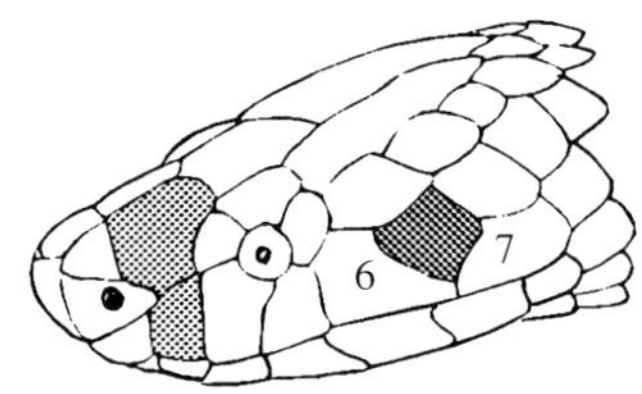

Fig 8: *Furina tristis* - a brown headed snake. Less typical elapid head scalation, loreal absent but prefrontal and 2nd supralabial scales in contact (light shaded), rather than preocular and nasal scales (Fig 7). Typical elapid large temporalabial scale present (dark shaded).

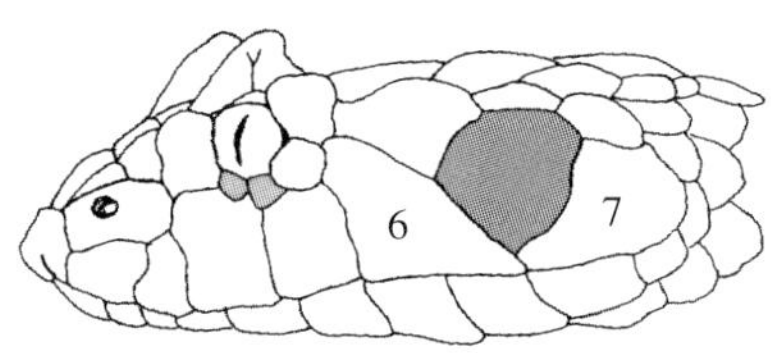

Fig 9: *Acanthophis praelongus* - a death adder. A typical elapid head scalation, large temporalabial scale present (dark shaded) between 6th and 7th supralabials and loreal absent but 'viperine' superoculars raised into 'horn like' projections over the eyes, subocular scales present (light shaded) beneath eye, and eye with vertically elliptical pupil.

SNAKE-LIZARD HEAD SCALATION

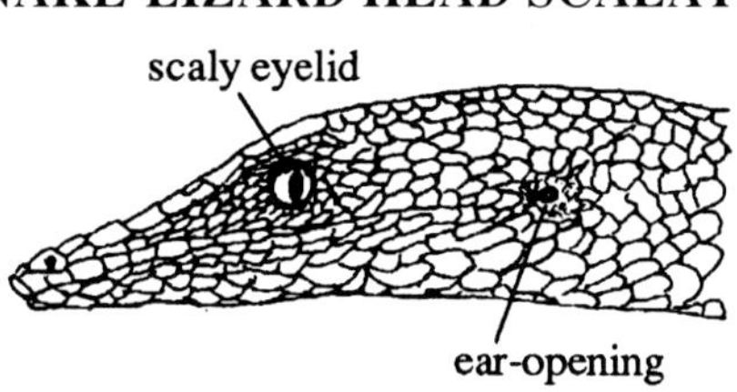

Fig 10: *Lialis burtonis* - a snake-lizard. Lizard characters include scaly eyelids and external ear-openings Pygopodids possess elongate jaws.

BLINDSNAKE HEAD SCALATION

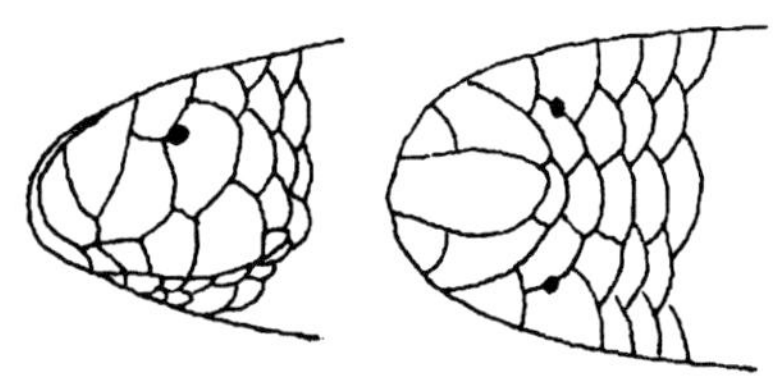

Fig 11: *Ramphotyphlops braminus* - a blindsnake. Notice large head scales followed by cycloid body scales and eyes reduced to pigmented areas beneath ocular scales.

PYTHON HEAD SCALATION

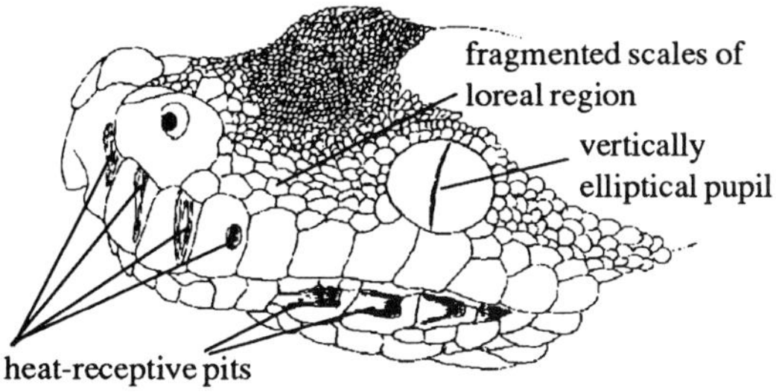

Fig 12: *Morelia viridis* - a python. Most head scales are fragmented and granular in the pythons *M. viridis* and *Morelia spilota*, and also the boas *Candoia* spp. The pupil of the eye is vertically ellipitical. Pythons also have distinctive heat-sensitive pits on the supralabials and infralabials.

FILE SNAKE HEAD SCALATION

Fig 13: *Acrochordus arafurae* - a file snake.
Nostrils forward facing and surrounded by numerous granular scales.

SEA KRAIT & SEA SNAKE HEAD SCALATION

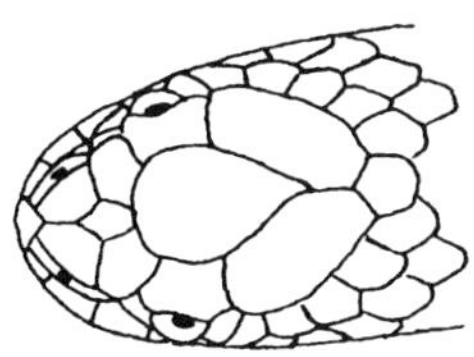

Fig 14: *Laticauda colubrina* - a sea krait.
Nasal scales (shaded) separated by internasal scales.

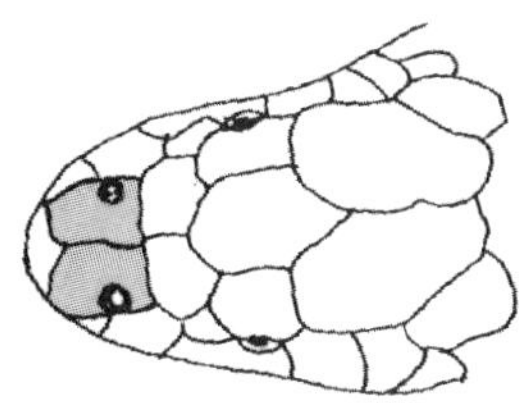

Fig 15: *Hydrophis elegans* - a sea snake.
Internasal scales absent, nasal scales (shaded) in contact.

BASIC SNAKE BODY & TAIL SCALATION

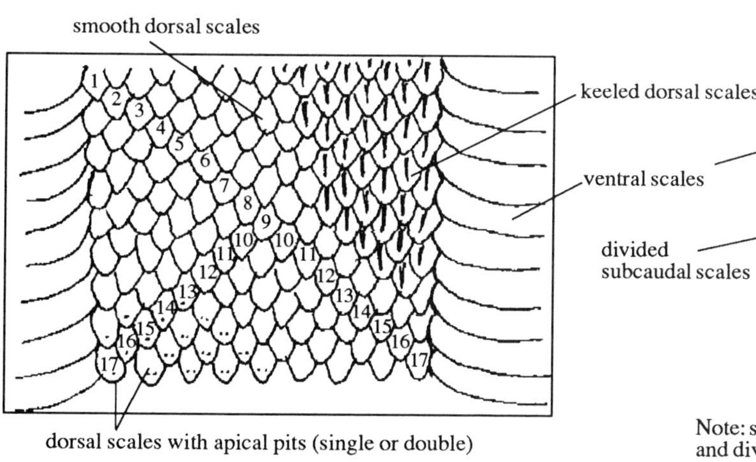

Fig 16: Basic snake body scalation illustrating two methods of counting the DMB (dorsals at midbody).

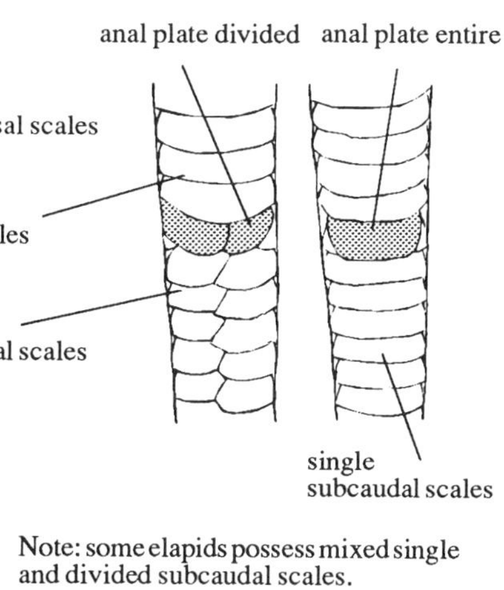

Fig 17: Basic snake belly and tail scalation (anal or cloacal plate shaded).

OTHER SNAKE TAILS (not to scale)

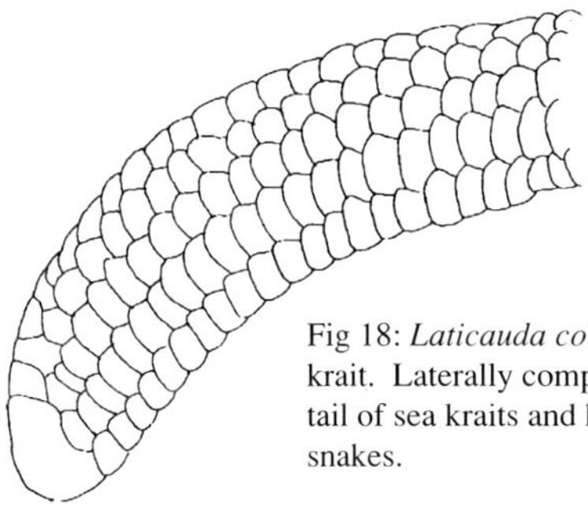

Fig 18: *Laticauda colubrina* - a sea krait. Laterally compressed paddle-like tail of sea kraits and hydrophiid sea snakes.

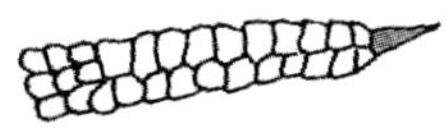

Fig 19: *Acanthophis* sp. a death adder.
Slightly compressed yellow or white tail tip of death adder exhibiting a short terminal spine (shaded).

Fig 20: *Typhlops* sp. - a blindsnake. Short rounded tail of blindsnake exhibiting a short terminal spine (shaded).

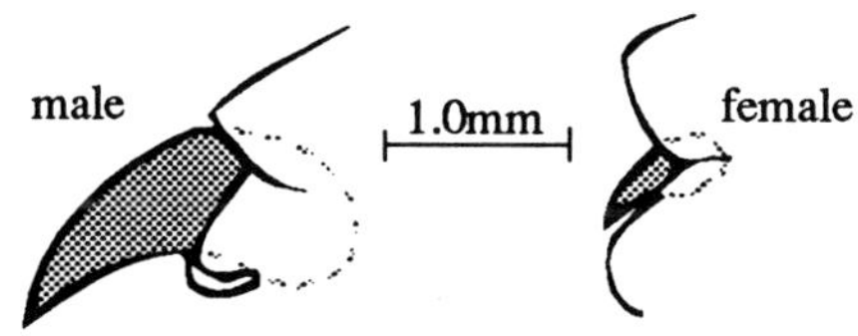

Fig 21: *Candoia carinata* - a boa (after Stickel & Stickel, 1945). Large cloacal spurs are present on either side of the cloacal plate in all male boas, *Candoia* spp., and pythons. Spurs are reduced in size in female pythons and Irian Jayan populations of *C. carinata*, reduced and confined to one side (usually the right) in female *C. carinata* from Bougainville and completely absent in all female *C. aspera*, and female *C. carinata* from mainland PNG and the Bismark Archipelago (McDowell, 1979).

KEY TO THE FAMILIES

1a. The head is very long, slender and pointed; ear openings are present (fig. 10); the tongue is not forked. **Snake-lizards:** Family Pygopodidae

1b. The head is not slender and pointed; ear openings are absent; the tongue is forked. 2

2a. A small, 'worm-like' snake, the body is slender and glossy; the belly scales are small and indistinguishable from those of the rest of the body; the tail terminates suddenly in a sharp, harmless spine (fig. 20); the head is small, rounded and not distinct from body; the eyes appear only as pigmented spots (fig. 11); burrowing terrestrial species. **Blindsnakes:** Family Typhlopidae

2b. Not worm-like in appearance, with discernible head and eyes and a gradually tapering or flattened tail. 3

3a. The skin of the body is loose and 'flabby', the scales of the head and body are numerous, raised and tuberculate; the ventral scales no larger than the scales on the rest of the body; the eyes are very small and the nostrils are positioned dorsally, on the top of the head (fig. 13). **File snakes:** Family Acrochordidae

3b. The skin of the body is not flabby and loose, the dorsal scales are normal, smooth or keeled, not tuberculate; the ventral scales are either much wider than the scales of the rest of the body (fig. 18) or, if they are the same size as the dorsal scales, the tail is 'paddle-shaped'; the eyes may be large or small and the nostrils positioned dorsally or laterally. 4

4a. The tail is laterally compressed, flattened and 'paddle-shaped' (fig. 18); the snake is marine, estuarine or occasionally freshwater in habit. 5

4b. The tail tapers gradually and is not 'paddle-shaped'. 7

5a. Ventral scales as broad as (or almost as broad as) ventral scales of terrestrial snakes (Colubridae or Elapidae). 6

5b. Ventral scales as broad as (or fractionally broader than) adjacent lateral scales of body. **Seasnakes:** Family Hydrophiidae

6a. Nasal scales in contact on top of head (fig. 15). **Seasnakes:** Family Hydrophiidae

6b. Nasal scales separated by internasal scales (fig. 14). **Sea kraits:** Family Elapidae

7a. The loreal scale is absent allowing contact between the preocular and nasal scales (fig. 7) or between the prefrontal and supralabial scales (fig. 8). 8

7b. A single loreal scale is present (figs. 2 & 5) or the loreal scale may be fragmented into numerous granular scales (fig. 12) which prevent contact between the prefrontal and supralabial scales and between the preoculars and nasal scales. 9

8a. Dorsal scales arranged in 23-29 rows at midbody (fig. 16); the eye is small; temporalabial scale absent (fig. 6) - *Fordonia* mangrove snake. **Harmless snakes:** Family Colubridae

8b. Dorsal scales arranged in 15-23 rows at midbody (fig. 16), if in 23 rows then either with supraocular scale raised into a 'horn-like' projection over the small eyes (fig. 17) or shelved supraocular over large eyes which are largely obscured from above; temporalabial scale present (figs. 7 - 9). **Venomous land snakes:** Family Elapidae

9a. Dorsal scale rows arranged in 15-29 rows at midbody; loreal scale single; scales on top of head always composed of enlarged 'scutes' (fig. 1); heat sensitive pits and cloacal spurs always absent; pupil of eye round or vertically elliptical. **Harmless snakes:** Family Colubridae

9b. Dorsal scale rows arranged in more than 30 rows at midbody; loreal scale either single or fragmented into granular scales (fig. 12); scales on top of head either enlarged 'scutes' or fragmented into numerous granular scales; heat sensitive pits may be present in the lip scales (fig. 12); cloacal spurs may be present either side of the cloaca (fig. 21); pupil of eye always vertically elliptical (fig. 12). **Boas & Pythons:** Family Boidae

Keys to the Families are to be found on the following pages:

CHECKLIST OF SNAKES FROM NEW GUINEA

including type specimen collection localities and museum accession numbers

A full checklist of the snakes from the island of New Guinea (including Irian Jaya) and the Admiralty Islands, Bismarck Archipelago, Louisiade and d'Entrecasteaux Archipelagoes and North Solomons Province to the east of the PNG mainland is provided below. One Solomons Island species from close proximity to Bougainville is also included.

Latin binomials, and subspecies trinomials, where recognised, are provided followed by the original author and publication date.

English common names are included to assist persons unfamiliar with Latin binomials. Common names in inverted commas are not formally recognised names but rather suggested names coined either from the distribution of the species or through the Anglicisation of the Latin binomial. Some of these names were arrived at in collaboration with P. David and I. Ineich (*pers. comm.* & *in prep*).

The type locality is the collection locality for the 'type' or first specimen/s of the species. Several of the older collection localities are fairly vague.

The type number is the museum accession number for the original type specimen. Some of the type specimens have been either lost or destroyed. Museum acronyms follow the checklist.

SPECIES AUTHOR AND COMMON NAME	TYPE LOCALITY	TYPE NO. [(?) = number not found]
FAMILY **TYPHLOPIDAE** (blindsnakes)		
Genus ***Ramphotyphlops*** (Australasian blindsnakes)		
R. braminus (Daudin, 1803) Brahminy blindsnake	Vizagapatam	P.43 in Russell, 1796
R. erycinus (Werner, 1901) 'Northern New Guinea blindsnake'	German New Guinea (= ne PNG)	(presumed lost)
R. flaviventer (W. Peters, 1865) 'Yellow-bellied blindsnake'	Ternate, Indonesia	ZMB 5029
R. leucoproctus (Boulenger, 1889) 'White-tailed blindsnake'	Murray Is., Torres St., Aust.	BMNH 1946.1.10.58-60
	Fly River, Western Province	BMNH 1946.1.11.84
	Qld, Australia	BMNH 1946.1.11.97
R. multilineatus (Schlegel, 1839) 'Hook-nosed blindsnake'	New Guinea (Irian Jaya)	MNHN 1067
R. polygrammicus (Schlegel, 1839) 'Southern New Guinea blindsnake'	Timor, Indonesia	RMNH 3712
R. subocularis (Waite, 1897) 'Bismark sharp-nosed blindsnake'	Duke of York Is., ENB. Prov.	AMS 2202
Genus ***Typhlops*** (Tropical blindsnakes)		
T. ater (Schlegel, 1839) 'Black blindsnake'	Java, Indonesia	RMNH 3714
T. depressiceps (Sternfeld, 1913) 'Lowland beaked blindsnake'	New Guinea (PNG no exact locality)	ZMB 23986
T. diardii (Schlegel, 1839) Diard's blindsnake	East Indies	MNHN 1065
T. inornatus (Boulenger, 1888) 'Montane blindsnake'	Sogere (=Sogeri), Central Province	BMNH 1946.1.11.80

[Note : Wallach 1995, *in press* & Wallach & Wynne, *in prep.* considerably increases the typhlopid fauna of New Guinea - see Addenda for explanations, descriptions, keys and checklist]

SPECIES AUTHOR AND COMMON NAME	TYPE LOCALITY	TYPE NO.
FAMILY **BOIDAE** (boas and pythons)		
Subfamily **Boinae** (boas)		
Genus ***Candoia*** (Pacific boas)		
C. aspera (Günther, 1877) New Guinea ground boa	Duke of York Is., East New Britain	BMNH 1946.1.10.33
C. a. 'aspera' (Günther, 1877) (New Britain race)	as above	as above
C. a. 'schmidti' (Stull, 1932) (mainland race)	Kaiserin-Augusta (=Sepik) River	MCZ 29778
C. carinata (Schneider, 1801) Pacific boa	no record of type locality	NMHG (?) (lost)
C. c. 'carinata' (Schneider, 1801) (tree boa phase)	as above	as above

Taxon	Type locality	Type specimen
C. c. 'paulsoni' (Stull, 1956) (ground boa phase)	Ugi Is., Solomon Islands	MCZ 14521
Subfamily **Pythoninae** (pythons)		
Genus ***Apodora*** (Papuan python)		
*A. papuana (*Peters & Doria, 1878) Papuan olive python	Ramoi, nr. Sorong, Irian Jaya	MSNG 29988
Genus ***Bothrochilus*** (ringed python)		
B. boa (Schlegel, 1837) Bismarck ringed python	New Ireland, no exact locality	MNHN 7172
Genus ***Leiopython*** (white-lipped python)		
L. albertisi (Peters & Doria, 1878) D'Albertis python	Andai, nr. Dorei, Irian Jaya	MSNG 29989
	Kapoar, Onin Peninsula, Irian Jaya	MSNG 29990
Genus ***Liasis*** (Australian water and olive pythons)		
L. fuscus (Peters, 1873) Brown water python	Port Clinton, Qld, Australia	ZMB 7840 (lost)
Genus ***Morelia*** (carpet and scrub pythons)		
M. amethistina (Schneider, 1801) Amethystine or scrub python		
	no type locality given	ZMU 1485
M. a. amethistina (Schneider, 1801) N.G. amethystine python	as above	as above
M. boeleni (Brongersma, 1953) Boelen's python	Dimija, Wissel Lakes, Irian Jaya	RMNH 9651
	Okaitadi, Irian Jaya	RMNH 9652
M. spilota (Lacépède, 1804) Carpet python	New Holland (=Australia)	MNHN 3272
M. s. varigata (Gray, 1842) Carpet python	Port Essington, N.T., Australia	BMNH iv.1.2a
M. viridis (Schlegel, 1872) Green tree python	Aroe (=Aru) Is., Indonesia	RMNH 4672

FAMILY **ACROCHORDIDAE** (filesnakes)

Taxon	Type locality	Type specimen
Genus ***Acrochordus*** (file or wart snakes)		
A. arafurae (McDowell, 1979) Arafura filesnake	Lake Daviumbo, Western Province	AMNH 59887
A. granulatus (Schneider, 1799) Little filesnake	not found	(presumed lost)

FAMILY **COLUBRIDAE** (treesnakes, watersnakes and ground snakes)

Taxon	Type locality	Type specimen
Subfamily **Boiginae** (rear-fanged venomous snakes)		
Genus ***Boiga*** (catsnakes)		
Boiga irregularis (Merrem, 1802) Brown catsnake	not found	(presumed lost)
Subfamily **Colubrinae** (typical snakes)		
Genus ***Dendrelaphis*** (treesnakes)		
D. calligastra (Günther, 1867) Green treesnake	Cape York, Qld, Australia	BMNH 67.5.6.71
D. gastrostictus (Boulenger, 1894) 'Montane treesnake'	Northwest New Guinea	BMNH 1946.1.23.20
D. lorentzi (van Lidth de Jeude, 1911) 'Lorentz River treesnake'		
	Sabang, Lorentz River, Irian Jaya	RMNH 4710
D. papuensis (Boulenger, 1895) Papuan treesnake	Trobriand Is., Milne Bay Prov.	BMNH 1946.1.6.57-61
D. punctulatus (Gray, 1827) Common or black treesnake	Careening Bay, Western Aust.	BMNH 1946.1.23.34
D. salomonis (Günther, 1872) 'Solomons treesnake'	Solomon Islands	BMNH 1946.1.5.97
	Solomon Islands	BMNH 1946.1.6.11
Subfamily **Homalopsinae** (water, mud & mangrove snakes)		
Genus ***Cantoria*** (watersnakes)		
Cantoria annulata (de Jong, 1927) Banded watersnake	Prinz Frederik Hendrik Is., (=Kolepom)	
	Irian Jaya	ZMA 11065
Genus ***Cerberus*** (watersnakes)		
Cerberus rynchops (Schneider, 1799) Bockadam	Ganjam, India (presumed lost)	(lost)
C. r. novaeguineae (Loveridge, 1948) New Guinea bockadam		
	Merauke, Irian Jaya	MCZ 22818
Genus ***Enhydris*** (watersnakes)		
Enhydris polylepis (Fischer, 1886) Smooth watersnake	Fly River, Western Province	MTKD D437 (lost)
Genus ***Fordonia*** (mangrove snakes)		
Fordonia leucobalia (Schlegel, 1837) White-bellied mangrove snake		
	Timor, Indonesia	RMNH 1161

Genus ***Heurnia*** (watersnakes)		
Heurnia ventromaculata (Jong, 1926) 'Mamberamo River watersnake'		
	Pionierbivak, Mamberamo River, Irian Jaya	ZMA 11066
Genus ***Myron*** (mangrove snakes)		
Myron richardsoni (Gray, 1849) Richardson's grey mangrove snake		
	Northwest Australia	BMNH 1946.1.2.43
Subfamily **Lycodontinae** (wolf-toothed snakes)		
Genus ***Stegonotus*** (ground snakes)		
S. cucullatus (D. B. & D., 1854) Slatey-grey snake	Dore, Japen Is., Indonesia	MNHN 3412
S. diehli (Lindholm, 1905) 'Diehl's little ground snake'	Bogadjim, Madang Prov.	MWNH 1244
S. guentheri (Boulenger, 1895) 'Milne Bay ground snake'	Fergusson Is., Milne Bay Prov.	BMNH 1946.1.11.37-9
S. heterurus (Boulenger, 1893) 'Bismarck ground snake'	New Britain	BMNH 1946.1.14.91
	Duke of York Is., East New Britain Prov.	BMNH 1946.1.14.95
	Duke of York Is., East New Britain Prov.	BMNH 1946.1.15.10
S. modestus (Schlegel, 1837) 'Northern New Guinea ground snake'		
	Ambon, Indonesia	RMNH 324
S. parvus (Meyer, 1875) 'Common ground snake'	Japen Is. (Jobi), Indonesia	(not found)
Subfamily **Natricinae** (keelback watersnakes)		
Genus ***Tropidonophis*** (Indo-Australian keelbacks)		
T. aenigmaticus (Malnate & Underwood, 1988) 'East Papuan keelback'		
	Fergusson Is., Milne Bay Prov.	AMNH 76633
T. dahlii (Werner, 1899) 'New Britain keelback'	Herbertshohe, Neu-Pommern (=N.Brit.)	NHMV 1899:373
T. doriae (Boulenger, 1897) Barred keelback	Moroka, Central Province	BMNH 97.12.10.116
	Haveri, Central Province	BMNH 97.12.10.117-118
T. elongatus (Jan, 1865) 'Moluccan keelback'	Ambon, Indonesia	RMNH 1085
T. hypomelas (Günther, 1877) 'Bismarck keelback'	Duke of York Is., East New Britain Prov.	BMNH 72.2.24.24
T. mairii (Gray, 1841) Common or Mair's keelback	New Holland (=Australia)	(lost)
T. m. mairii (Gray, 1841) Common keelback	as above	as above
T. m. plumbea (Macleay, 1877) Common keelback	Katow (=Mawatta), Western Prov.	AMS 31898-31901
T. mcdowelli (Malnate & Underw'd, 1988) 'Northern New Guinea keelback'		
	Camp 3, Otakwa River, Irian Jaya	BMNH 1913.11.1.100
T. montanus (van Lidth de Jeude, 1911) 'North Irian montane keelback'		
	Bivak V., Helwig Gebirge, Irian Jaya	RMNH 4704
T. multiscutellatus (Brongersma, 1948) 'Many-scaled keelback'		
	Alkmaar, Lorentz River, Irian Jaya	RMNH 8669
T. novaeguineae (van Lidth de Jeude, 1911) New Guinea keelback		
	Alkmaar, Lorentz River, Irian Jaya	RMNH 4702
T. parkeri (Malnate & Underwood, 1988) 'Highland keelback'		
	Waghi River, Eastern Highlands Prov.	CAS 139588
T. picturatus (Schlegel, 1837) Painted keelback	Lobo Bay, Irian Jaya	RMNH 1076
T. statisticus (Malnate & Underwood, 1988) 'PNG montane keelback'		
	Orumba, Eastern Highlands Prov.	MCZ 142422

FAMILY **ELAPIDAE** (front-fanged venomous terrestrial snakes)

Subfamily **Oxyuraninae** (Australasian venomous snakes)		
Genus ***Acanthophis*** (death adders)		
A. antarcticus (Shaw & Nodder, 1802) Common death adder	Australasia	(lost)
A. praelongus (Ramsay, 1877) Northern death adder	Somerset, Cape York, Qld, Australia	AMS 451
Genus ***Aspidomorphus*** (New Guinea crowned snakes)		
A. lineaticollis (Werner, 1903) 'Striped crowned snake'	New Guinea	(not found)
A. muelleri (Schlegel, 1837) Müller's crowned snake	Lobo, Triton Bay, Irian Jaya	(not found)
A. schlegeli (Günther, 1872) Schlegel's crowned snake	Misool Is., Indonesia	BMNH 1946.1.18.43

Taxon	Locality	Specimen
Genus ***Demansia*** (Australian whip snakes)		
D. atra (Macleay, 1884) Black whip snake	Ingham, Qld, Australia	AMS 5941 & 31920
D. papuensis (Macleay, 1877) Papuan whip snake	Hall Sound, PNG	AMS 31919
D. papuensis papuensis (Macleay, 1877) Papuan whip snake	as above	as above
Genus ***Furina*** (nape-banded snakes)		
F. tristis (Günther, 1858) Brown-headed or grey-naped snake	NE coast of Australia	BMNH 1946.1.18.99
Genus ***Loveridgelaps*** ('Solomons small-eyed snake')		
L. elapoides (Boulenger, 1890) 'Solomons small-eyed snake'	Florida Is., Solomon Is.	BMNH 1946.1.18.98
Genus ***Micropechis*** (New Guinea small-eyed snakes)		
M. ikaheka (Lesson *in* Duperrey, 1829) New Guinea small-eyed or ikaheka snake	Dore, Irian Jaya	MNHN 7669
Genus ***Oxyuranus*** (taipan)		
O. scutellatus (Peters, 1867) Coastal taipan	Rockhampton, Qld, Australia	ZMB 5883
O. scutellatus canni (Slater, 1956) Papuan taipan	Napa Napa, Port Moresby, NCD	NMV D8614
Genus ***Pseudechis*** (black snakes)		
P. australis (Gray, 1842) King brown or mulga snake	Port Essington, N.T., Aust.	BMNH 1946.1.20.39
P. papuanus (Peters & Doria, 1878) Papuan blacksnake	Yule Is., Central Province	MSNG 8018
Genus ***Pseudonaja*** (brown snakes)		
P. textilis (Duméril, B. & D., 1854) Eastern brown snake	Australia	MNHN 3944 (lost)
Genus ***Rhinoplocephalus*** (Australian small-eyed snakes)		
R. boschmai (Brongersma & Knaap van Meeuven, 1961) Carpentaria whip snake	Merauke, Irian Jaya	RMNH 10874
R. nigrostriatus (Krefft, 1864) Black-striped snake	Australia	ZMH 479 (lost)
Genus ***Salomonelaps*** ('Solomons coral snake')		
S. par (Boulenger, 1884) 'Solomons coral snake'	Faro Is. (=Fauro Is.), Solomon Is.	BMNH 1946.1.20.66
Genus ***Toxicocalamus*** ('New Guinea forest snakes')		
T. buergersi (Sternfeld, 1913) 'Buergers's forest snake'	German New Guinea	ZMB 25232
T. grandis (Boulenger, 1914) 'Setekwa River forest snake'	Launch camp, Setekwa R., Irian Jaya	BMNH 1946.1.18.34
T. holopelturus (McDowell, 1969) 'Mt Rossel forest snake'	Mt. Rossel, Rossel Is., Milne Bay Prov.	AMNH 76660
T. longissimus (Boulenger, 1896) 'Woodlark or Fergusson Island forest snake'	Woodlark Is., Milne Bay Prov.	BMNH 1946.1.18.92-93
T. loriae (Boulenger, 1898) 'Loria forest snake'	Haveri, Central Province	MSNG 29141
T. misimae (McDowell, 1969) 'Misima Island forest snake'	Misima Is., Milne Bay Prov.	AMNH 76684
T. preussi (Sternfeld, 1913) 'Preuss's forest snake'	Seleo Is., West Sepik Prov.	ZMB 23948
T. p. preussi (Sternfeld, 1913) 'Preuss's Sepik forest snake'		as above as above
T. p. angusticintus (Bogert & Matalas, 1945) 'Fly River forest snake'	Palmer Junct., Fly R., Western Prov.	AMNH 57512
T. spilolepidotus (McDowell, 1969) 'Spotted forest snake'	Purosa, nr. Okapa, Eastern Highlands	AMNH 85745
T. stanleyanus (Boulenger, 1903) 'Owen Stanley Range forest snake'	Dinawa, Owen Stanleys, Central Prov.	BMNH 1946.1.7.55
Subfamily **Elapinae** (Oriental coral snakes)		
Genus ***Parapistocalamus*** ('Bougainvillian coral snake')		
P. hedigeri (Roux, 1934) 'Bougainvillian or Hediger's coral snake'	Buin, s. Bougainville Is., N.Solomons Prov.	NMBA 11967
Subfamily **Laticaudinae** (Sea kraits)		
Genus ***Laticauda*** (sea kraits)		
L. colubrina (Schneider, 1799) Colubrine or yellow-lipped sea krait	not specified	ZMB 9078
L. laticaudata (Linnaeus, 1754) Common or blue-lipped sea krait	"in Indiis"	NHRM 87-88

FAMILY **HYDROPHIIDAE** (seasnakes)		
Subfamily **Ephalopiinae** (thick seasnakes)		
Genus ***Aipysurus*** (mighty-tailed seasnakes)		
A. duboisii (Bavay, 1869) Reef shallows seasnake	Lifou (= Lifu), Loyalty Is., New Caledonia	(not found)
A. eydouxii (Gray, 1849) Spine-tailed seasnake	Indian Ocean	BMNH 1946.1.6.86
A. laevis (Lacépède, 1804) Olive-brown seasnake	Locker Is., Western Australia	WAM 22384
Genus ***Hydrelaps*** (mangrove seasnakes)		
H. darwiniensis (Boulenger, 1896) Port Darwin seasnake	Port Darwin, NT., Australia	BMNH 1946.1.1.91-92
Genus ***Emydocephalus*** (turtle-headed seasnakes)		
E. annulatus (Krefft, 1869) 'Turtle-headed seasnake'	Australian Seas	AMS 454 & 6633
Genus ***Parahydrophis*** (mangrove seasnakes)		
P. mertoni (Roux, 1910) Arafura smooth or northern mangrove seasnake	Sungai Waskei, Wokam, Aru Is.	NMBA 6246
Subfamily **Hydrophiinae** (flat seasnakes)		
Genus ***Acalyptophis*** (spiny-headed seasnake)		
A. peronii (Dumeril, 1853) Spiny-headed seasnake	New Holland (= Australia)	MNHN 7711
Genus ***Astrotia*** (large-headed seasnakes)		
A. stokesi (Gray, 1846) Stokes's seasnake	Australian seas	BMNH 1946.1.17.12
Genus ***Disteria*** (seasnakes)		
D. kingi (Boulenger, 1896) Spectacled or King's seasnake	North Australia	BMNH 1946.1.10.10
D. major (Shaw, 1802) Olive-headed or greater seasnake	Indian seas (Indian Ocean)	BMNH 1946.1.9.24
Genus ***Enhydrina*** (common seasnakes)		
E. schistosa (Daudin, 1803) Common or beaked seasnake	Tranquebar, India	BMNH 1946.1.10.7
E. zweifeli (Kharin, 1985) 'Sepik or Zweifel's beaked seasnake'	off mouth of Sepik River, PNG	AMNH 104340
Genus ***Hydrophis*** (banded seasnakes)		
H. atriceps (Günther, 1864) Black-headed seasnake	Siam (= Thailand)	BMNH 1946.1.2.62
	Siam (= Thailand)	BMNH 63.9.29.5
H. elegans (Gray, 1842) Elegant or bar-bellied seasnake	Port Essington, N.T., Australia	BMNH 1946.1.3.89
H. gracilis (Shaw, 1802) Graceful small-headed or slender seasnake	not specified	BMNH 1946.1.17.37
H. macdowelli (Kharin, 1983) Small-headed or McDowell's seasnake	Northern Australian shelf	ZIL 19678
H. melanosoma (Günther, 1864) Black-banded or robust seasnake	not specified	BMNH 1946.1.10.6
H. ornatus (Gray, 1842) Ornate reef seasnake	not specified (Indian Ocean ?)	BMNH 1946.1.23.72
H. o. ornatus (Gray, 1842) Ornate reef seasnake	as above	as above
H. o. ocellatus (Gray, 1849) Reef seasnake	Australia	BM III.6.9.a
H. pacificus (Boulenger, 1896) Pacific seasnake	New Britain, PNG	BMNH 1946.1.10.14
H. vorisi (Kharin, 1984) Estuarine seasnake	New Guinea	AMNH 58869
Genus ***Lapemis*** (spine-bellied seasnakes)		
L. hardwickii (Gray, 1834) Hardwicke's spine-bellied seasnake	India	BMNH 1946.1.18.39
Genus ***Pelamis*** (pelagic seasnake)		
P. platurus (Linneaus, 1766) Yellow-bellied or pelagic seasnake	not specified	(lost)

Museum collection acronyms (from Leviton et al 1985):

AMNH	=	American Museum of Natural History, New York, USA
AMS	=	Australian Museum, Sydney, Australia
BMNH	=	British Museum (Natural History), London, UK
CAS	=	California Academy of Sciences, San Francisco
IRSNB	=	Institut Royal des Sciences Naturales de Belgique, Brussels, Belgium
MCZ	=	Harvard University, Museum of Comparative Zoology, Massachusetts, USA
MNHN	=	Museum National d'Histoire Naturelle, Paris, France
MSNG	=	Museo Civico di Storia Naturale di Genova, Italy
MTKD	=	Staatliches Museum für Tierkunde, Dresden, Germany
MWNH	=	Museum Wiesbaden, Natural History, Wiesbaden, Germany
NHMG	=	Naturhistoriska Riksmuseet, Goteborg, Sweden
NHMV	=	Vienna Museum (*see* NHMW)
NHMW	=	Naturhistorisches Museum, Wien, Austria
NHRM	=	Naturhistoriska Riksmuseet, Stockholm, Sweden
NMBA	=	Natuurhistorisches Museum, Basel, Switzerland
NMV	=	National Museum of Victoria, Melbourne, Australia
RMNH	=	Rijksmuseum van Natuurlijke Historie, Leiden, Holland
WAM	=	Western Australian Museum, Perth, Australia
ZIL	=	Zoological Institute, Academy of Sciences, St Petersburg (Leningrad), Russia
ZMA	=	Zoologisch Museum, Universität van Amsterdam, Holland
ZMB	=	Zoologisches Museum, Universität Humboldt, Berlin, Germany
ZMH	=	Zoologisches Institut und Museum, Universitat Hamburg, Germany
ZMU	=	Berlin Museum (*see* ZMB)

AUSTRALIAN SNAKE LIZARDS
Family Pygopodidae

The lizard family Pygopodidae is a family of approximately thirty species of legless, or almost legless, lizards endemic to the Australasian region. Most of the species are confined to Australia but the genus *Lialis* (two species) also occurs in New Guinea where one species, *L. jicari*, is an endemic and therefore the only pygopodid which does not occur in Australia. Pygopodids are believed to have evolved from the geckoes. Their limblessness is considered by many scientists to be an advanced trait which is mirrored by similar evolutionary transitions in a) the skink lizards from long-legged species to more advanced reduced limbed or limbless species, and possibly b) from the monitor lizards to the snakes.

SNAKE-LIZARDS or SCALY-FEET
Lialis spp.

(2 species) Map: 01

NON-VENOMOUS
Average length: 0.5-0.8m.
Description:
a) physique: Extremely elongate 'snake-like' lizards with tails which are, if complete, 1-1.5x longer than the body; **hindlimbs present as vestigal 'flaps' or 'scaly feet';** forelimbs absent; head only slightly indistinct from neck; snout pointed and elongate, especially in *L. jicari,* with fine 'forceps-like' jaws; eyes small; **ear opening present.**
b) colouration: Variable: grey, brown, buff or even pinkish with, or without, darker longitudinal stripes and fine dark speckling; lighter below.
c) scalation: Highly variable and species specific (*see* key to the species and individual accounts below).

Habitat: All habitats with loose soil and ground vegetation, especially common in gardens.
Habits: Snake-lizards are easily distinquished from snakes through the presence of a) external ear openings; b) vestigal hindlimbs reduced to small scaly flaps; c) elongate forceps-like jaws and d) a short notched, rather than forked tongue. Unlike any species of New Guinea snakes they also practice caudal autotomy (voluntary tail loss) when molested and specimens with truncated or regenerating tails are common. Their inoffensive nature earns them the colloquial name of 'friendly snakes' and being more rigid than snakes they are also called 'pencil snakes'. They swallow their prey whole, usually head first, and are capable of taking larger prey animals, in relation to their own body size, than similarly sized snakes. Scientifically referred to as pygopodids, snake-lizards are considered to be closely related to geckos. They are crepuscular or nocturnal and they can vocalise. These lizards are oviparous with a clutch size of two.
Diet: Small lizards such as skinks of genus *Carlia*, which are killed and consumed whole. Other prey recorded, from Australian *L. burtonis*, includes agamid lizards, geckos, a snake and pygopodid lizards of genus *Delma*. *L. jicari* has been recorded taking reptile eggs.
Distribution within PNG: Throughout the coastal provinces of PNG and also a few highland and insular areas.
Distribution outside PNG: Irian Jaya and neighbouring islands, and Australia.

Key to the species of pygopodid lizard genus *Lialis* of New Guinea

1a. 19-22 dorsal scale rows at midbody; 4 preanal pores present, 3-5 anal scales (fig. 3); head truncated with 13-17 supralabials (fig. 1). *Lialis burtonis*
1b. 22 dorsal scale rows at midbody; 6 - 8 preanal pores, 5-6 anal scales (fig. 4); head especially elongate with 17- 22 supralabials (fig. 2). *Lialis jicari*

BURTON'S SNAKE-LIZARD
Lialis burtonis
Average length: 500-600mm.
Description: scalation: DMB 19- 22, all smooth; V 70-105; 4 preanal pores; anal scales 3-5; supralabials 13-17.
Distribution within PNG: WEST: Oriomo Plateau and Daru Is. NCD. CENT: Mekeo, Yule Is. and low southern Owen Stanley Mt. foothills. ORO: Owen Stanley Mts. and Popondetta. M BAY: Alotau.
Distribution outside PNG: Widespread throughout most of mainland Australia, southern Irian Jaya and the Aru Islands.

JICAR'S or NEW GUINEA SNAKE-LIZARD
Lialis jicari
Average length: 750-810mm.
Description: scalation: DMB 22, all smooth; V 90-114; 6-8 preanal pores; anal scales 5-6; supralabials 17-22.
Distribution within PNG: Altitude records 0 - 1600m. WEST: Daru Is., e. Oriomo Plateau, Fly R. and Lake Murray. NCD: Port Moresby. MOR: Lae, Garaina and Markham R. MAD: Astrolabe Bay, Madang, Kar Kar Is., and Ramu R. delta. E.SPK: Sepik R. and Wewak. W.SPK: Aitape and Torricelli Mts. W.HIGH: Baiyer R. ENB: Rabaul.
Distribution outside PNG: Irian Jaya only, an endemic to the island of New Guinea.
Extralimital Irian Jayan species yet to be reported from PNG

DIBAMID
Dibamus novaeguineae
D. novaeguineae is a member of the unique lizard family Dibamidae, sometimes termed 'blind skinks' which occur from Malaysia and the Philippines to western Irian Jaya. It can be distinguished from the pygopodid snake-lizards by its rounded head. Of uncertain relationships within the lizards, legless *Dibamus* has reduced eyes and lacks ears. Reduced flap-like hind limbs are present only in males. The head is armoured for burrowing. Little is known of the ecology of *Dibamus*.

Burton's snake-lizard – *Lialis burtonis* (Port Moresby)

Jicari snake-lizard – *Lialis jicari* (Madang Province)

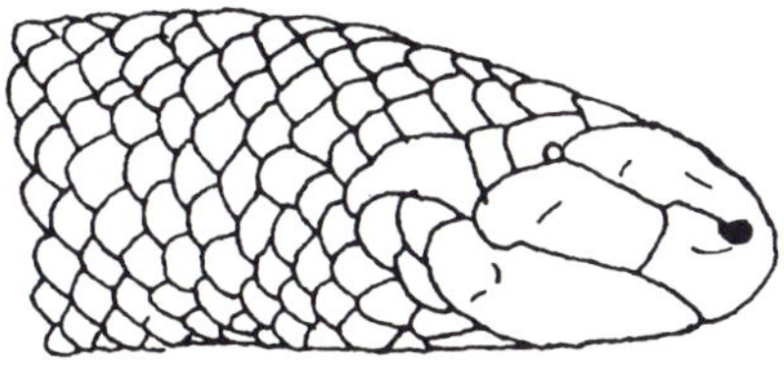

Dibamus novaeguineae
(after de Rooji 1915)

Figures 1 - 4 : page 48

SNAKE-LIZARD HEADS AND PREANAL PORES

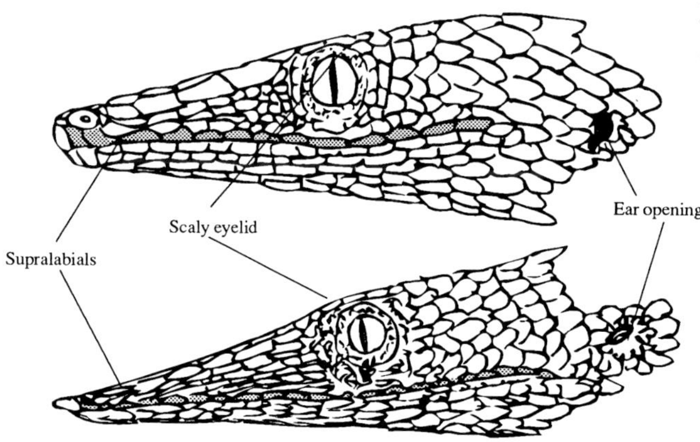

Fig 1: *Lialis burtonis* - Burton's snake-lizard. Truncated head with 13-19 supralabial scales (shaded).

Fig 2: *Lialis jicari* - Jicar's snake-lizard. Elongated head with 17-22 supralabial scales (shaded).

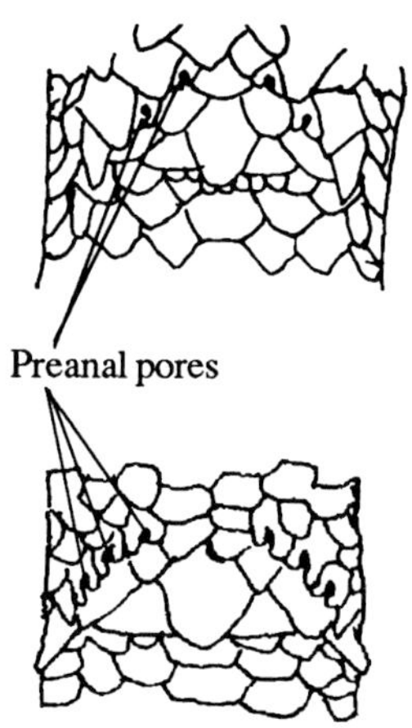

Fig 3: *Lialis butronis* - Burton's snake-lizard. 4 preanal pores.

Fig 4: *Lialis jicari* - Burton's snake-lizard. 6-8 preanal pores.

Blindsnakes
Family Typhlopidae

The family Typhlopidae is a pantropical family of small to medium-sized slender, shiny-scaled snakes with degenerate eyes and a small mouth containing teeth in the upper jaw only, which prey on soft-bodied invertebrates. The family is very well represented in the Australasian region.

NEW GUINEA BLINDSNAKES
Ramphotyphlops & *Typhlops* spp.

(Recent name changes, *see* Taxonomic note)

(11 species) Maps: 02a-c

NON VENOMOUS

Average/Maximum length: 142mm / 325mm.

Description:

a) physique: extremely slender shiny, 'worm-like' snakes (3-8mm diameter) with short tails (max. 13mm *R. polygrammicus*) terminating in a sharp spine; head indistinct from body; eyes degenerate, covered by large transparent ocular and preocular scales.

b) colouration: head and upper body unicolour dark or light brown; underside usually yellow, whitish or dark brown.

c) scalation: Scales at midbody 18-36, ventrals not differentiated from dorsals, all smooth; scales of head enlarged, comprising large rostral (rounded or keeled and 'beak-like') and nasal (may be divided into nasal and supranasal by nasal cleft) extending onto dorsum of head, preocular, ocular, prefrontal, supraoculars and frontal, 3-5 supralabials (1st smallest, 3rd or 4th largest), 2-7 infralabials, subocular present (*Typhlops*), absent (*Ramphotyphlops*) or numerous (*R. subocularis*), postoculars and parietals (*see* fig. 1 and key to the species and individual accounts below).

Habitat: All habitats with loose soil, especially cultivated coastal ground.

Habits: Very secretive totally fossorial (burrowing) snakes which inhabit soil. Usually only seen above the surface nocturnally after rain although may be encountered during digging or when moving rocks, paving slabs, oil drums or other debris, even in built-up areas. Inoffensive. Probably all oviparous. *R. braminus* is the only known parthenogenetic snake - the species is asexual existing in female only populations. Parthenogenetic species are good colonisers and *R. braminus* has colonised most of the tropical world. Clutch sizes have not been recorded for New Guinea but Australian blindsnakes produce between 2 - 8 eggs, although much higher clutches have been recorded.

Diet: Soft-bodied invertebrates, primarily ant larvae and pupae and, to a lesser degree termites and their winged alates, adult ants and their eggs. *R. polygrammicus* also feeds on adult ants and worker termites while *R. subocularis* is the only species known to feed on earthworms (Webb & Shine 1993).

Taxonomic note: *Typhlops* and *Ramphotyphlops* are separated on the structure of the male genetalia. Since *braminus* appears to exist only in parthenogenic all-female populations its exact generic status is difficult to determine. However taxonomists generally consider *braminus* to be closest to *Ramphotyphlops* (*Typhlina* of McDowell, 1974) and McDowell noted similarities in scalation and head architecture between this species, *R. erycinus* and other members of the *R. polygrammicus* group. Recent and forthcoming papers (Wallach, 1995, *in press* & Wallach & Wynn, *in prep.*) expand the typhlopid fauna of New Guinea to three genera and 19 species (*see* Addenda).

Distribution within PNG: Throughout the coastal mainland and islands with a few high altitude locations (*see* individual accounts following).

Distribution outside PNG: Six species of PNG *Ramphotyphlops* also occur in Irian Jaya, Queensland or the Solomon Islands (*see* individual accounts below). In addition Irian Jaya contains a further species of *Ramphotyphlops* and three of *Typhlops* (listed below); the Solomon Islands three *Ramphotyphlops*; and northern Australia a further nine species of *Ramphotyphlops*.

Key to the blindsnakes, family Typhlopidae: genera *Ramphotyphlops* and *Typhlops* of New Guinea
This is a difficult group of snakes to identify requiring the examination of head scalation characteristics and exact body scale counts and is therefore beyond the scope of many workers without access to a microscope. Evenso our herpetofaunal knowledge would benefit considerably from collections made in remote regions and the enthusiastic naturalist is referred to "A Plea for Information" later in this guide. A tentative key is provided and readers requiring additional data are referred primarily to McDowell, 1974. [For Australo-Papuan species *see* also Cogger (1992) and Ehmann (1992)].

1a. Ocular and preocular not fragmented, a single subocular present (fig. 1) or absent (fig. 3); rostral rounded if viewed from above (may be keeled and hooked ventrally); scales around midbody 24 or fewer. 2
1b. Lower surfaces of ocular and preocular fragmented to form numerous suboculars (fig. 2); rostral not rounded but angular and projecting forward if viewed from above; scales around midbody greater than 24. *Ramphotyphlops subocularis*

2a. Subocular scale present (fig. 1). 3
2b. Subocular scale absent (fig. 3). 5

3a. Subocular between preocular and 3rd supralabial (fig. 4); scales around midbody 18. *Typhlops ater*
3b. Subocular between ocular and 4th supralabial (fig. 5); scales around midbody 20 - 24. 4

4a. Rostral strongly keeled to form a downward pointing 'beak-like' projection (fig. 6); scales around midbody 20 - 24. *Typhlops depressiceps*
4b. Rostral rounded without keel or downward projection (fig. 5); scales around midbody 20 - 22; small elongate scale may be present between preocular and nasal. *Typhlops inornatus*

Brahminy blindsnake – *Ramphotyphlops braminus* (NCD)

5a. Scales around midbody 24 - 28; (fig. 3). *Typhlops diardii*
5b. Scales around midbody 20 - 22. 6

6a. Nasal completely divided by nasal cleft (fig. 1). 7
6b. Nasal incompletely divided by nasal cleft (figs. 7). 8

7a. Nasal cleft contacts preocular (fig. 8) allowing contact between preocular and anterior half of nasal but excluding posterior (upper) half of nasal from contact with supralabials. *Ramphotyphlops braminus*
7b. Nasal cleft contacts 1st supralabial (fig. 9) preventing contact between preocular and anterior half of nasal but allowing posterior half of nasal to contact 1st and 2nd supralabials. *Ramphotyphlops erycinus*

8a. Upward projecting keel on rostral (fig. 10). *Ramphotyphlops multilineatus*
8b. Rostral rounded without keel. 9

9a. Scales around midbody 20; underside dark brown; (fig. 7). *Ramphotyphlops leucoproctus*
9b. Scales around midbody 22; underside yellow. 10

10a. Nasal cleft short, barely extending beyond nostril (fig. 11). *Ramphotyphlops flaviventer*
10b. Nasal cleft long, extending dorsally almost entire length of nasal (fig. 12). *Ramphotyphlops polygrammicus*

Fig 1: **Basic Typhlopid Head Scalation**

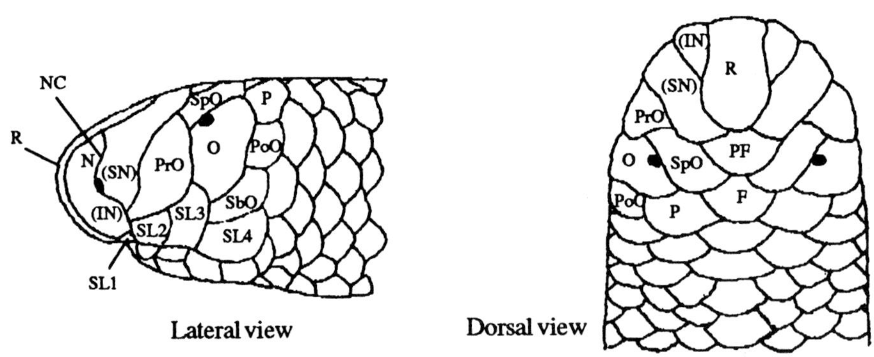

KEY: F - frontal; (IN) - infranasal; N - nasal; NC - nasal cleft: P - parietal; PF - prefrontal; PoO - postocular; PrO - preocular; PF - prefrontal; R- rostral; SL - supralabials; (SN) - supranasal; SbO - subocular; SpO - pupraocular (after McDowell 1974 *J.Herp. 8*(1) reproduced with the kind permission of the author and SSAR)

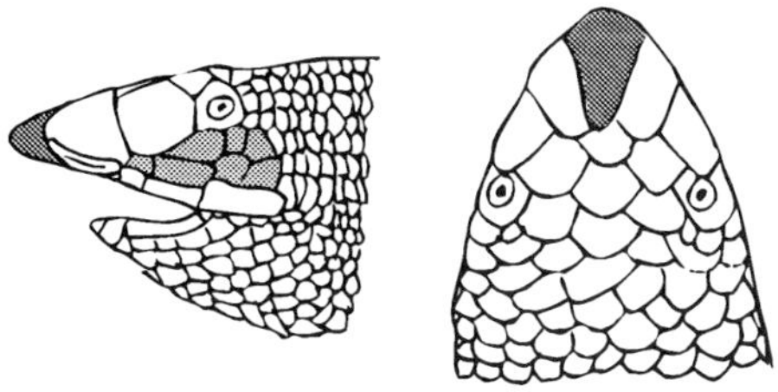

Ramphotyphlops subocularis BBM 5457
Awelkom, Umboi Is., Morobe Prov.
Fig 2: Rostral angular and pointed (dark shaded); lower ocular and preocular scales fragmented to form numerous suboculars (light shaded). (after McDowell 1975 *J.Herp. 9*(1) reproduced with the kind permission of the author and SSAR)

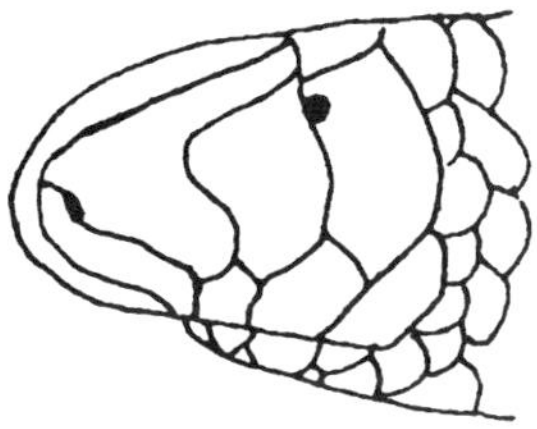

Typhlops diardii RMNH 6328 Doré, West Irian.
Fig 3: Subocular scale absent, ocular and preocular in contact with supralabials. (after McDowell 1974 *J.Herp. 8*(1) reproduced with the kind permission of the author and SSAR).

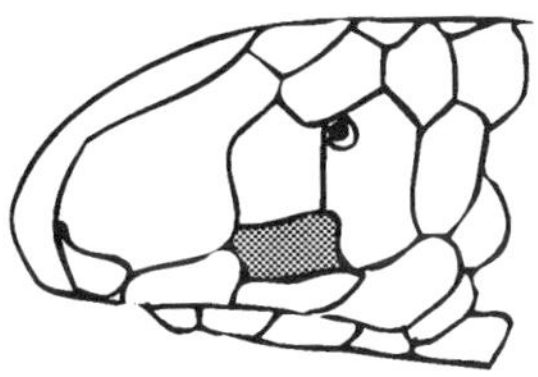

Typhlops ater RMNH 11436 Java, Indonesia.
Fig 4: Subocular scale present (shaded) between preocular and 3rd supralabial. (after McDowell 1974 *J.Herp. 8*(1) reproduced with the kind permission of the author and SSAR).

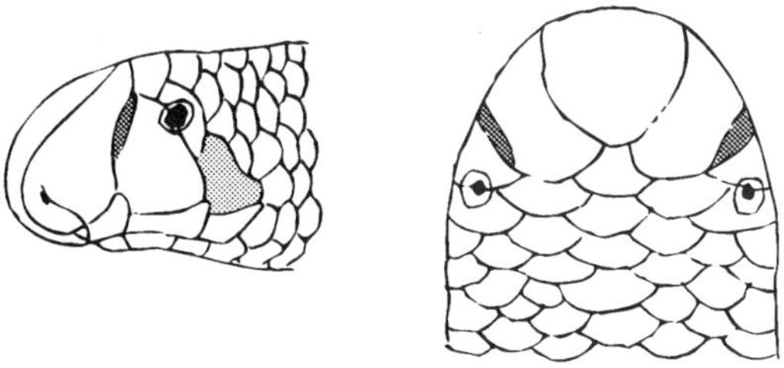

Typhlops inornatus BBM 2772 Sogere (=Sogeri), Central Prov.
Fig 5: Subocular scale present (light shaded) between ocular and 4th supralabial; small, elongate scale may be present (dark shaded) between preocular and nasal. (after McDowell 1974 *J.Herp. 8*(1) reproduced with the kind permission of the author and SSAR).

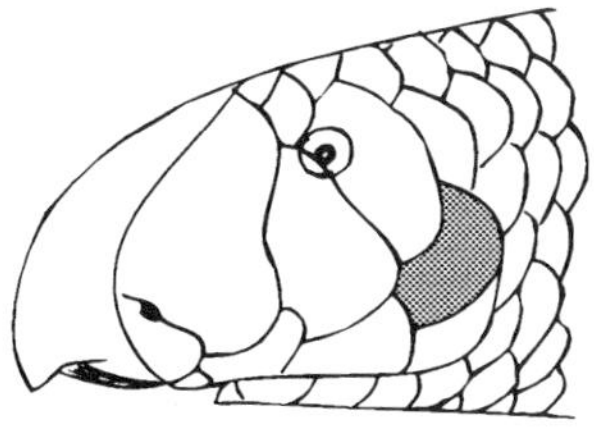

Typhlops depressiceps Alpa Helix no: YPM 107 Omuru, Madang Prov.
Fig 6: Subocular present (shaded) between ocular and 4th supralabial; rostral keeled to form a downward projecting 'beak'.

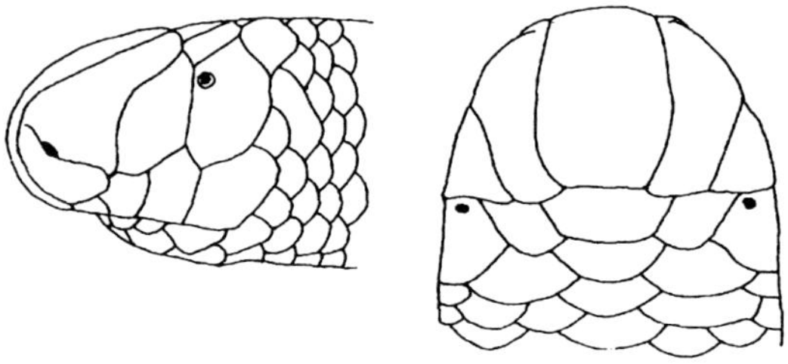

Ramphotyphlops leucoproctus BM 1946.1.11.84
Fly River, Western Prov.
Fig 7: Nasal cleft not completely dividing nasal.

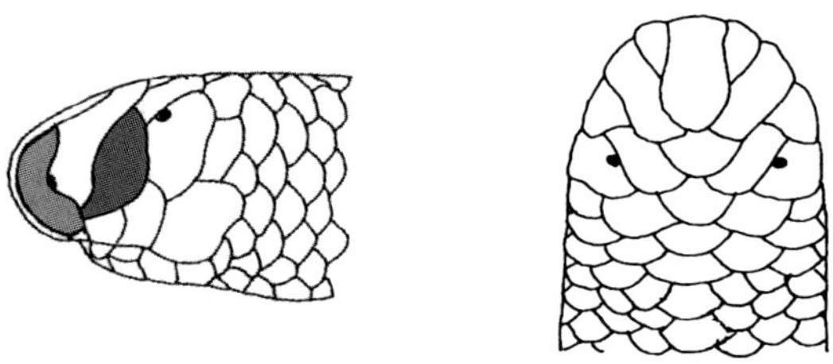

Ramphotyphlops braminus BBM 27752
Bulolo, Morobe Prov.
Fig 8: Nasal cleft contacts preocular allowing contact between preocular (dark shaded) and anterior half of nasal (or infranasal, light shaded), preventing contact between posterior half of nasal (or supranasal) and supralabials.

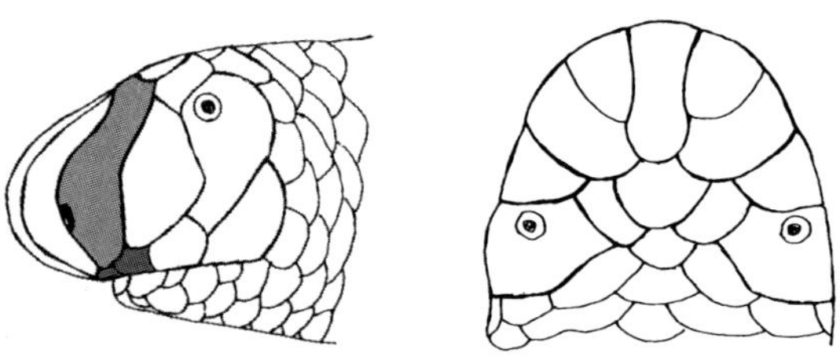

Ramphotyphlops erycinus MCZ 49396 Toem, W. Irian.
Fig 9: Nasal cleft contacts 1st supralabial allowing contact between posterior half of nasal (or supranasal, light shaded) and 1st and 2nd supralabials (dark shaded), preventing contact between anterior half of nasal (or infranasal) and preocular.

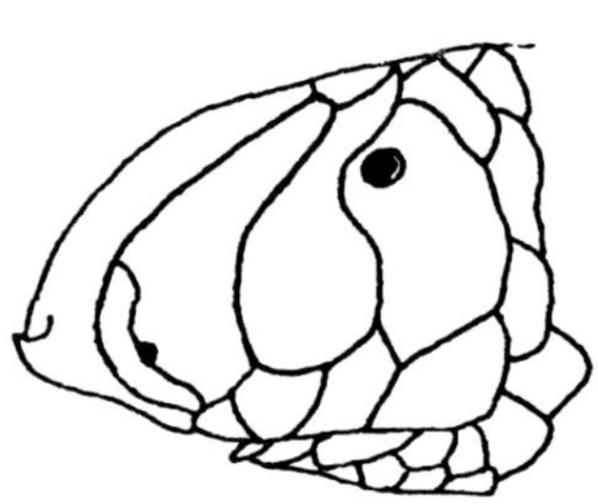

Ramphotyphlops multilineatus AMNH 62468
Bernhard Camp, Idenburg R., W. Irian.
Fig 10: Rostral with upward projecting keel.

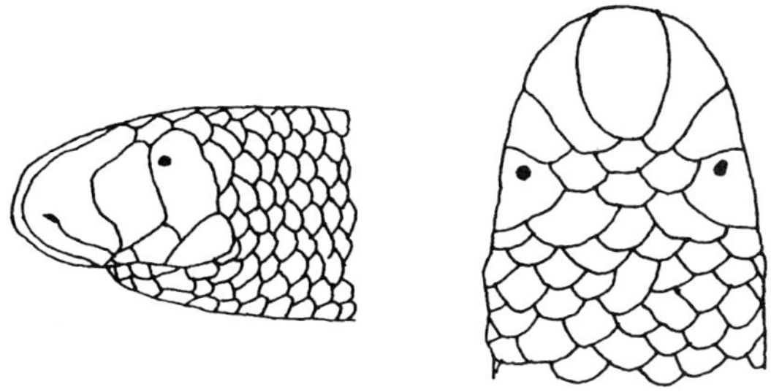

Ramphotyphlops flaviventer BBM 3127
sw. Manokwari, Kebar Valley, Arfak Mts., W. Irian.
Fig 11: Nasal cleft short, barely extending beyond nares (nostril).

Ramphotyphlops polygrammicus AMNH 59875
Sturt Is., Fly River, Western Prov.
Fig 12: Nasal cleft long, extending upwards and backwards.

BRAHMINY BLINDSNAKE or FLOWERPOT SNAKE

Ramphotyphlops braminus

Average/Maximum length: 142 / 170mm.

Description: scalation: Scales at midbody 20; dorsal scales from rostral to above cloaca 302-310; dorsal scales above tail 14, terminal scale usually with a sharp spine; nasal completely divided by nasal cleft; 4 supralabials; 3 infralabials; suboculars absent.

Distribution within PNG: Altitude records 0-2600m. E.SPK: Wewak. MAD: Alexishafen and Kar Kar Is. MOR: Lae and Bulolo. ORO: Lego. NCD: Port Moresby. GULF: Kikori. W.HIGH: Tomba. ENB: Rabaul. N.SOL: s. Bougainville.

Being parthenogenetic this species probably colonised New Guinea from Asia, McDowell suggests it may have been introduced by the Japanese during WWII, a factor which could account for the northern distribution pattern which is centred on former areas of Japanese occupation.

Distribution outside PNG: Irian Jaya (Geelvink Bay to Merauke), Solomon Islands, Indonesia and northern Australia, also many Pacific Islands (Hawaii, Pelau, Mariana Is., Guam, Okinawa etc.), Asia, Madagascar and Mexico.

'NORTHERN NEW GUINEA BLINDSNAKE'

Ramphotyphlops erycinus

Maximum length: 275mm.

Description: scalation: Scales at midbody 20; dorsal scales from rostral to above cloaca 300-320; dorsal scales above tail 16-18, terminal scale usually with a sharp spine; nasal completely divided by nasal cleft; 4 supralabials; 3 infralabials; suboculars absent.

Distribution within PNG: MAD: Ramu R. MOR: Lae and Huon Pen. WEST: upper Fly R.

Distribution outside PNG: Irian Jaya (Mamberamo River).

'YELLOW-BELLIED BLINDSNAKE' **(see Addenda)**

Ramphotyphlops flaviventer

Maximum length: 265mm.

Description: scalation: Scales at midbody 22; dorsal scales from rostral to above cloaca 264 - 326; dorsal scales above tail 14-15, terminal scale usually with a sharp spine; nasal incompletely divided by nasal cleft; 4 supralabials; 3 infralabials; suboculars absent.

Distribution within PNG: Altitude record 550m. MAN: Lorengau and Derimbat; ENB: Gazelle Pen., Nakanai Mts. and Duke of York Is. N.SOL: Nissan Atoll and s. Bougainville.

Distribution outside PNG: Irian Jaya (Vogelkop and Arfak Mts.), North Moluccan Is. and Solomon Islands (incl. Shortlands Island).

'WHITE-TAILED BLINDSNAKE'

Ramphotyphlops leucoproctus

Average/Maximum length: 180mm / 250mm

Description: scalation: Scales at midbody 20; dorsal scales from rostral to above cloaca 372 - 409; dorsal scales above tail 14-17, terminal scale usually with a sharp spine; nasal almost completely divided by nasal cleft; 4 supralabials; 3 infralabials; subocular absent.

Distribution within PNG: Low-lying elevations. WEST: Lower Fly R. and Mata.

Distribution outside PNG: Torres Strait Islands (Murray Is.) and Queensland, Australia.

'SOUTHERN NEW GUINEA BLINDSNAKE'

Ramphotyphlops polygrammicus

Average/Maximum length: 250mm / 400mm.

Description: scalation: Scales at midbody 22; dorsal scales from rostral to above cloaca 346; dorsal scales above tail 19, terminal scale usually with a sharp spine; nasal incompletely divided by nasal cleft; 4 supralabials; 3 infralabials; suboculars absent.

Distribution within PNG: Low-lying elevations. WEST: Fly R., Oriomo Plateau, Morehead and Daru Is.

Distribution outside PNG: Lesser Sunda Islands, southern Irian Jaya (Digul R.), Torres Strait Islands (Murray Is.) and Queensland, Victoria and New South Wales, Australia.

'SHARP-NOSED BLINDSNAKE' **(see Addenda)**

Ramphotyphlops subocularis

Maximum length: 245mm.

Description: scalation: Scales at midbody 26 - 36; dorsal scales from rostral to above cloaca >360; dorsal scales above tail 15, terminal scale usually with a sharp spine; snout rounded or angular in dorsal profile, usually sharp in lateral profile and projecting forward; nasal completely divided by nasal cleft; 4 supralabials; 3 infralabials; subocular scales may be numerous.

Distribution within PNG: Altitude records 25m-1000m. M.BAY: Alotau; MOR: Umboi Is. WNB: Whiteman Mts. ENB: Jacquinot Bay, Gazelle Pen., Nakanai Mts. and Duke of York Is. N.IRE: Fissoa and Lelet Plateau. N.SOL: Bougainville.

Distribution outside PNG: Solomon Islands.

'LOWLAND or BEAKED BLINDSNAKE'

Typhlops depressiceps

Known length: 328 mm

Description: scalation: Scales at midbody 22 - 24; dorsal scales from rostral to above cloaca 628; dorsal scales above tail 25, terminal scale usually with a sharp spine; rostral sharply keeled and projecting downward to form a 'beak-like' projection; nasal incompletely divided by nasal cleft; 4 supralabials; 2 infralabials; suboculars present between ocular and 4th supralabial.

Distribution within PNG: All low elevations. E.SPK: Sepik R. MAD: Madang. MOR: Lae. ORO: Afore. M.BAY: Trobriand Is. - Kiriwina Is.
Distribution outside PNG: No records, believed a PNG endemic.

'MONTANE BLINDSNAKE'
Typhlops inornatus
Average length: 194mm.
Description: scalation: Scales at midbody 20 - 22; all smooth; dorsal scales from rostral to above cloaca 371; dorsal scales above tail 17, terminal scale usually with a sharp spine; nasal incompletely divided by nasal cleft; 4 supralabials; 3 infralabials; suboculars present between ocular and 4th supralabial; small elongate scale present between nasal and preocular.
Distribution within PNG: Altitude records 580-2000m. CENT: Sogeri, (580m). ORO: Albert Edward Mts. (2000m). MOR: Bulolo and Huon Pen.
Distribution outside PNG: No records, believed a PNG endemic.

Extralimital Irian Jayan species yet to be reported from PNG:
'HOOK-NOSED BLINDSNAKE'
Ramphotyphlops multilineatus - North Irian Jaya (Vogelkop, Angadi Is., Lake Jamur, Idenberg R.), Kei Islands and Salawati.
DIARD'S BLACKSNAKE
Typhlops diardii - Doré, Vogelkop? (questionable), S.E. Asia and India.
'BLACK BLINDSNAKE'
Typhlops ater - Andai, Irian Jaya, Waigeo, Salawati Islands, Java, Ternate, Sulawesi and Halmahera.

ADDENDA: TYPHLOPID BLINDSNAKES IN THE NORTH SOLOMONS PROVINCE AND BISMARCK ARCHIPELAGO, PNG

An important new paper on Bismarck/Solomons typhlopid blindsnake taxonomy [Wallach V. 1995 A new genus for the *Ramphotyphlops subocularis* species group (Serpentes: Typhlopidae), with description of a new species. *Asiatic Herpetological Research* **6**: 132-150] which was received too late for incorporation into the main body of this book, is here summarised. In addition Wallach has provided unpublished data from the descriptions of two new species from PNG that are currently *in press*.

I. *Ramphotyphlops subocularis*

This species of typhlopid blindsnake differs from all other Australo-Papuan-Solomon species through its internal anatomy, external morphology and scale counts, and also in its dietary preferences (Webb & Shine 1993). Wallach addressed the situation by a) creating a new genus, *Acutotyphlops*, for this sharp-snouted species; b) resurrecting two species from synonymy within the former *R. subocularis* and c) describing a new species as the type species for *Acutotyphlops*.

Acutotyphlops of Wallach therefore contains four species: *Acutotyphlops subocularis*; *A. infralabialis* and *A. solomonis* (removed from synonymy) and *A. kunuaensis* (the newly described type species). Brief notes follow on the diagnostic characteristics and distribution of each species together with a field key adapted from Wallach (p. 143). Wallach contains full descriptions of these species and also detailed maps of their known ranges.

Acutotyphlops subocularis

Description: scalation: Scales at midbody 34 - 36; dorsal scales from rostral to the terminal spine 363 - 472; rostral scale rounded; supranasals present; ocular single; preocular divided into dorsal and ventral preoculars; suboculars 4 - 7.

Distribution within PNG: Altitude records 25m - 1065m. MOR: Awelkon, Umboi Is.; WNB: Whiteman Mts. - Iambon, Williaumez Pen. - Talasea and Mosa, w.Nakanai; ENB: Jacquinot Bay, Gazelle Pen. - Yalom, Keravat, Toma and Kokopo, and Duke of York Is.; N.IRE: Fissoa/Madina district & Lemkamin.

Distribution outside PNG: None, a PNG endemic.

Acutotyphlops solomonis

Description: scalation: Scales at midbody 30 - 34; dorsal scales from rostral to the terminal spine 334 - 424; rostral scale rounded; supranasals absent, fused with nasal; ocular divided into 1 anterior and 1 - 3 posterior oculars; preocular divided into dorsal and ventral preoculars; subocular divided into 1 - 3 anterior suboculars and 2 - 3 posterior suboculars.

Distribution within PNG: Altitude records near sea level. N.SOL: Bougainville - Kunua, Melelup, Mutuhai, Torokina, Empress Augusta Bay, Kieta & Buin; M.BAY: Alotau.

Distribution outside PNG: No records but potentially from the western Solomons Islands.

Acutotyphlops infralabialis

Description: scalation: Scales at midbody 26 - 28; dorsal scales from rostral to the terminal spine 418 - 526; rostral scale pointed; supranasals absent, fused with nasal; ocular divided into anterior and posterior oculars; preocular single; subocular divided into 1 anterior and 2 posterior suboculars.

Distribution within PNG: Altitude records 15m - 245m. N.SOL: s. Bougainville - Buin district.

Distribution outside PNG: Solomons Islands - New Georgia, Guadalcanal and Malaita Islands.

Acutotyphlops kunuaensis

Description: scalation: Scales at midbody 30 - 36; dorsal scales from rostral to the terminal spine 360 - 542; rostral scale pointed; supranasals absent, fused with nasal; ocular divided into 1 anterior and 1 postocular oculars; preocular single; subocular divided into 1 anterior and 2 posterior suboculars.

Distribution within PNG: Altitude records near sea level - 915m. N.SOL: Bougainville - Kunua, Melelup, Mutuhai, Topanos, Torokina, Empress Augusta Bay, North Nasioi, Kieta & Buin.

Distribution outside PNG: No records, a Bougainville endemic.

Key to the sharp-snouted blindsnakes, genus *Acutotyphlops*, of New Guinea
(field key adapted from Wallach 1995)

1a. Preocular single, anterior subocular paired, dorsal snout profile rounded. 2
1b. Preocular divided, anterior subocular usually single, dorsal snout profile pointed. 3

2a. Midbody scale rows 34 - 36, supranasals present, ocular single. *A. subocularis*
2b. Midbody scale rows 30 - 34, supranasals absent, ocular divided, anterior and posterior. *A solomonis*

3a. Midbody scale rows 26 - 28. *A. infralabialis*
3b. Midbody scale rows 30 - 36. *A. kunuaensis*

II. ***Ramphotyphlops flaviventer*** (North Solomons and Bismarck populations)

Ramphotyphlops flaviventer (fide McDowell 1974) is distributed through the north Moluccas, Irian Jaya, Manus Province, the Bismarck Archipelago, North Solomons Province and the Solomon Islands. In his paper Wallach restricts *R. flaviventer* to Indonesia and western Irian Jaya, resurrecting *R. depressus* from synonymy within *R. flaviventer* for the Bougainvillian and Bismarck populations. There are no records of either *R. flaviventer* or *R. depressus* from mainland PNG. Wallach also resurrects *R. similis*, from Manokwari, western Irian Jaya, and *R. supranasalis* from Salawati Is., far western Irian Jaya, from synonymy within *R. flaviventer*. *Ramphotyphlops supralabialis* can be differentiated from *R. flaviventer* by the presence of supranasal shields. Both of these species are known only from their type localities. Of the three species, *R. flaviventer*, *R. similis* and *R. depressus*, only the latter occurs in Papua New Guinea.

Key to the yellow-bellied blindsnakes, *Ramphotyphlops flaviventer* species group, of New Guinea

1a. Scales around midbody 20; dorsal scales from rostral to the terminal spine fewer than 240; subcaudals 9 - 12; yellow caudal ring present; postocular single; superior nasal suture incomplete, nasal not completely divided, but suture visible from above. *R. similis*

1b. Scales around midbody 22 - 24; dorsal scales from rostral to the terminal spine more than 290; subcaudals 13 - 25; yellow caudal ring absent; postoculars 2 - 3; superior nasal suture complete or incomplete but invisible from above. 2

2a. Broad transition zone, 3 - 5 scales rows wide, between dark dorsum and light venter; light venter 3 - 9 scale rows wide; superior nasal suture completely dividing nasal. *R. depressus*

2b. Sharp demarkational line between dark dorsum and light venter; light venter 11 - 13 scale rows wide; superior nasal suture incomplete, nasal not completely divided. 3

3a. Supranasal shields present; rostral notched posteriorly. *R. supranasalis*

3b. Supranasal shields absent; rostral not notched posteriorly. *R. flaviventer*

Ramphotyphlops depressus

Description: b) colouration: Dorsally reddish-brown, ventrally cream, separated by narrow light brown lateral transitional zone 3 - 5 scales wide; c) scalation: Scales at midbody 22 - 24; dorsal scales from rostral to the terminal spine 292 - 440; nasal usually completely divided by nasal cleft; postoculars 2 - 3.

Distribution within PNG: Altitude records near sea level - MAN: Lorengau, Manus Is., Bundrahei; ENB: Matupi & Duke of York Is.; WNB: Hoskins & Lagunen Point; New Brit: Forsayth & Hanam Hafen; N.IRE: Kavieng, St Matthias Group - Ekalin & Enaie, & Ambitle Is.; N.SOL: Bougainville Is. - Torokina, Empress Augusta Bay, Toriboru & Kieta; MOR: Siassi Is..

Distribution outside PNG: Solomon Islands Shortlands, Santa Ysabel, New Georgia, Olu Malau, San Cristobal, Guadalcanal, Malaita, Florida, Russell Is. groups and possibly Fiji.

Extralimital Irian Jayan species yet to be reported from PNG:

'Yellow-bellied blindsnake; *Ramphotyphlops flaviventer* - Makulu, Salawati & Batanta Is. & Doberai & Onin Pens. w. Irian Jaya.

'Twenty-rowed blindsnake' *Ramphotyphlops similis* - Manokwari, Doberai Pen., w. Irian Jaya only.

'Salawati blindsnake' *Ramphotyphlops supranasalis* - Salawati Is., w. Irian Jaya only.

III. *Typhlops* spp.

Wallach also provided me with data pertaining to a number of new species of the genus *Typhlops* from Papua New Guinea which are mentioned briefly here. Two belong to the *T. ater* species group. *Typhlops fredparkeri* Wallach, from Korobosea, Port Moresby, NCD, is characterised by possessing 16 scale rows at midbody. *Typhlops mcdowelli* Wallach, from Port Moresby and Hombrom's Bluff, resembles *T. depressiceps* except for its low scale count from rostral to caudal spine (431 - 464), thicker body, shape of rostral scale and small apical spine. A third undescribed species, known from Manus Island and Melilup and Buin on Bougainville Island resembles *R. affinis* and is characterised by 18 midbody scale rows.

Wallach (1995 and *in press*) and Wallach & Wynn *(in prep.)* potentially raise the New Guinea typhloid fauna from 11 to 19 species and the known snake fauna of Papua New Guinea to 98 species. Wallach urges further collecting of typhlopids, especially in the eastern provinces of PNG in view of the collection of the Bougainvillean species, *Acutotyplops solomonis*, from Alotau, mainland Milne Bay Province. This Bougainville/Milne Bay link may be mirrored by the distribution of the colubrid treesnake *Dendrelaphis salomonis* which occurs on Bougainville, and Misima Island, Milne Bay Province.

IV. Checklist to New Guinea typhlopid blindsnakes taking into consideration recent data from Wallach

English common names in inverted commas were coined in collaboration with Van Wallach (*pers. comm.*).

Species, author and common name	Type locality	Type no. {(?) = number not found]
FAMILY **TYPHLOPIDAE** (blindsnakes)		
Genus ***Acutotyphlops*** (Sharp-nosed blindsnakes)		
A. infralabialis (Waite, 1918) 'Solomon Is. sharp-nosed blindsnake'	Malaita, Solomon Islands	AMS 4609
A. kunuaensis (Wallach, 1995) 'Bougainville sharp-nosed blindsnake'	Kunua, nw Bougainville, N. Sol. Prov.	MCZ 76964
A. solomonis (Parker, 1939) 'Bougainville blunt-nosed blindsnake'	Bougainville, N. Solomons Prov.	IRSNB 2029
A. subocularis (Waite, 1897) 'Bismarck sharp-nosed blindsnake'	Duke of York Is., ENB Prov.	AMS 2202
Genus ***Ramphotyphlops*** (Australasian blindsnakes)		
R. braminus (Daudin, 1803) Brahminy blindsnake	Vizagapatam	Pl. 43 in Russel, 1796
R. depressus (Peters, 1881) 'Solomon Is. round-nosed blindsnake'	Duke of York Is., E.N.B. Prov.	ZMB 9667
R. erycinus (Werner, 1901) 'N. New Guinea blindsnake'	German New Guinea (= ne. PNG)	(presumed lost)
R. flaviventer (W. Peters, 1865) 'Yellow-bellied blindsnake'	Ternate, Indonesia	ZMB 5029
R. supranasalis (Brongersma, 1934) 'Salawati blindsnake'	Salawati Is., w. Irian Jaya	RMNH 6291-92
R. leucoproctus (Boulenger, 1889) 'White-tailed blindsnake'	Murray Is., Torres St.	BMNH 1946.1.10.58-60
	Fly River, Western Prov.	BMNH 1946.1.11.84
	Queensland, Australia	BMNH 1946.1.11.97
R. multilineatus (Schlegel, 1839) 'Hook-nosed blindsnake'	New Guinea (Irian Jaya)	MNHN 1067
R. polygrammicus (Schlegel, 1839) 'S. New Guinea blindsnake'	Timor, Indonesia	RMNH 3712
R. similis (Brongersma, 1934) 'Twenty-rowed blindsnake'	Doré (= Manokwari), Irian Jaya	RMNH 6295-96

Genus ***Typhlops*** (Tropical blindsnakes)

T. ater (Schlegel, 1839) 'Black blindsnake'	Java, Indonesia	RMNH 3714
T. depressiceps (Sternfeld, 1913) 'Lowland beaked blindsnake'	New Guinea (PNG no exact locality)	ZMB 23986
T. diardii (Schlegel, 1839) Diard's blindsnake	East Indies	MNHN 1065
T. fredparkeri (Wallach, *in prep.*) 'Fred Parker's blindsnake'	Korobosea, Port Moresby, NCD	MCZ 142651
T. inornatus (Boulenger, 1888) 'Montane blindsnake'	Sogere (= Sogeri), Cent. Pr.	BMNH 1946.1.11.80
T. mcdowelli (Wallach, *in prep.*) 'McDowell's blindsnake'	Hombrom's Bluff, Central Prov.	UPNG 7502

Boas and Pythons

Family Boidae

The family Boidae is extremely well represented in the Australasian region. Worldwide only Australia and Indonesia exceed PNG in the number of python species they contain: 15, 13 and 8 respectively. Africa and mainland Asia only contain four and three species respectively. New Guinea and Indonesia also contain two of the three species of Pacific boas, a taxa absent from Australia. Only the Solomon Islands possesses all three species.

Key to the Boas and Pythons of New Guinea: Family Boidae

Most specimens from this family will be easily identified by reference to the photographs accompanying the individual species accounts.

1a. Deep sensory pits present on supralabials, infralabials or both (fig. 1); dorsal scales smooth, with or without apical pits (fig. 2); subcaudal scales mostly divided. Subfamily Pythoninae 4

1b. Sensory pits absent (fig. 3); dorsal scales keeled (fig. 4); subcaudal scales single. Subfamily Boinae 2

2a. Body extremely slender with a long etiolated (fig. 5), prehensile tail; head, anterior of eyes, long, squarish, narrow and flattened; head and body usually grey or brown with a single cream, pinkish or light brown saddle across the back level with the base of the tail. *Candoia carinata carinata*

2b. Body relatively or very stout, head not narrow; tail not slender; body ground colour brown, red, grey or off-white; tail without single pale saddle level with base of tail. 3

3a. Tail very short (fig. 6), fewer than 25 subcaudals; body colour light brown with a dorsal pattern of darker square saddles which may or may not be joined to form an irregular vertebral pattern. *Candoia aspera*

3b. Tail fairly short (fig. 7), more than 35 subcaudals; body colour colour brown, red, grey or white with a dorsal pattern of darker diamonds which are usually linked vertebrally to form a zig-zag pattern. *Candoia carinata paulsoni*

4a. Apical pits present on dorsal scales, either single or as a pair (fig. 2); tail not prehensile; 1 or 2 loreal scales (figs. 12 - 15). 8

4b. Apical pits absent from dorsal scales (fig. 2); tail strongly prehensile; more than 2 loreal scales, often 7 - 10 (figs. 8 - 11). 5

5a. Majority of dorsal head scales, excluding anterior most scales, fragmented into much smaller scales (figs. 8 - 9). 6

5b. Head covered in large regular scales (figs. 10 - 15). 7

6a. Body and head either bright green, rarely blue (adults) with white and yellow or light blue vertebral patterning, or yellow, occasionally orange, (juveniles) with vertebral patterning of black and white; underside yellow (adults) or white (juveniles); all dorsal scales posterior to nasal scales fragmented and granular (fig. 8). *Morelia viridis*

6b. Body and head yellow to light brown with patterning of darker brown saddles and irregular stripes; on dorsal of head, only nasals, internasals, 1 - 2 pairs of scales following internasals and small frontal scale not fragmented (fig. 9). *Morelia spilota*

7a. Head and body deep blue-black throughout; head markings confined to a white throat and 'piano-key bars' of black and white, or yellow, on the lip scales; body markings confined to the anterior ventral surfaces which are white or yellow with diagonal 'fingers' of white or yellow protruding onto the anterior body; fewer than 300 ventrals; fewer than 12 supralabials, 3 are pitted; usually two pairs of parietal scales, interparietal absent; two pairs of prefrontal scales, anterior and posterior with both contacting loreal scales (fig. 10). *Morelia boeleni*

7b. Head and body dark or light brown above; scales of mouth without 'piano-key' markings, body markings dark brown or black reticulations; tail with red-brown rings; more than 300 ventrals; more than 10 supralabials, 4 - 5 are pitted; usually three pairs of parietal scales with interparietal usually present between 2nd and 3rd pairs; two pairs of prefrontals, occasionally with a small azygous shield contacting all four prefrontals and frontal scale (fig. 11). *Morelia amethistina*

8a. Two pairs of prefrontal scales present, median and lateral pairs (figs. 12 - 13). 9

8b. One pair of prefrontal scales present, median pair only (figs. 14 - 15). 10

9a. Scales of lips with 'piano-key bars' of black and white, head black in contrast to rest of body which is brown above, yellow lower down and white below (northern provinces) or blue-brown above and white below (southern provinces); fewer than 284 ventrals; more than 11 supralabials; loreal scale large and elongate (fig. 12); mainland New Guinea and St Matthias Group, Manus. *Leiopython albertisii*

9b. Lips scales not barred black and white; head black or brown with body patterning either orange and black rings or unicolour brown; fewer than 40 dorsal scale rows at midbody; loreal scale small and discrete (fig. 13); Bismarck Arch., excluding St Matthias group. *Bothrochilus boa*

10a. Head light grey with scales edged by black interstitial sutures; body generally brown with flanks lighter than mid-dorsal region; more than 60 dorsal scale rows at midbody; more than 350 ventrals; more than 80 subcaudals; usually one pair of parietals with an interparietal scale present; two pairs of prefrontals, elongate median pair and small, discrete lateral pair which do not contact the loreal scale (fig. 14). *Apodora papuana*

10b. Head and body dark brown to black; fewer than 50 dorsal scale rows at midbody; fewer than 300 ventrals; fewer than 80 subcaudals; two pairs of parietals with an interparietal scale present; two pairs of prefrontals with both median and lateral pairs fairly large and in contact with loreal scale (fig. 15). *Liasis fuscus*

PYTHON AND BOA LABIAL AND APICAL PITS

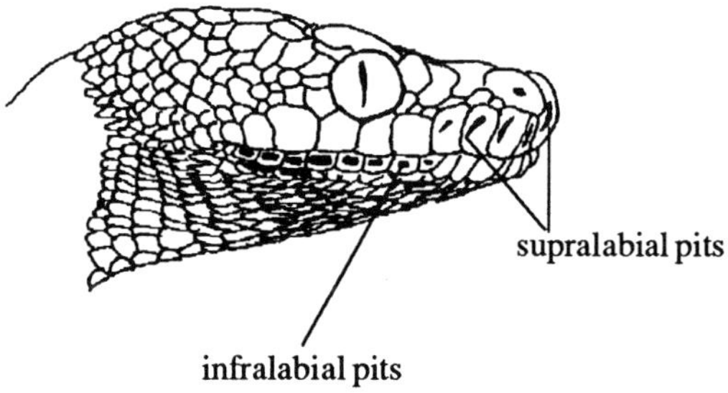

Fig 1: *Morelia amethistina* - amethystine python. (after de Rooji 1917)
Heat-sensitive pits present in supralabials and infralabials.

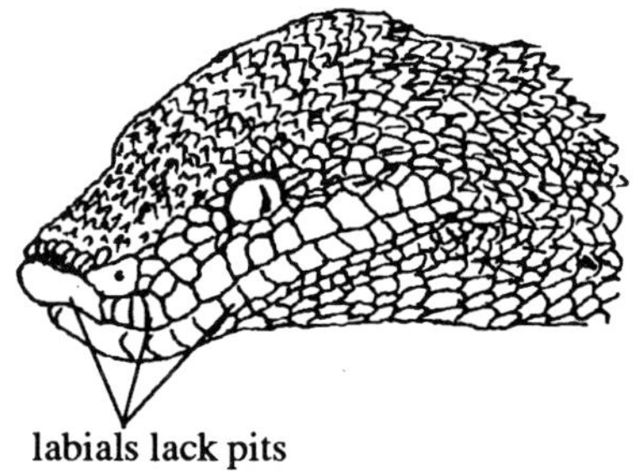

Fig 3: *Candoia aspera* - ground boa. (after McDowell 1979)
Heat-sensitive pits absent from labial scales.

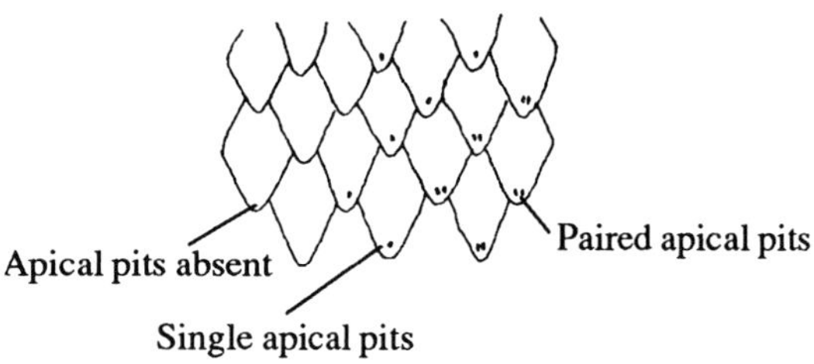

Fig 2: Smooth dorsal scales (pythons) with apical pits single, paired or absent.

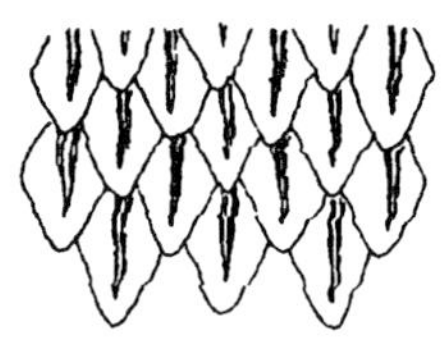

Fig 4: Keeled dorsal scales (boas).

PACIFIC BOAS
(not to scale)

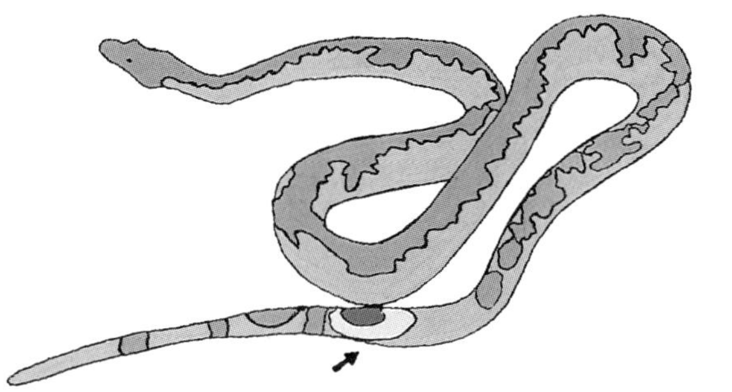

Fig 5: *Candoia carinata 'carinata'* - Pacific tree boa (Jais Aben, Madang). Long-tailed *Candoia carinata* with slender arboreal body and a narrow head; markings consist of an irregular reticulated zig-zag; (arrow indicates position of pale saddle over cloaca and base of tail).

Fig 6: *Candoia aspera* - New Guinea ground boa (Kar Kar Is., Madang). Stout terrestrial bodied boa, broad head and very short tail; markings consist of brown saddles; (arrow indicates position of cloaca and base of tail).

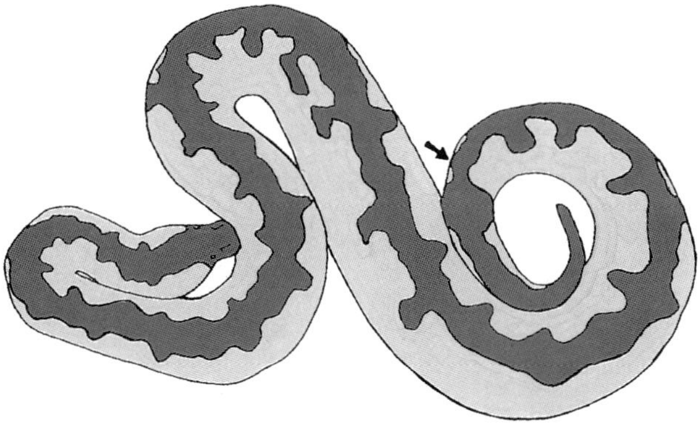

Fig 7: *Candoia carinata 'paulsoni'* - Solomons ground boa (Bougainville Is., North Solomons). Short-tailed *Candoia carinata* with stout semi-terrestrial body and a broad head; markings consist of a broad irregular reticulated zig-zag; (arrow indicates position of cloaca and base of tail).

PYTHON HEAD SCALATION 1

(not to scale)

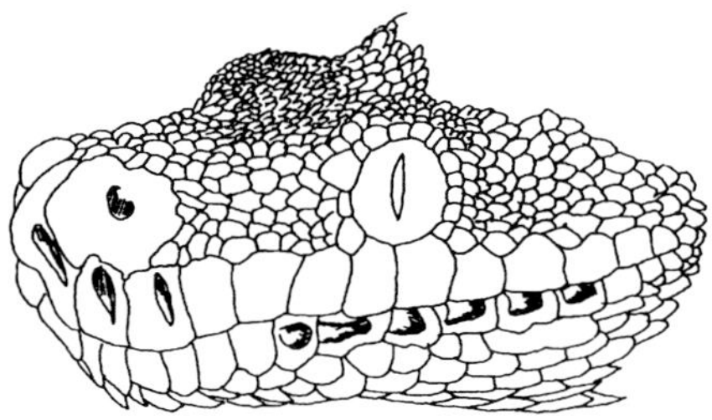

Fig 8: *Morelia viridis* - green tree python (Jais Aben, Madang).
All dorsal head scales, excluding nasal and supralabials, fragmented and granular.

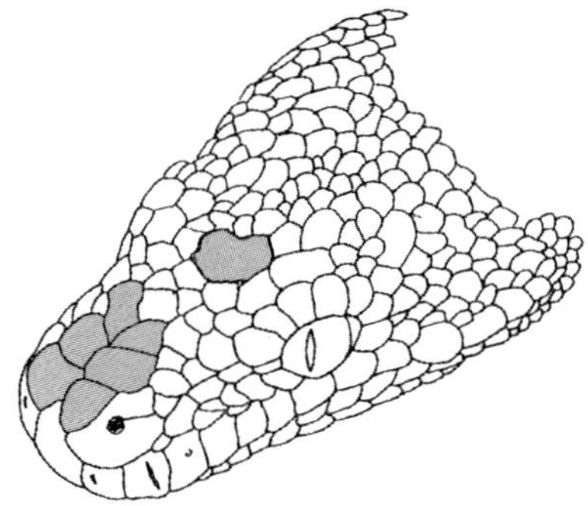

Fig 9: *Morelia spilota* - carpet python (Brown River, Central).
All dorsal head scales posterior of nasal scales, excluding internasals, 1-2 scale pairs following internasals and frontal scale (all shaded), fragmented and granular.

Pythons with unfragmented scutes:
light shading - loreal scales;
anterior dark shading - prefrontal scales;
posterior dark shading - parietal scales.

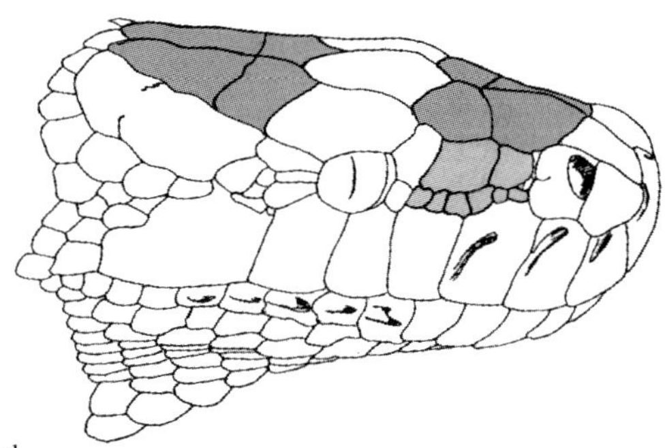

Fig 10: *Morelia boeleni* - Boelen's python (Tari, Southern Highlands).
Two pairs of parietals followed by a post-parietal; two pairs of prefrontals, anterior and posterior; loreals fragmented and numerous.

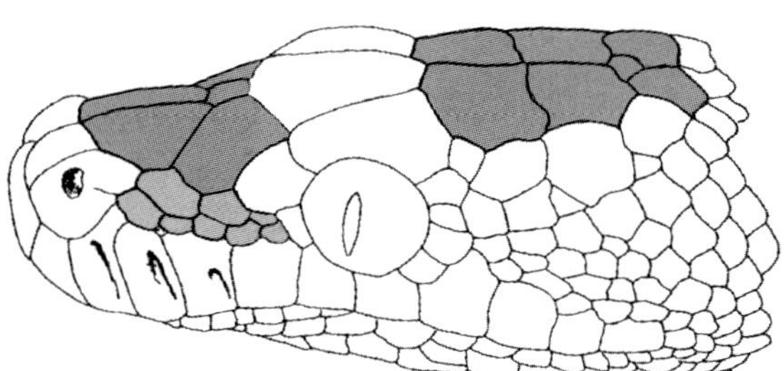

Fig 11: *Morelia amethistina* - amethystine python. (Jais Aben, Madang).
Three pairs of parietals with a post-parietal between 2nd and 3rd pairs; two pairs of prefrontals with median prefrontals either in contact with the frontal scute or separated by a small azygous scale and both median and lateral prefrontals in contact with loreals; loreals fragmented and numerous.

PYTHON HEAD SCALATION 2

(not to scale)

Pythons with unfragmented scutes:
Light shading - loreal scales;
anterior dark shading - prefrontal scales;
posterior dark shading - parietal scales.

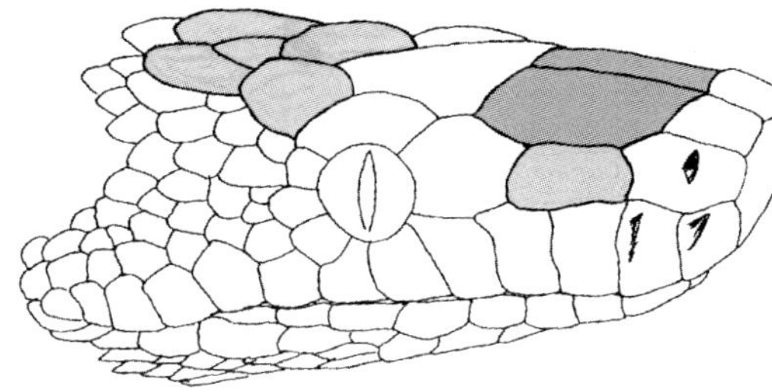

Fig 12: *Leiopython albertisi* - D'Albertis python (Siar, Madang).
Two pairs of parietals with an interparietal; one pair of elongated prefrontals; a single large loreal.

Fig 13: *Bothrochilus boa* - Bismarck ringed python
PNGM 22012
Two pairs of parietals with an interparietal; one pair of prefrontals; a single small loreal.

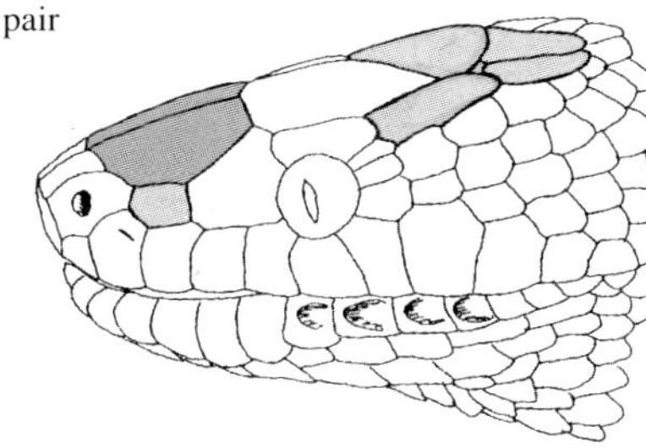

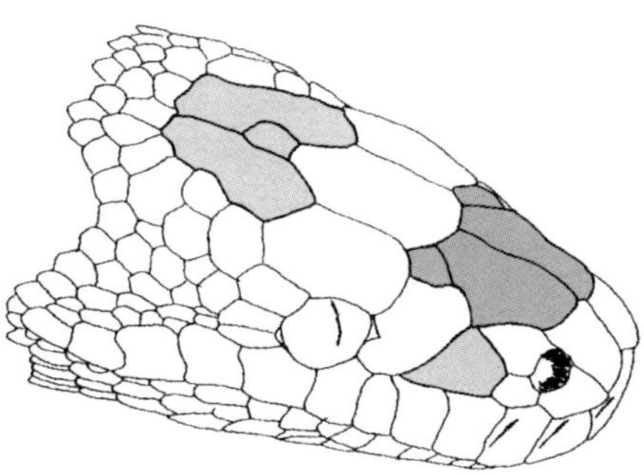

Fig 14: *Apodora papuana* - Papuan olive python (Inawabui, Mekeo, Madang Prov.)
One pair of parietals with an interparietal; two pairs of prefrontals with lateral prefrontals reduced and separated from loreal by an enlarged preocular; a single loreal.

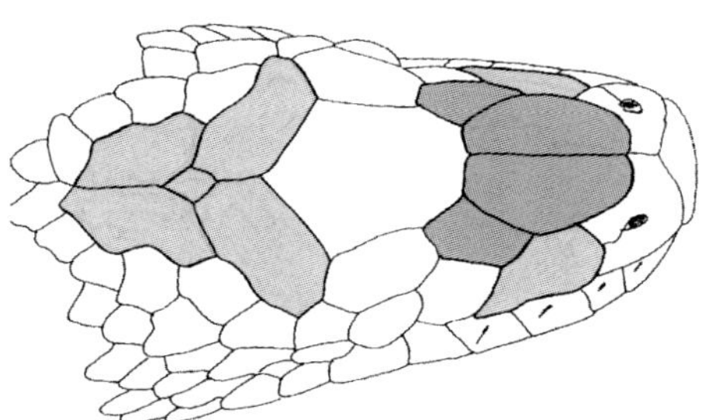

Fig 15: *Liasis fuscus* - Brown water python
PNGM 22408
Two pairs of parietals with an interparietal; two pairs of prefrontals with smaller lateral prefrontals in contact with single loreal.

PACIFIC GROUND AND TREE BOAS

Candoia carinata

Map: 03

NON VENOMOUS

Average/Maximum length: 0.6 / 0.72m (*C. c.'carinata'*); 0.75 / 1.0m (*C. c.'paulsoni'*).

Description:

a) physique: Two phases exist: a) slender bodied *'carinata'* arboreal phase with a long prehensile tail (fig. 5) b) stout bodied *'paulsoni'* phase with a short tail (fig 7) resembling *C. aspera*. Pelvic spurs are present in males on either side of cloaca, absent in females from mainland PNG and the Bismarcks, but present on one side (usually the right) in females from Bougainville (Irian Jayan females have a pair of small spurs); head flattened, concave above, angular and chunky in *'paulsoni'* phase, elongate in *'carinata'*, distinct from the neck; eyes very small with vertical pupils.

b) colouration: Highly variable not only within a small area but even in individual snakes which have the ability to lighten or darken in colour. Body grey, dark brown or red-brown, red, orange, light brown or cream with patterning ranging from square dorsolateral blotches to complete or broken longitudinal vertebral zig-zags in a variety of colours. Often a light coloured vertebral saddle just anterior to the cloaca in long-tailed *'carinata'* phase. Underside yellow with dark spotting.

c) scalation: DMB 31 - 46, all strongly keeled except ventrolateral rows; V 160 - 216; SC 35 - 60, all single; anal scale single with pelvic spurs present in males and absent or reduced in females; loreal, preocular, postocular and most of dorsal head scales numerous and granular; supralabial scales 10-14, 2nd and 3rd touching the eye; infralabials 11-18.

Habitat: A wide variety of habitats from rainforests to coffee, cocoa and coconut plantations. Also known to enter and live in caves.

Habits: The long-tailed phase occurs in sympatry with *C. aspera* where it adopts primarily an arboreal habit whilst the short-tailed terrestrial phase occurs in areas where *C. aspera* is absent. However, both phases may occur in low vegetation or forage nocturnally on the ground. Viviparous exhibiting considerable differences in clutch size i.e. 40 - 50 in the *'paulsoni'* phase but only six in the *'carinata'* phase. The biogeography of *C. carinata, C. aspera* and a third species distributed from the Solomon Islands to Samoa (*C. bibroni* which also occurs in two phases) is extremely complex (*see* Taxonomic note below and McDowell 1979).

Diet: *C. carinata 'paulsoni'* - primarily lizards but also frogs and small mammals; *C. carinata 'carinata'* - primarily skinks but also gecko eggs and frogs.

Taxonomic note: Some authors recognised the existence of two subspecies of *C. carinata* and although the status of these populations as distinct subspecies has not been proven, the names *'carinata'* and *'paulsoni'* are used to separate the two primary phases or 'morphs' of *C. carinata* - the long-tailed and short-tailed forms respectively. McDowell (1979) made a detailed study of the genus *Candoia* and plotted the distributions of these two morphs and also noted that an independent 'long-tailed' form also occurred on Misima Island in Milne Bay Province whilst apparent intergrades between the *'carinata'* and *'paulsoni'* occurred in Morobe and New Ireland Provinces.

Distribution within PNG: Altitude records 0-915m, highest 1525m (Telefomin, W.SPK).

C. carinata 'carinata' ('long-tailed' Pacific tree boa phase) – WEST: Upper and lower Fly R., Aramia R. and Lake Murray. GULF: Baimuru and Purari R. deltas. W.SPK: Aitape. E.SPK: Wewak, Kairiru Is. and upper Sepik R., Ama. MAD: Ramu R. delta, Madang, Astrolabe Bay, Kar Kar and Manam Is. WNB: south coast and Aware Is. ENB: Gazelle Pen., Duke of York and Mioko Is. N.IRE: New Ireland, New Hanover, Mussau and Tench Is. MAN: Manus, Lou, Baluan and Rambutyo Is.

C. carinata 'paulsoni' ('short-tailed' Pacific ground boa phase) – W.SPK: Telefomin. WEST: Kiunga, upper Fly R. and s.Trans-Fly (rare). MOR: Markham R., Huon Pen., Garaina and Umboi Is. ORO:

Popondetta, Lego and Owen Stanley Mts. M.BAY: Cape Vogel, Biniguni, Samarai Is., d'Entrecasteaux Is., Goodenough, Fergusson, Dobu and Normanby Is., Trobriand Is., Kiriwina and Kitava Is., Woodlark Is., Rossel and Tagula Is. N.SOL: Buka and Bougainville Is., Tinputz.
Possible intergrade populations - MOR: west of Lae. N.IRE: Tabar Is.
C. carinata (long-tailed phase) - M.BAY: Misima Is.
C. carinata (undetermined phase) - MAD: Crown and Long Is. N.IRE: Babase Is.
Distribution outside PNG: *C. carinata 'carinata'* occurs in Irian Jaya and on the Indonesian islands to the southwest eg. Ceram and Tanimbar Islands. *C. carinata 'paulsoni'* occurs in the Solomon Islands and Santa Cruz Islands and on the Indonesian islands to the west of Irian Jaya eg. Sulawesi and Halmahera.

Pacific tree boa – *Candoia carinata 'carinata'*
(Jais Aben, Madang Province)

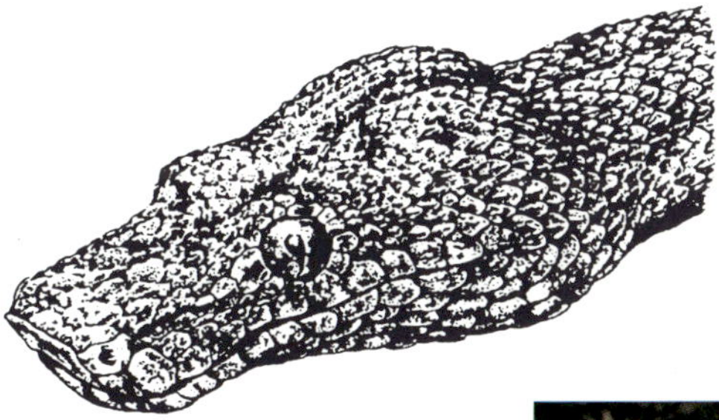

Candoia carinata carinata
BBM 2942 Gaulim, ENB Prov.
(after McDowell 1979 *J.Herp 13*(1))
reproduced with the kind permission of the author and SSAR.

Pacific ground boa – *Candoia carinata 'paulsoni'*
(North Solomons Province).

NEW GUINEA GROUND BOA

Candoia aspera

Map: 04

NON VENOMOUS

Average/Maximum length: 0.6 - 0.9m / 1.0m.

Description:

a) physique: A very stout bodied snake with an extremely short, non-prehensile tail less than 2x length of the head (fig 6); pelvic spurs present in males on either side of the cloaca; head flattened, concave above, angular and chunky, distinct from the neck; eyes very small with vertical pupils. Specimens from the Bismarck Archipelago appear less robustly built than those from the mainland (*see* Taxonomic note below).

b) colouration: Body dark brown or red-brown with square dark brown or black, occasionally reddish, blotches which may become united to form a vertebral zig-zag pattern at midbody; underside yellow to light brown with black or red mottling.

c) scalation: DMB 30 - 45, all strongly keeled except ventrolateral rows; V 127-153; SC 11 - 22, all single; anal plate single with pelvic spurs present in males; loreal, preocular, postocular and most of dorsal head scales numerous and granular; supralabial scales 9-14; infralabials 12-17.

Habitat: A wide variety of damp wooded or forested habitats but particularly common in discarded husk piles in coconut plantations.

Habits: A thick bodied terrestrial boa which superficially resembles the highly venomous death adder (*see* death adder account), *C. aspera* is sometimes called the 'viper boa'. Rarely encountered in the open, unless after heavy rain, this is a nocturnal and secretive species which inhabits leaf-litter and decaying vegetation. Although also known as the 'sleepy snake' *C. aspera* has enlarged front teeth and can deliver a painful bite. Specimens do not always strike or bite, many taking up a defensive 'balling' position (rolling into a ball with the head in the centre) if molested. Ground boas are viviparous.

Diet: Frogs, small mammals and lizards.

Taxonomic note: Stull (1932) and Loveridge (1948) recognised two subspecies in the former species *Enygrus asper: E. asper asper* from the Bismarck Archipelago and *E. a. schmidti* from mainland New Guinea, based on the scalation and patterning characteristics listed below:

	DMB	V	SC	SL	IL	circumorbital scales	dorsal spots
E. a. asper	37-41	149-150	18-19	10	12-13	13-15	24-25
E. a. schmidti	34-37	127-138	15-21	10-12	12-16	11-16	15-22

NB: *Enygrus* was the old generic name for *Candoia*.

SL (supralabials or scales of upper lip); IL (infralabials or scales of lower lip); circumorbital scales (minute scales surrounding the eye).

Comparison of a juvenile specimen from New Ireland with similar sized specimens from Kar Kar Island certainly indicated differences. The New Ireland specimen appeared much more slender with a higher vertebral ridge, a longer tail and much lighter coloured eyes than the Kar Kar juveniles.

Distribution within PNG: Altitude records 0-1000m highest 1300m (Wabag, ENGA). WEST: upper Fly R. and Lake Murray. GULF: Omati. CENT: Owen Stanley Mts., Brown R., Subitana and Mekeo. M.BAY: Biniguni. ORO: Owen Stanley Mts. and Popondetta. MOR: Garaina, Markham R., Lae, Huon Pen. and Umboi Is. MAD: Adelbert Mts., Kar Kar Is., Awar and Astrolabe Bay. E.SPK: Sepik R., Narimok and Wewak. W.SPK: Sepik R., Leonard Schultze R., Torricelli Mts., Aitape and Seleo Is. SIM: Karimui. ENGA: Wabag. WNB: Whiteman Mts., Cape Hoskins and Garu. ENB: Gazelle Pen. and Duke of York Is. N.IRE: New Ireland and New Hanover Is. MAN: Manus, Los Negros, Lou, Baluan and Rambatyo Is. Recorded in error from Bougainville.

Distribution outside PNG: Irian Jaya and offshore Indonesian islands.

New Guinea ground boa – *Candoia aspera 'schmidti'*
(Kar Kar Is., Madang Prov.).

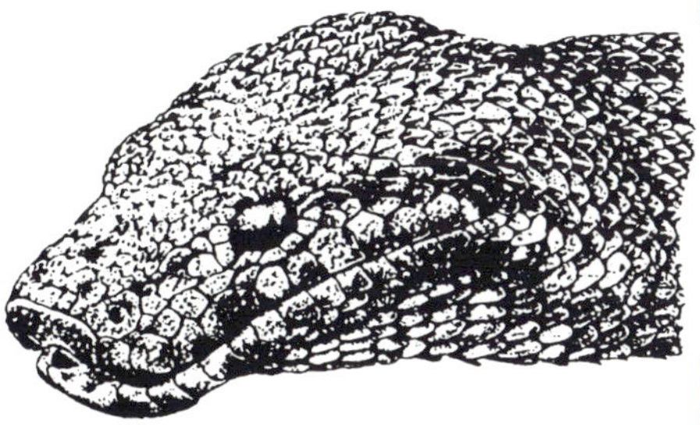

Candoia aspera
BBM 1658
Los Negrol Is., Manus Prov.
(after McDowell 1979 *J.Herp 13*(1))
reproduced with the kind permission of the author and SSAR.

New Guinea ground boa – *Candoia aspera 'aspera'* (New Ireland Prov.)
Photographs: above M. O'Shea below Mike McCoy

GREEN TREE PYTHON

Morelia viridis

(recent name change, *see* Taxonomic note)

Map: 05

NON VENOMOUS

Average/Maximum length: 1.45m / 1.5m.

Description:

a) physique: A slender python with a fairly short, chunky head, which is distinct from the neck; tail strongly prehensile; eyes large with vertical pupils.

b) colouration: Adults are bright green above, occasionally blue, and either white or yellow below and on the lower flanks with a suffusion of green spots invading from the dorsum. Dorsal patterning may consist of a continuous or punctuated white, occasionally light blue, vertebral stripe. Alternatively markings may be confined to a few scattered white or light blue spots or completely absent. The head is coloured as the body, with yellow or white lips, chin and throat. Juveniles are usually yellow above, although some populations may be brick red, reportedly those from West Sepik Province, orange or even green, with white or yellow underbellies. Markings consist of small white vertebral bars and lateral spots which may be black or brown edged. A white, dark edged, stripe runs along from the snout through the eye to the angle of the jaw. The eye is white in the juvenile and yellow in the adult. The tail in the juvenile is black and white but a less obvious yellow in the adult.

c) scalation: DMB 55-67, all smooth; V 219-250; SC 68-129, almost all paired; anal plate single; all dorsal head scales, excluding large nasal scutes, numerous and granular; supralabial scales 12-16 with 6th and 7th or 7th and 8th touching the eye; infralabials 14-18 (fig 8).

Habitat: A wide variety of wooded or forested habitats but also encountered in gardens and hedges around buildings.

Habits: Although this is a highly agile species with a prehensile tail there is evidence to suggest that this python forages on the ground at night. Sleeping specimens may be located within 0.5m of the ground in sheltered tree clefts, epiphytic plants and on low branches during the day. Juvenile green tree pythons are generally yellow or orange, only changing to their adult green livery at about 15 months. This process may take a few days or several months and occasional specimens retain their juvenile colouration throughout life. The enlarged front teeth of this species can deliver a painful but non-venomous bite. Oviparous with a clutch size of 8-12.

Diet: Primarily small mammals when adult but feeding on lizards, especially skinks, as juveniles. There does not appear to be any evidence that birds form a significant part of the diet.

Taxonomic note: The green tree python is considered closely related to the carpet python, *Morelia spilota* and recent workers (Underwood & Stimson 1990; Kluge 1993) have therefore placed it in the genus *Morelia*. Although the resultant loss of the long-established generic name *Chondropython* is likely to be unpopular with herpetoculturists and zoos, this taxonomic change is likely to be generally accepted by taxonomists and herpetologists.

Distribution within PNG: Altitude reports 0-2000m in almost all mainland provinces. WEST: Trans-Fly, Sturt Is., Lake Murray, Aramia R. and upper Fly R. GULF: Omati and Kerema. NCD. CENT: Brown R., Sogeri, Mt Diamond and Owen Stanley Mts., Woitape. M.BAY: Sinaeada, Biniguni, Fergusson and Normanby Is. ORO: Popondetta, Mt. Lamington, Albert Edward Mts. and Owen Stanley Mts. MOR: Garaina, Wau, Bulolo, Markham R., Mt. Rawlinson and Huon Pen. MAD: Astrolabe Bay, Adelbert Mts., Ramu R. and Schrader Mts., Simbai. E.SPK: Sepik R. W.SPK: Aitape, Torricelli Mts. and Telefomin. W.HIGH: Waghi, Baiyer and Jimi R. E.HIGH: Goroka, Kratke Mts. and Aiyura. SIM: Chimbu R. and Kubor Mts.

Distribution outside PNG: Most of Irian Jaya, the islands of Geelvink Bay, Misool and the Aru Islands, Indonesia. Also Cape York Peninsula, far-northern Queensland, Australia.

Green tree python (juv.) – *Morelia viridis*
(Goroka, E. Highlands Prov.) Photo: Mike McCoy

Green tree python (adult) – *Morelia viridis*
(Jais Aben, Madang Prov.)

Green tree python (juv..) – *Morelia viridis*
(Jais Aben, Madang Prov.)

Green tree python (adult) – *Morelia viridis* (Riwo, Madang)

CARPET PYTHON

Morelia spilota variegata

(*see* Taxonomic note)

Map: 06

NON VENOMOUS

Average/Maximum length: 1.0-2.0m / (4.0m Australia).

Description:

a) physique: A heavily built python with a moderately short head, 'chunky' in adults, which is distinct from the narrow neck; tail long and prehensile; eye large with a vertical pupil.

b) colouration: Generally a yellow to light brown ground colour with a highly variable series of broad, irregular, dark brown edged, chestnut-brown bars, saddles and blotches along the body and tail. Top of head as body with a chestnut-brown central arrow extending from the angle of the jaw to above the eyes and the tip of the snout and brown stripes above the white lips from the jaw to the snout. Undersides off-white with fainter representations of the dorsal saddles.

c) scalation: DMB 45-51, all smooth; V 254-270; SC 71-81, almost all paired; anal plate single; most dorsal head scales (excluding nasal and internasal scales, two pairs of enlarged scales following internasals, and frontal scale) small and granular; supralabial scales 12-13 with 6th & 7th touching the eye; infralabials 18-20. (fig. 9)

Habitat: Confined to southern Papuan coastal eucalypt savannas where hollow logs and abandoned buildings are also inhabited.

Habits: The carpet python is the most commonly encountered nocturnal python in and around Port Moresby and the species most frequently killed on NCD roads. Australian carpet pythons are reported to attain twice the size recorded for New Guinea specimens. Although Australian carpets may also be encountered in tropical rainforest and desert, the New Guinea carpet python is a relatively small savanna woodland species although occasional larger specimens around 2m in length are found. Since this species is well known for its willingness to hiss and bite if handled, all carpet pythons should be treated with respect. Despite their irrasibility, this python is a beneficial devourer of rats and mice and should be preserved rather than persecuted. Carpet pythons are oviparous and may lay clutches of up to 20 eggs.

Diet: Small mammals, mice and rats, possibly ground nesting birds and lizards, agamas (dragon lizards) and skinks.

Taxonomic note: The widely distributed Australian carpet python populations, formerly enveloped under the single subspecific name *M. spilota variegata*, have seen revision in recent years. Wells and Wellington (1984 & 1986) described three new species but Barker and Barker (1995) took the more conservative approach of recognising these populations as subspecies of *M. spilota*. In Barker and Barker *M. s. variegata* is therefore confined to the a northern strip extending from the Gulf of Carpentaria region of northern Queensland and Northern Territory to the Kimberley region of northern Western Australia. This would still support the recognition of the name *M. s. variegata* for southern New Guinea populations of *Morelia spilota*, at least those found in the broad, seasonally flooded Merauke (Irian Jaya) /Western Province (PNG) savanna belt. However, the population occupying the majority of northern and eastern Queensland may be recognised as a separate subspecies, *M. spilota mcdowelli,* and whether this population has any links with the carpet python population of Central Province, PNG, has yet to been determined. A potential connection is already suggested by the presence of elements of the New Guinea ophiofauna in northern Queensland and on off-shore islands of the Torres Straits i.e. *Morelia amethistina* and *M. viridis*, and *Leiopython albertisi* and *Stegonotus parvus* respectively.

Distribution within PNG: Altitude records 0-300m. WEST: S. Trans-Fly. NCD. CENT: Sogeri Plateau, Brown R., Laloki R., Goldie R. and Yule Is.

Distribution outside PNG: Across the border into the Irian Jayan savannas as far as Merauke and throughout mainland Australia where six subspecies are recognised (Barker & Barker 1995), most notably the diamond python *Morelia spilota spilota*.

Carpet python – *Morelia spilota variegata* (Bluff Inn, Central Province)

Carpet python – *Morelia spilota variegata* (16 mile, Central Province)

AMETHYSTINE PYTHON

Morelia amethistina amethistina

Map: 08

NON VENOMOUS

Average/Maximum length: 5.5-6.0m / 8.5m.

Description:

a) physique: A large slender python with an elongate head which is distinct from the narrow neck; tail long and prehensile; eye large with vertical pupil.

b) colouration: Highly variable throughout its range. New Guinea specimens usually either dark grey-brown (forest phase) or sandy brown (savanna phase) with darker reticulate markings either throughout the body or confined to the mid and posterior sections of the body. (Populations from the Moluccas, Indonesia may be dark green with fine black markings). The entire dorsal pattern is overlaid by a purple-blue iridescence which shimmers in sunlight and earns the species the name 'amethystine python'. Head as body but often lighter with dark pigment etching edges of scales. Undersides cream to off-white. Juveniles more vividly patterned often with red-brown rings on the tails.

c) scalation: DMB 34-57, all smooth: V 308-346; SC 85-122, almost all paired; anal plate single; loreal scales small and numerous; supralabial scales 11-14 with 6th and 7th or 7th and 8th touching the eye; infralabials 17-24; preoculars 2-3; postoculars 3-5; two pairs of prefrontals, all contacting loreals but may be separated from frontal scale by small azygous scale; three pairs of parietals with a small post-parietal between 2nd and 3rd pairs (fig 11).

Habitat: Occuring in a wide variety of habitats from savanna to rainforest to an altitude of 1600m. Frequently encountered along river banks, around buildings or crossing roads.

Habits: Possibly the longest Australasian snake species and in New Guinea rivalled only by the more heavily built Papuan olive python, *Liasis papuanua*. Secretive by day and often only observed crossing roads at night or entering out-buildings. Generally irritable and inclined to bite if provoked but a valuable rat-eater if left alone. Very large specimens may possibly represent a threat to unattended pets but it is unlikely that this python represents any threat to humans. Amethystine pythons are oviparous.

Diet: Mammals, from rats to wallabies and pigs; birds and lizards, goannas and skinks.

Distribution within PNG: Altitude records 0-1600m. WEST: Oriomo Plateau, Trans-Fly, Daru and Bobo Is., lower and upper Fly R., Lake Murray and Aramia R. GULF: Omati. NCD. CENT: Mekeo, Bereina, Laloki, Goldie and Brown R., Owen Stanley Mts. and Yule Is. M.BAY: Cape Vogel, Sinaeada, Trobriand Is., Normanby Is. and Rossel Is. ORO: Popondetta. MOR: Garaina, Wau, Lae, Markham R., Mt. Rawlinson, Huon Pen., and Umboi Is. MAD: Astrolabe Bay, Adelbert Mts. and Bogadjim. SIM: Kundiawa. WNB: Whiteman Mts., Mt. Vale, Garu and Fullerborn Harbour. ENB: Gazelle Pen., Rabaul and Duke of York Is. N.IRE: Kavieng, New Ireland and St Matthias Group, Mussau Is.

Distribution outside PNG: The New Guinea subspecies is also recorded from Irian Jaya and the satellite islands to the north, west and south in addition to the Aru and Kei Islands and Moluccan archipelagos. Australian populations from Cape York Peninsula, Queensland and the Torres Straits are recognised as a separate subspecies, *Morelia amethistina kinghorni*.

Amethystine python – *Morelia amethistina amethistina* (Alexishafen, Madang Prov.)

Amethystine python – *Morelia amethistina amethistina*
(Siar, Madang Prov.)

Amethystine python – *Morelia amethistina amethistina* (Central Prov.)

BOELEN'S PYTHON

Morelia boeleni

Map: 07

NON VENOMOUS

Average/Maximum length: 1.83-2.44m / max. length not known.

Description:

a) physique: A moderately stocky built python with a broad, thick-set head which is distinct from the narrow neck; tail fairly long; eye large with vertical pupil.

b) colouration: A striking species. The upper surfaces are deep black with an overlying purple-blue iridescence. The only markings are a series of irregular forward-facing yellow or white diagonal 'finger-markings' which extend from the white anterior ventral surface onto the lower flanks of the first third of the body. Head as body with broad white barring on the lip scales (labials) and black mottling on the white surfaces of the chin and throat. Underside off-white or yellow anteriorly, becoming darker by midbody and totally black posteriorly. Juveniles are red-brown with faint representions of the 'finger-marks' which become more obvious as the dorsum darkens.

c) scalation: DMB 44-51, all smooth; V 282-298; SC 57-64+, almost all paired; anal plate single; loreal scales small and numerous, 6-12; supralabial scales 8-11 with 5th & 6th touching the eye; infralabials 14-17; preoculars 2; postoculars 2-4; two pairs of prefrontals, all contacting loreals; two pairs of parietals followed by a small post-parietal (fig 10).

Habitat: A mid-montane rainforest species occuring above 1000m, but below the treeline, and appearing to favour humid, low-light conditions.

Habits: Almost nothing is known about the habits of Boelen's python from wild observations. It seems likely that the species requires a high degree of humidity to survive. Although probably a forest floor dweller, it is agile and locals from the Fane region report that it climbs well. The Boelen's python is oviparous with clutch sizes of at least 14 eggs.

Diet: Little known but probably small mammals and ground nesting birds. Captive specimens at the National Museum of PNG feed on day old chicks.

Distribution within PNG: Altitude records 1000-2000+m. CENT: Woitape and Mt. Brown. MOR: Wau and Mts. behind Lae, (no precise locality data). E.HIGH: Kainantu and Goroka. MAD: above Bundi, (no precise locality data). S.HIGH: Tari. M.BAY: Goodenough Is. Anecdotally also WEST: Star Mts. (no precise locality data). A New Guinea endemic. This disjunct distribution is probably more a reflection of our limited knowledge concerning the montane herpetofaunas of PNG than any rarity on the part of the python. **However, this species is the most protected reptile in PNG, receiving protection similar to that afforded to the Birds of Paradise.** It is therefore illegal for a non-national person to kill or possess this species or any derivative such as its skin etc.

Distribution outside PNG: Only known from scattered localities in Irian Jaya, most notably the area around Wissel Lakes around 1750m in the far west.

Boelen's python – *Morelia boeleni* (Tari, S. Highlands Prov.)

Boelen's Python – *Morelia boeleni* (Woitape, Central Province)

Boelen's python – *Morelia boeleni* (Woitape, Central Prov.)

Boelen's python – *Morelia boeleni* (Woitape, Central Prov.)

D'ALBERTIS or WHITE-LIPPED PYTHON

Leiopython albertisii

(recent name change, *see* Taxonomic note)

Map: 09

NON VENOMOUS

Average/Maximum length: 1.0-1.8m (northern) / 2.44m (southern).

Description:

a) physique: A moderately heavily built python with a fairly long head, which is distinct from the neck; tail not prehensile; eyes with vertical pupils.

b) colouration: At least two distinct forms appear to exist on mainland New Guinea which in PNG may be termed the Northern or Momase and Southern or Papuan races, although the actual ranges of the two morphs have not been fully studied. Northern specimens are smaller in size, light brown to copper above, yellow-brown on the lower flanks and immaculate white below in stark contrast to their dark glossy black heads and black and white barred lips scales. Southern specimens appear much larger but much less brightly marked with an iridescent, dark blue-brown body, which contrasts much less with the black and white head markings, and a cream underbelly.

c) scalation: DMB 44-55, all smooth; V 259-283; SC 62-72, almost all paired; anal plate single; usually one large loreal scale occasionally 2-3; supralabial scales 12-14 with 5th to 7th touching the eye; infralabials 14-18; single preocular and 3-4 postoculars; one pair of elongate prefrontals; two pairs of parietals, interparietal present (fig 12).

Habitat: Encountered in a wide variety of habitats but showing a preference for rainforest or monsoon forests in close proximity to freshwater. Around Port Moresby the southern race of the D'Albertis python replaces the more common carpet python in damp, lowlying habitats.

Habits: When alarmed specimens of the smaller northern race will 'ball' (roll into a loose ball shape), burying their heads in the centre of the coil in the manner of West African Calabar and Royal pythons. This defensive behaviour was not noted in the larger southern race. These pythons are nocturnal and commonly seen on roads after rain, especially in Madang Province. D'Albertis pythons are oviparous with clutch sizes between eight and fifteen.

Diet: Mammals from mice to bandicoots and probably also skinks, when juvenile.

Taxonomic note: The *Liasis* group of pythons have undergone considerable changes in nomenclature in recent years. Revisions have resulted in the resurrection of *Bothrochilus* (Cogger *et al,* 1983) for these non-prehensile-tailed Australo-Papuan pythons, the transfer of all species to *Morelia* (Underwood & Stimson, 1990) and the reinstatement of *Liasis* (Cogger 1992, Ehmann 1992). The most recent published change (Kluge, 1993) recognises the considerable diversity existing within the *Liasis-Bothrochilus* group and resurrects the old generic name *Leiopython* for *albertisii*, thereby creating a new monotypic genus.

Distribution within PNG: Altitude records 0-1600m. WEST: Oriomo Plateau, Trans-Fly, Aramia and Bensbach R., Lake Murray and upper Fly R. GULF: Omati. NCD. CENT: Veimauri, Laloki, Brown and Goldie R., Sogeri, Mekeo, Subitana and Rigo. M.BAY: Normanby Is. ORO: Girua. MOR: Garaina, Wau-Bulolo, Lae and Finschhafen. SIM: Karimui. W.HIGH: Baiyer R. MAD: Ramu R. delta, Astrolabe Bay and Adelbert Mts. E.SPK: Wewak. W.SPK: Torricelli Mts. and Telefomin. N.IRE: St. Matthias group - Mussau Is. The isolated population on the remote Mussau Island in the St. Matthias Group, to the north of New Ireland, is interesting since *L. albertisi* does not occur elsewhere in the Bismarck Archipelago (*see* Bismarck ringed python *Bothrochilus boa*). This disjunct distribution is mirrored by the little file snake *Acrochordus granulatus.*

Distribution outside PNG: Throughout northern Irian Jaya as far west as Biak and Salawati Islands and Fak Fak and in southern Irian Jaya around Merauke. Also reported for Australia from the northern Torres Strait islands, near the PNG coast. Formerly considered a subspecies of the Australian species *Liasis fuscus* (*see* Brown water python *Liasis fuscus).*

Northern D'Albertis python
– *Leiopython albertisi*
(Siar, Madang Prov.)

Northern D'Albertis python
– *Leiopython albertisi*
(Siar, Madang Prov.)

Southern D'Albertis python
– *Leiopython albertisi*
(Goldie River, Central Prov.)

BISMARCK RINGED PYTHON

Bothrochilus boa

(recent name change, *see* Taxonomic note)

Map: 10

NON VENOMOUS

Average/Maximum length: 0.95-1.54m / 1.73m.

Description:

a) physique: A slenderly built python with a fairly short, pointed head, which is barely distinct from the neck; tail not prehensile; eyes small with vertical pupils.
b) colouration: At least two colour phases exist; a unicolour brown phase with a light yellow underbelly, which appears commonest in the western part of the range on Umboi Island and New Britain (although a specimen purportedly from New Ireland is known) and an orange/brown and black phase. In this phase the black markings take the form of either bands or rings around the body or a series of irregular broken spots, blotches and stripes. Specimens from both phases possess dark brown or black heads, a constant characteristic in a variable species. A light spot may occur behind the eye. In the unicolour brown phase there may be faint suggestions of darker blotches or crossbands.
c) scalation: DMB 34-39, all smooth; V 245-267; SC 45-54, almost all paired; anal plate single; single small loreal scale; supralabial scales 9-12 with 4th to 6th touching the eye; infralabials 12-13; single preocular and 2-3 postoculars; one pair of prefrontals; two pairs of parietals, a small interparietal may be present. (fig 13).

Habitat: Forests, open land and coconut plantations.

Habits: Little is known about the ecology of this species. It appears to be a nocturnal, secretive, terrestrial, possibly semi-fossorial species. Oviparous.

Diet: Primarily small mammals and ground dwelling lizards, especially skinks. There is also a possibility that this species may exhibit occasional ophiophagous habits.

Taxonomic note: The *Liasis* group of pythons have undergone significant nomenclatural changes in recent years. Recent revisions have resulted in the resurrection of *Bothrochilus* (Cogger *et al.,* 1983) for these non-prehensile-tailed Australo-Papuan pythons, the transfer of all species to *Morelia* (Underwood & Stimson, 1990) and the reinstatement of *Liasis*. The most recent published change (Kluge, 1993) re-establishes *Bothrochilus* as a monotypic genus.

Distribution within PNG: MOR: Umboi Is. WNB: Whiteman Mts., Stettin Bay, Garu, Arawe and Gasmata Is. ENB: Rabaul, Gazelle Pen., Duke of York Is. and Mioko Is. N.IRE: New Ireland, Kavieng & Namatanai, Tatau, Djaul and New Hanover Is. N.SOL: Nissan Atoll. *B. boa* is notably absent from the St. Matthias Island group where the largest island, Mussau, is inhabited by an isolated population of the otherwise non-Bismarck species *Leiopython albertisii*.

Distribution outside PNG: Not recorded elsewhere, the ringed python is a Papua New Guinea endemic.

Bismarck Ringed Python (adult)
– *Bothrochilus boa*
(New Ireland Province)

Bismarck Ringed Python (juv.)
– *Bothrochilus boa*
(US captive bred)

Bismarck Ringed Python (adult)
– *Bothrochilus boa*
(New Hanover Is.,
New Ireland Prov.)
Photo: Mike McCoy

PAPUAN OLIVE PYTHON

Apodora papuana

(recent name change, *see* Taxonomic note)
Map: 11

NON VENOMOUS
Average/Maximum length: 1.4-3.6m / 4.27m.
Description:

a) physique: A particularly heavily built python with a fairly short, stumpy head, which is distinct from the neck; tail not prehensile; eyes with vertical pupils.

b) colouration: Olive brown to yellow brown on the flanks, somewhat darker dorsally presenting a 'two-tone' pattern, in contrast to the black interstitial skin between the scales. Underbelly light green to buff with lighter interstitial skin, head scales brown or light grey with contrasting black edging, lips and throat also light with darker speckling, though not as obviously marked as D'Albertis or Boelen's pythons. A short dark stripe often present behind the eye.

c) scalation: DMB 63-72, all smooth; V 358-374; SC 82-88, almost all paired; anal plate single; single large loreal scale; supralabial scales 10-11 with 5th to 7th touching the eye; infralabials 16-20; 1-2 preoculars and 2-3 postoculars; two pairs of prefrontals, lateral prefrontals small and separated from loreal scale by large preocular; one pair of parietals, interparietal present (fig 14).

Habitat: Lowland monsoon forests, savanna-woodlands and savannas.

Habits: This is an extremely powerful python species well capable of killing and devouring large prey (*see* below) and although rivalled by the amethystine python for the title of longest New Guinea snake, it is without doubt the heavier species. Large specimens are fairly frequently encountered on roads at night and 3-4m 'road kills' are not uncommon. Papuan olive pythons are oviparous.

Diet: Mammals to the size of 22.7kg wallabies (Parker, 1982) but also a snake-eater. A 2.0m specimen from Kunini, southern Western Province, contained a slightly longer amethystine python, *Morelia amethistina*. Ophiophagy is a rare trait in pythons. Papuan olive pythons may also be potentially cannibalistic since a specimen in the National Museum exhibited keratophagy, the eating of its own shed skin, unusual behaviour which may lead to preying on smaller specimens of the same species.

Taxonomic note: The *Liasis* group of pythons have seen considerable changes in nomenclature in recent years and revisions have resulted in the resurrection of *Bothrochilus* (Cogger *et al.*, 1983) for these non-prehensile-tailed Australo-Papuan pythons, the transfer of all species to *Morelia* (Underwood & Stimson, 1990) and the reinstatement of *Liasis*. Underwood and Stimson (1990) considered the Papuan olive python *papuanus* close to the Australian olive python *olivaceus*. The most recent published research (Kluge, 1993) recognises the considerable diversity existing within the *Liasis - Bothrochilus* group and proposes the creation of a new monotypic genus *Apodora* for *papuanus* which would become *Apodora papuana*. Kluge placed the Australian olive python species, *olivaceus* in *Liasis* with *mackloti*. McDowell (1975) suggests that the giant python measuring 4.3m, reportedly collected from Astrolabe Bay and described as *Liasis maximus* by Werner (1936), but which is now lost, may be an aberrant specimen of *A. papuana*. The largest existing specimen of *A. papuana*, and the largest known New Guinea python, is the type of *A. papuana* (AMNH 73993) from Biniguni, Milne Bay Province, which measures 4.78m.

Distribution within PNG: WEST: S. Trans-Fly, Aramia and Fly R., Lake Murray and Lake Daviumbu. GULF: Omati. NCD. CENT: Mekeo, Laloki, Goldie and Brown R. M.BAY: Biniguni, and Fergusson Is. ORO: Girua. MOR: Garaina, Lae and Markham R. MAD: Bogadjim, Astrolabe Bay, Maclay Coast, Astrolabe Bay and Adelbert Mts.

Distribution outside PNG: Probably widespread in Irian Jaya, especially in the north and west, and also recorded from the offshore islands of Biak and Misool. Formerly considered a subspecies of the Australian olive python (*Liasis olivaceus*) which is distributed widely along the western and northern coastlines of Western Australia, Northern Territory and Queensland.

Papuan Olive Python
– *Apodora papuana*
(Inauabui, Mekeo,
Central Province)

Drowned 2m Papuan olive python which had predated a 2m. Amethystine python (Kunini, Western Province)

Papuan Olive Python
– *Apodora papuana*
(Inauabui, Mekeo,
Central Province)

BROWN WATER PYTHON

Liasis fuscus

(recent name change: ***Liasis mackloti.*** *See* Taxonomic note)
Map: 12

NON VENOMOUS
Average/Maximum length: 1.5-2.0m / 3.0m (Australia).
Description:

a) physique: A slender python with a fairly long head, which is barely distinct from the neck; tail not prehensile; eyes with vertical pupils.
b) colouration: Western Province specimens are glossy black or dark brown. The chin and throat are white, the underside of the body is yellow and the underside of the tail is grey-green or yellow.
c) scalation: DMB 45-49, all smooth; V 291-285; SC 65-76, almost all paired; anal plate single; single large loreal scale; supralabial scales 10-12 with 5th and 6th touching the eye; infralabials 14-17; single loreal; single preocular and 2 postoculars; two pairs of prefrontals, lateral prefrontals in contact with loreal; two pairs of parietals, interparietal present (fig. 15).

Habitat: In PNG this species is believed to be confined to damp habitats such as swamps or lagoons.
Habits: Thought to spend prolonged periods of time in water, diving into submerged vegetation to escape capture. Probably the most aquatic python in New Guinea and rivalling the Boelen's python for the least documented. Nocturnal and oviparous.
Diet: Small mammals, wading birds, lizards and hatchling crocodiles.
Taxonomic note: The *Liasis* group of pythons have undergone considerable nomenclatural changes in recent years. These revisions have resulted in the resurrection of *Bothrochilus* (Cogger *et al,* 1983) for these non-prehensile-tailed Australo-Papuan pythons, the transfer of all species to *Morelia* (Underwood & Stimson, 1990) and the reinstatement of *Liasis.* Kluge (1993) also synonymised *Liasis fuscus* of Australasia with *Liasis mackloti* of Indonesia and referred to the Sawu Island population (*see* below) as a varient rather than a valid subspecies. He commented that although this species does exhibit considerable geographical variation in both colouration and scalation, there did not appear to be sufficient evidence "to suggest the presence of two or more historical entities". However, American herpetoculturists who have examined large numbers of live specimens of *L. mackloti* from the Lesser Sunda Islands, *L. fuscus* from northern Australia and water pythons from Irian Jaya, maintain that the New Guinea snakes are much closer morphologically to the Australian *L. fuscus* than they are to the Indonesian *L. mackloti* (Dave Barker, Boerne, Texas *pers. comm.*). I am not currently in a position to examine or compare living specimens of any of these populations, so for the purpose of this Guide, and in recognition of the established herpetological links between the Western Province/Merauke region of New Guinea and northern Australia, the name *Liasis fuscus* has been applied for the water python.
Distribution within PNG: WEST: s.Trans-Fly from Bensbach R. to Oriomo R., Lake Murray, Lake Daviumbu, Aramia R. and Sigabaduru.
Distribution outside PNG: The Irian Jayan populations of water python are located in the southeast around Merauke whilst Australian populations of *L. fuscus* occur in the north of Western Australia, Northern Territory, Queensland and the Torres Strait islands. Macklot's water python, *L. mackloti,* occurs on the Lesser Sunda Islands of Timor, Wetar and Samao to the West of New Guinea. A former subspecies of *L. mackloti, L. m. savuensis,* from Sawu Island in the Lesser Sundas, is sometimes recognised as a valid species *L. savuensis.* Kluge (1993) synonymises both *L. m. savuensis,* and *L. fuscus* with *L. mackloti* as a single species.

Brown water python
– *Liasis fuscus*
(Western Province PNGM 22408)

Brown water python
– *Liasis fuscus*
(Western Province PNGM 22408)

Macklot's water python
– *Liasis mackloti mackloti*
(UK captive from Indonesia)

FILESNAKES
Family Acrochordidae

The family Acrochordidae is a small family containing only three species of fully aquatic snakes, which is confined to the Indo-Australian region. The common names of 'wart snake' or 'filesnake' refer to the rough texture of their tuberculate skin, whilst the third commonly applied name, 'elephants-trunk snake', is a reference to the loose, 'baggy' nature of the skin on the body. Two species occur in the Australasian region, one primarily in freshwater and one in marine conditions.

WART or FILESNAKES
Acrochordus spp.

(2 species) Map: 13

NON-VENOMOUS
Average/Maximum length: 0.5-1.35m / 2.5m.
Description:
a) physique: Snakes with extremely loose and warty 'baggy' skins (especially *A. arafurae*); tail slightly compressed and prehensile; head distinct from neck, largely covered in granular scales as body; eyes dorsolateral, very small with round pupils; nostrils facing forwards (*A. arafurae*) or upwards (*A. granulatus*).
b) colouration: (*see* key to the species and individual accounts below.)
c) scalation: (*see* key to the species and individual accounts below.)

Habits: Excellent swimmers but helpless on land. Generally inoffensive but large specimens of *A. arafurae* may deliver a painful bite. Generally nocturnal and inactive for long periods, resting on the muddy bottom or anchored to submerged vegetation by their prehensile tails. File snakes are viviparous and clutch sizes range from 6-12 (*A. granulatus*) to 17-32 (*A. arafurae*).

Key to the species of filesnakes genus *Acrochordus* in New Guinea
1a. Dorsal scale rows 120-180 at midbody; loreal scales 11-14; supralabials 9-11; habitat freshwater swamp or riverine; average length exceeds 1.0m. *A. arafurae*
1b. Dorsal scale rows 90-160 at midbody; loreal scales 5-7; supralabials 5-7; habitat coastal waters, mangrove swamps or estuaries; average length less than 1.0m. *A. granulatus*

ARAFURA WART or FILESNAKE
Acrochordus arafurae
Average/Maximum length: 1.2- 1.5m / 2.5m.
Description:
b) colouration: Unicolour drab grey or brown or with a broad darker vertebral band and faint crossbands; underside lighter than above.
c) scalation: DMB 120-180, all small, keeled, granular or tuberculate; enlarged ventral scales absent; subcaudal scales and anal scale absent; loreal scales 11-14; supralabials 9-11.

Habitat: *A. arafurae* occurs in slow-moving freshwater rivers, lakes and swamps, including isolated ponds, which are reached during monsoon floods, and also occasionally in brackish or inshore marine situations.
Diet: Entirely fish.
Taxonomic note: Formerly recognised as part of the S.E.Asian species *A. javanicus* from which it was separated by McDowell (1979) largely using characters of the skull, dentition and hemipenes.

Distribution within PNG: WEST: Fly-Strickland R. system, Lake Daviumbo, Lake Murray, Bensbach, Morehead, Binaturi, Pahoturi, Oriomo and Aramia R., Balimo.
Distribution outside PNG: Southern Irian Jaya (Mimika and Lorentz R.) and northern Australia.

LITTLE WART or FILESNAKE

Acrochordus granulatus
Average/Maximum length: 0.5- 0.6m / 1.2m.
Description:

b) colouration: Grey brown or black with numerous lighter grey, brown or orange bands around the body; underside lighter than above.
c) scalation: DMB 90-160, all extremely small, keeled, granular or tuberculate; enlarged ventral scales absent; subcaudal scales and anal scale absent; loreal scales 5-7; supralabials 5-7.

Habitat: A marine species usually found in the intertidal zone in mangrove swamps or along inshore coral reefs but also in brackish estuaries and inland freshwater grass swamps. This species has been collected at sea 3 km from land in 18m of water.
Diet: Mainly marine or estuarine fish but also crabs.
Distribution within PNG: WEST: Daru and Bobo Is., Oriomo, Fly, Aramia and Wawao R. and Lake Murray. GULF: Kerema, Kikori, Uramu Is. and Gulf of Papua. CENT: Yule Is. M.BAY: Kiriwina Is., Trobriand Is. MAD: Nagada Harbour. WNB: Willaumez Pen. and Garu. E.SPK: Wewak, Sepik R. and Kairiru Is. N.IRE: Mussau Is. N.SOL: Buka and n. peninsula Bougainville.
Distribution outside PNG: India to Indonesia, coastal Irian Jaya, Solomon Islands and northern coastal Australia.

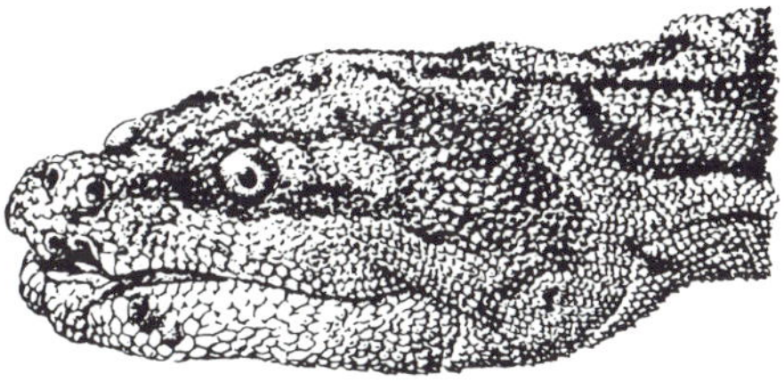

Acrochordus arafurae (juv) BBM3092
Balimo, Western Prov.
(after McDowell 1979 *J. Herp.13*(1))
reproduced with the kind permission of the author and SSAR.

Left: Arafura filesnake – *Acrochordus arafurae*
(Oriomo, Oriomo R., Western Province)

Below: Little filesnake – *Acrochordus granulatus*
(Nagada Harbour, Madang Province)

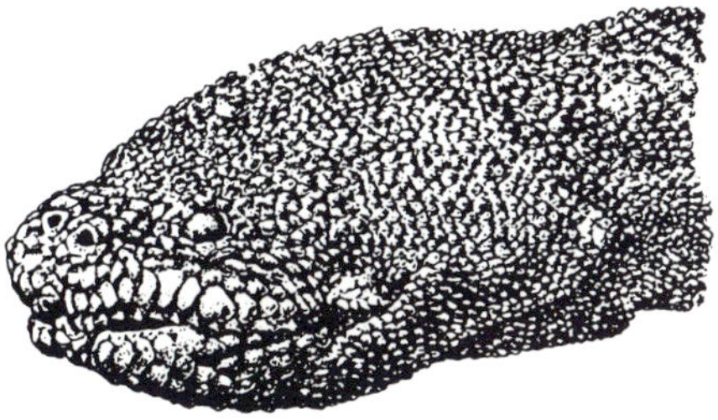

Acrochordus granulatus BBM 3384
Japen Is., Geelvink Bay, West Irian
(after McDowell 1979 *J.Herp.13*(1))
reproduced with the kind permission of the author and SSAR

Tree, Ground and Watersnakes
Family Colubridae

The family Colubridae is the most diverse and widely distributed snake family worldwide containing some 2000 species, 60% of the known snake fauna. In the Australasian region the family is generally poorly represented. Australia possesses only 10 species, approximately 7.5% of the snake fauna, (excluding seasnakes and sea kraits), but New Guinea contains 32 species which constitutes almost 37% of its snake fauna.

Key to the Harmless Terrestrial and Freshwater Snakes of New Guinea: Family Colubridae

1a. Loreal scale absent allowing point contact between preocular and nasal scales and prefrontal and 2nd supralabial scales (fig. 1). *Fordonia leucobalia*
1b. Loreal scale present (fig. 2) preventing contact between preocular and nasal scales or prefrontal and supralabial scales. 2

2a. Dorsal scales strongly keeled. 3
2b. Dorsal scales smooth or weakly keeled. 4

3a. Dorsal scales in 17 or fewer rows at midbody; parietal scales entire (fig. 3). *Tropidonophis* spp.
3b. Dorsal scales in 23 or more rows at midbody; parietal scales fragmented (fig. 4). *Cerberus rhynchops*

4a. Dorsal scales arranged in oblique rows (fig. 5). 5
4b. Dorsal scales arranged in normal, non-oblique rows (fig. 6). 6

5a. Dorsal scale in 13 rows at midbody; eye with round pupil; anal plate divided (fig. 7). *Dendrelaphis* spp.
5b. Dorsal scales in 19 - 23 rows at midbody; eye with vertically elliptical 'cat-like' pupil; anal plate entire (fig. 7). *Boiga irregularis*

6a. Anal plate entire (fig. 7); 19 or fewer dorsal scale rows at midbody, all smooth. *Stegonotus* spp.
6b. Anal plate divided (fig. 7); 19 or more dorsal scale rows at midbody, if 19 then dorsal scales weakly keeled. 7

7a. Nasal scales in contact (fig. 8). *Enhydris polylepis*
7b. Nasal scales separated by single internasal scale (fig. 9). 8

8a. Three light bands across rear of head; nasal cleft absent, nasal scale entire. *Cantoria annulata*
8b. No bands across neck or head; nasal cleft present, dividing nasal scale into two separate halves. 9

9a. Dorsal scales weakly keeled, in 19 - 21 rows at midbody; fewer than 150 ventral scales; fewer than 45 subcaudal scales. *Myron richardsoni*
9b. Dorsal scales smooth, in 27 rows at midbody; more than 150 ventral scales; more than 60 subcaudal scales. *Heurnia ventromaculata*

Keys to the large colubrid genera are to be found on the following pages:

Genus	Page
Tropidonophis	96
Dendrelaphis	102
Stegonotus	108

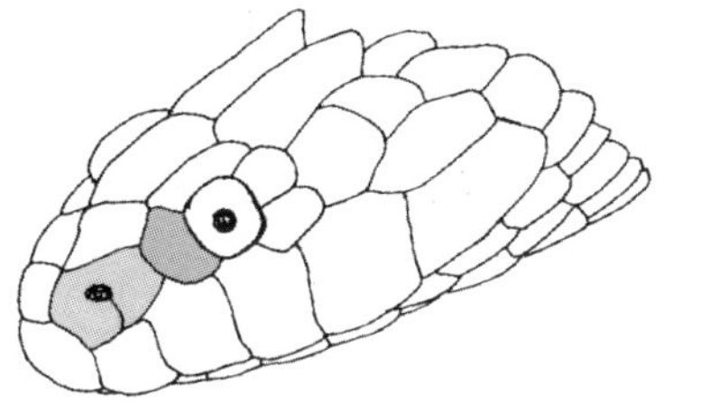

Fig 1: *Fordonia leucobalia* - a mangrove snake. Loreal scale absent, preocular scale (dark shaded) in contact with nasal scale (light shaded).

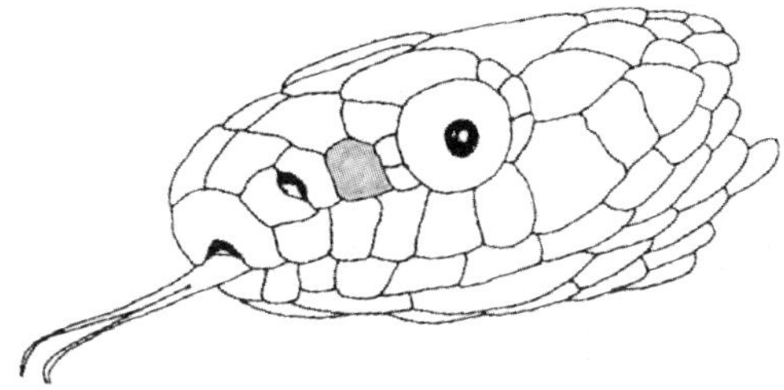

Fig 2: *Tropidonophis statisticus* - a keelback. Loreal scale present (shaded) separating preocular scales and nasal scales.

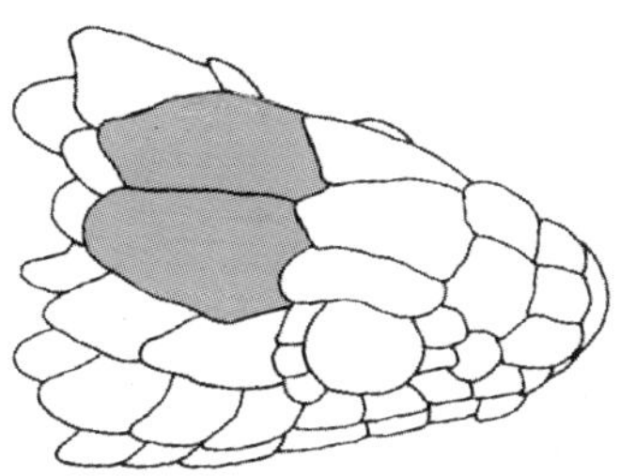

Fig 3: *Tropidonophis mairii* - a keelback. Parietal scales (shaded) entire.

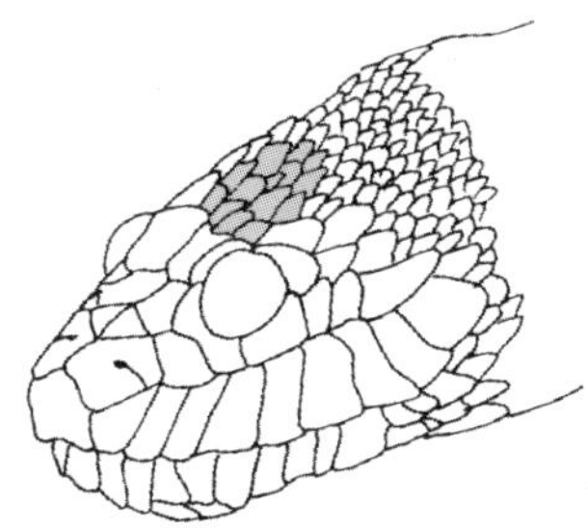

Fig 4: *Cerberus rynchops* - a watersnake. Parietals scales (shaded) fragmented.

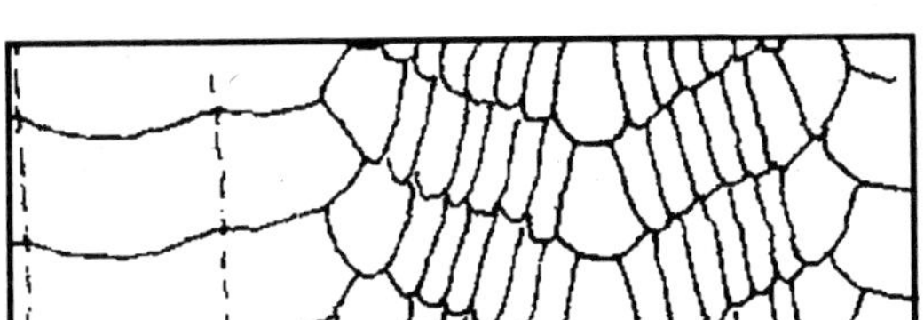

Fig 5: *Dendrelaphis* or *Boiga* - treesnakes. Dorsal scales arranged obliquely; ventral scales with a longitudinally keeled edge; body 'arch-shaped' in cross-section.

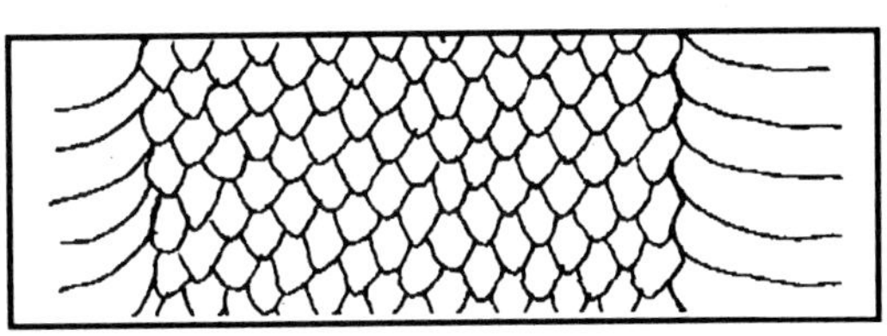

Fig 6: Dorsal scales arranged normally; ventral scales without a longitudinally keeled edge; body rounded in cross-section.

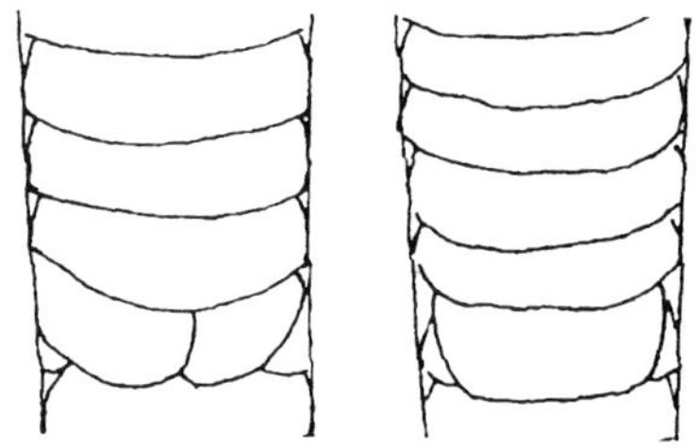

Fig 7: Anal plate divided (left) or entire (right).

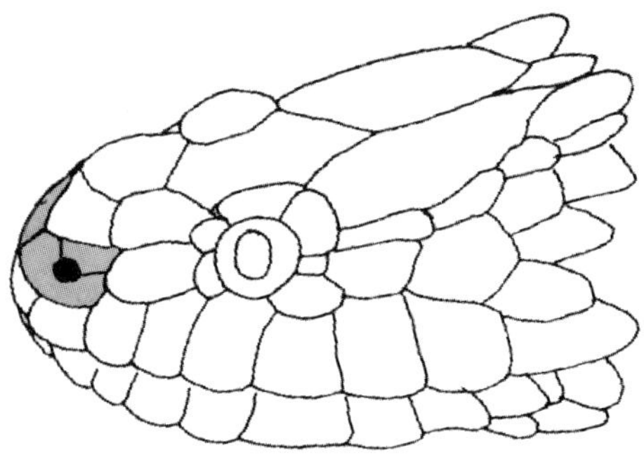

Fig 8: *Enhydris polylepis* - a watersnake.
Nasal scales (shaded) in contact.

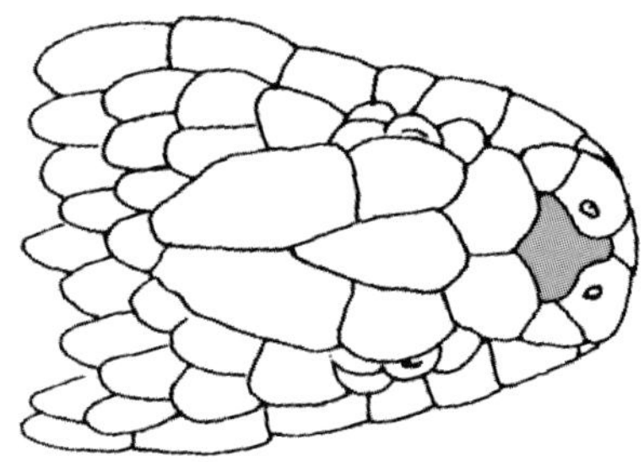

Fig 9: *Myron richardsoni* - mangrove snake.
Nasal scales separated by single internasal (shaded).

KEELBACKS

Tropidonophis spp.

(13 species) Maps: 14a-d

NON VENOMOUS

Average/Maximum length: 0.3-0.8m / 1.25m.

Description:

a) physique: Snakes of moderately slender build with elongate tails; head slightly distinct from the neck, small in *T. picturatus*; eye moderately large with a round pupil.

b) colouration: Usually dark grey, brown, olive or red-brown dorsally with darker speckling or barring or light stripes anteriorly, especially across the nape of the neck, in some species the stripes become more evident posteriorly and on the tail. Head generally as body or darker although snout tip and labials are lighter, the latter often yellow or white with dark surtures, black postocular stripes are present in most species (*T. aenigmaticus, T. dahlii, T. elongatus, T. hypomelas, T. mairii plumbea, T. mcdowelli, T. multiscutellatus, T. picturatus* and *T. statisticus*). Underside yellow to off-white with or without invasions of dorsal pigmentation as speckling, spotting or longitudinal stripes which become more apparent on the tail (*see* Malnate and Underwood 1988 for fuller species descriptions.)

c) scalation: (*see* key to the species and individual accounts below.) In all species the dorsal scales are keeled (weakly in *T. novaeguineae*), the subcaudals are paired, the anal plate is divided and the loreal is present.

Habitat: In close proximity to water in rainforest, open or cultivated country and towns. Inhabiting pool edges, slow rivers, swamps, creeks and drainage ditches.

Habits: Keelbacks are semi-aquatic in habit being equally at home on land as in the water but are usually encountered in or around freshwater environments. They may be diurnal or nocturnal and are generally inoffensive, though they are inclined to actively 'bluff' aggression if threatened. In this way dark specimens encountered around villages are often mis-identified as Papuan blacksnakes and killed. Oviparous with clutch sizes ranging from one to ten.

Diet: Primarily frogs but fish, tadpoles, frogs eggs and, less frequently, small skinks (*Cryptoblepharus*) and geckoes (*Nactus*) may be taken. Keelbacks appear to be the only amphibiophagous (frog-eating) snakes in New Guinea capable of preying on small specimens of the highly toxic, introduced S. American cane toad, *Bufo marinus*. This may be due to their ancestral origins, since the keelbacks are an Oriental group which spread southwards into Australasia and modern-day Asian species commonly feed on bufonid toads. Cane toads are believed to be one of the contributary factors in the demise of many true Australasian frog-eating animals, including snakes, which lack the keelback's defences against the powerful bufotoxin.

Taxonomic note: The New Guinea keelbacks have undergone a number of taxonomic changes in recent years and the generic names *Natrix, Amphiesma, Macropophis* and *Styporhynchus* may still be encountered in the literature. Malnate and Underwood (1988) carried out an intensive study of the Australasian members of this taxa with the result that several formerly recognised, wide-ranging species were split into more localised species i.e. the PNG populations of *T. montanus* (now confined to Irian Jaya) which became *T. aenigmaticus, T. statisticus* and *T. parkeri,* and also *T. mairii* which is now recognised as two separate subspecies occurring in the lowlands of Central and Western Provinces respectively. An older name exists for Western Province populations of *T. mairii* and therefore *T. m. brongersmai* (Malnate and Underwood 1988) is included here as *T. m. plumbea* (Shea 1990).

Distribution within PNG: All mainland and island provinces excluding North Solomons and Manus Provinces.

Distribution outside PNG: Several species inhabit Irian Jaya and one (*T. m. mairii*) occurs in northern and eastern Australia.

Keelback - *Tropidonophis multiscutellatus*
(Kar Kar Island, Madang Province)

Common keelback
– *Tropidonophis mairii mairii*
(Inauabui, Mekeo,Central Province)

Montane keelback
– *Tropidonophis statisticus*
(Tapini-Avios, Central Province)

Barred keelback
– *Tropidonophis doriae*
(Trifas, West Sepik Province)
Photo: Allen Allison

Key to the keelbacks, genus *Tropidonophis*, of New Guinea*
(adapted from Malnate and Underwood 1988) *See* Figure 1 head scalation characters.

1a. 17 dorsal scale rows at midbody. 2
1b. 15 dorsal scale rows at midbody. 4

2a. 8 supralabials with 3rd and 4th or 4th and 5th in contact with the eye. Mainland New Guinea and Aru Is. *Tropidonophis doriae*
2b. 9 supralabials with 5th and 6th in contact with the eye. 3

3a. Ventrals 170-185; subcaudals 101-114, subcaudal ratio** 0.369-0.396. Bismarck Archipelago. *Tropidonophis dahlii*
3b. Ventrals 184-196; subcaudals 103-104, subcaudal ratio** 0.344-0.350. Bismarck Archipelago. *Tropidonophis hypomelas*

4a. 3 posterior temporals, rarely 2, 4 or more. 5
4b. 2 posterior temporals, rarely 1, 3 or more. 7

5a. Ventrals 155-175; subcaudals 85-108. Extreme western Irian Jaya and Moluccas. *Tropidonophis elongatus*
5b. Ventrals 117-143; subcaudals 38-68. 6

6a. Fewer than 70 subcaudals; ventrals 117-140; 8 infralabials, rarely 7, 9 or more; dorsal pattern distinct, wide spaced narrow bands or dark spots on neck. Mainland New Guinea. *Tropidonophis picturatus*
6b. Fewer than 60 subcaudals; ventrals 128-143; 9 infralabials, rarely 8 or 10; dorsal pattern distinct, stripes on head and neck or unicolour. Mainland New Guinea. *Tropidonophis novaeguineae*

7a. 9, rarely 8, supralabials. West Sepik and Irian Jaya. *Tropidonophis mcdowelli*
7b. 8 supralabials, rarely 7 or fewer, or 9. 8

8a. 2, rarely 1 or 3, preoculars. 9
8b. 1, rarely 2, preoculars. 13

9a. 1 anterior temporal (fig. 1), rarely 2 or more. Mainland New Guinea and Milne Bay islands. *Tropidonophis statisticus*
9b. 2 anterior temporals, rarely 1, 3 or more. 10

10a. 8 infralabials, rarely 7 or 9; ventrals 152-171; subcaudals 79-89. Irian Jaya. *Tropidonophis montanus*
10b. 9 infralabials, rarely 8 or fewer or 10; ventrals 136-169; subcaudals 62 - 103. 11

11a. Ventrals 148-169; subcaudals 80-100. Highland provinces of PNG. *Tropidonophis parkeri*
11b. Ventrals 136-158; subcaudals 62-103. 12

12a. Ventrals 136-158; subcaudals 74-100. Mainland New Guinea and satellite islands. *Tropidonophis multiscutellatus*
12b. Ventrals 140-152; subcaudals 62-86. Eastern PNG and Milne Bay Archipelago. *Tropidonophis aenigmaticus*

13a. 1, rarely 2, anterior temporals; ventrals 140-166; subcaudals 58-76; dark postocular stripe usually absent. Central Province and northern Australia. *Tropidonophis mairii mairii*
13b. 1-2, rarely 3, anterior temporals; ventrals 133-152; subcaudals 70-80; dark postocular stripe usually present. Western Province and southeastern Irian Jaya. *Tropidonophis mairii plumbea*

* Note: Other than the number of dorsal scales at midbody, many of the scale counts recorded for *Tropidonophis* spp. show considerable overlap. Also dorsal colouration or patterning can be quite variable intraspecifically. Therefore it is suggested that the reader takes into consideration the altitudinal and geographical location of the collection site and refers to Malnate and Underwood (1988) when attempting to identify difficult specimens from the genus.

** Subcaudal ratio $= \frac{sc}{v + sc}$ where sc = no. subcaudals; v = no. ventrals

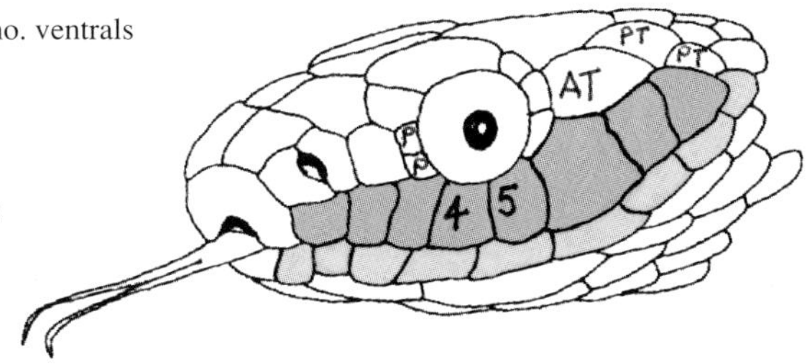

Fig 1: *Tropidonophis statisticus* - a keelback.
Dark shading - supralabials, 4th & 5th in contact with the eye; light shading - infralabials; P - preoculars; AT - anterior temporals; PT - posterior temporals.

'EAST PAPUAN MONTANE KEELBACK'

Tropidonophis aenigmaticus

Maximum length: 0.93m

Description: scalation: DMB 15; V 140-152; SC 62-86; 2 preoculars; 3, rarely 2 or 4, postoculars; 8, rarely 7 or 9, supralabials with 3rd-5th or 4th-5th in contact with eye; 9, rarely 8 or fewer, infralabials; 2-3, rarely 1 or 4, anterior temporals; 2-3, rarely 1 or 5, posterior temporals.

Distribution within PNG: Altitude records 700-1700m. E.HIGH: Aiyura and Kratke Mts. MOR: Huon Pen. and Garaina. M.BAY: Mt Dayman, Goodenough, Fergusson, Normanby, Trobriand and Misima Is.

Distribution outside PNG: None, an apparent PNG endemic.

'NEW BRITAIN KEELBACK'

Tropidonophis dahlii

Maximum length: 1.245m

Description: scalation: DMB 17; V 170-185; SC 101-114; 2 preoculars; 3, rarely 4, postoculars; 9, rarely 8, supralabials with 5th-6th in contact with eye; 9, rarely 8 or 10, infralabials; 2, rarely 3, anterior temporals; 3, rarely 2, posterior temporals.

Distribution within PNG: Altitude records 760-1070m. WNB: Whiteman Range and Willaumez Pen. ENB: Gazelle Pen. and Rabaul.

Distribution outside PNG: None, a PNG endemic.

BARRED KEELBACK

Tropidonophis doriae

Maximum length: 1.05m.

Description: scalation: DMB 17; V 134-159; SC 71-90; 2, rarely 1 or 3, preoculars; 3, rarely 2 or 4, postoculars; 8, rarely 7 or 9, supralabials with 3rd-5th or 4th-6th in contact with eye; 9, rarely 8 or 10, infralabials; 2, rarely 1, 3 or 4, anterior temporals; 2-3, rarely 4, posterior temporals.

Distribution within PNG: Altitude records sea level to 1545m. WEST: Lower and upper Fly R., Oriomo and Aramia R. GULF: Kikori and Purari R. CENT: Owen Stanley Mts. and Agaun. M.BAY: Mt Dayman. W.HIGH: Jimi R. MOR: Markham R. MAD: Teptep. E.SPK: Sepik R. W.SPK: Torricelli Mts.

Distribution outside PNG: Irian Jaya, east of Vogelkop, and the Aru Islands.

'BISMARCK KEELBACK'

Tropidonophis hypomelas

Maximum length: 0.95m.

Description: scalation: DMB 17; V 184-196; SC 103-104; 2 preoculars; 3, rarely 4, postoculars; 9, rarely 8, supralabials with 5th and 6th in contact with eye; 9, rarely 10, infralabials; 2, rarely 1, anterior temporals;

3, rarely 2 or 5, posterior temporals.
Distribution within PNG: Altitude record 950m. WNB: Cape Hoskins. ENB: Duke of York Is. N.IRE: Lelet Plateau.
Distribution outside PNG: None, a PNG endemic.

'NORTHERN NEW GUINEA or McDOWELL'S KEELBACK'
Tropidonophis mcdowelli
Maximum length: 0.65m.
Description: scalation: DBM 15; V 135-143; SC 79-95; 2, rarely 3, preoculars; 3 postoculars; 9, rarely 8, supralabials with 4th-6th or 3rd-5th in contact with eye; 9, rarely 8 or 10, infralabials; 3, rarely 1, 2 or 4, anterior temporals; 2, rarely 1, 3 or 4, posterior temporals.
Distribution within PNG: Altitude records 580-1885m. W.SPK: (Torricelli Mts).
Distribution outside PNG: Eastern Irian Jaya.

COMMON KEELBACK
Tropidonophis mairii
Maximum length: 0.85m.
Description: scalation: DMB 15; V 140-166 *(T. m. mairii*), 133-152 (*T. m. plumbea*); SC 58-76 (*T. m. mairii*), 70-80 (*T. m. plumbea*); 1, rarely 2, preoculars; 3, rarely 2 or 4, postoculars; 8, rarely 7 or 9, supralabials with 3rd-5th or 4th-6th in contact with eye; 9, rarely 7, 8 or 10, infralabials;1 - 2, rarely 3, anterior temporals; 2, rarely 1 or 3, posterior temporals.
Distribution within PNG: Altitude records sea level to 20m
T. m. mairii NCD. CENT: Brown and Laloki R., Subitana and Mekeo.
T. m. plumbea WEST: S. Trans-Fly from Aramia and Bamu R. to Bensbach R., Lake Murray and upper Fly R.
Distribution outside PNG: *T. m. mairii,* northern and eastern Australia. *T.m. plumbea,* se Irian Jaya (Merauke R.).

'MANY-SCALED KEELBACK'
Tropidonophis multiscutellatus
Maximum length: 0.95m.
Description: scalation: DMB 15; V 136-158; SC 74-103; 2, rarely 1 or 3, preoculars; 3, rarely 2 or 4, postoculars; 8, rarely 7 or 9, supralabials with 3rd-5th or 4th-6th in contact with eye; 9, rarely 8 or 10, infralabials; 2, rarely 1, 3 or 4, anterior temporals; 2, rarely 1, 3 or 4, posterior temporals.
Distribution within PNG: Altitude records 15m-1440m. WEST: s. Trans-Fly, upper Fly, Aramia and Bamu R. GULF: Pio R., Baimuru and Omati. CENT: Owen Stanley Mts and Mt Victoria. ORO: Owen Stanley Mts. and Popondetta. M.BAY: Fife Bay. MOR: Huon Pen., Gusiko, Bulolo, Wau, Lau and Garaina. MAD: Ramu R., Astrolabe Bay, Kar Kar Is. and Manam Is. W.HIGH: Baiyer R. and Tigi Plantation. SIM: Nondugl and Kubor Mts. E.SPK: Sepik R. and Prince Alexander Mts. W.SPK: Torricelli Mts. and coast.
Distribution outside PNG: Irian Jaya and islands to north and east.

NEW GUINEA KEELBACK
Tropidonophis novaeguineae
Maximum length: 0.82m.
Description: scalation: DMB 15; V 128 -143; SC 38-59; 2 preoculars; 3-4, rarely 4, postoculars; 8, rarely 7 or 9, supralabials with 3rd-5th in contact with eye; 9, rarely 8 or 10, infralabials; 2, rarely 1, 3 or 4, anterior temporals; 3, rarely 2, 4 or 5, posterior temporals.
Distribution within PNG: Altitude records 75-1067m.WEST: Lake Murray and Boboa Is. GULF: Purari R. and Omati. CENT: Owen Stanley Mts. SIM: Tive Plateau.
Distribution outside PNG: Central and western Irian Jaya.

'HIGHLAND' or 'PARKER'S KEELBACK'
Tropidonophis parkeri
Maximum length: 0.93m
Description: scalation: DMB 15; V 148-169; SC 80-100; 2, rarely 3, preoculars; 3, rarely 2, postoculars; 8, rarely 9, supralabials with 3rd-5th or 4th-5th in contact with eye; 9, rarely 8, infralabials; 2, rarely 3, anterior temporals; 2, rarely 3, posterior temporals.
Distribution within PNG: Altitude records 1070-2130m. W.HIGH: Waghi R. SIM: Kundiawa to Karimui. E.HIGH: Kratke Mts.
Distribution outside PNG: None, an apparent PNG endemic.

PAINTED KEELBACK
Tropidonophis picturatus
Maximum length: 0.58m.
Description: scalation: DMB 15; V 117-140; SC 38-68; 2, rarely 1 or 3, preoculars; 3, rarely 4 or 5, postoculars; 8-9, rarely 7, supralabials with 4th-5th or 4th-6th in contact with eye; 8-9, rarely 7 or 10, infralabials; 2-3, rarely 1 or 4, anterior and posterior temporals.
Distribution within PNG: Altitude records 200-1545m. WEST: Lake Murray and Aramia R. GULF: Omati. CENT: Owen Stanley Mts. M.BAY: Mt Dayman. SIM: Tive Plateau. MOR: Garaina. W.SPK: Torricelli Mts.
Distribution outside PNG: Irian Jaya, Misool, Salwatti and Waigeau.

'PNG MONTANE KEELBACK'
Tropidonophis statisticus
Maximum length: 0.87m.
Description: scalation: DMB 15; V 151-174; SC 65-87; 2, rarely 3, preoculars; 3, rarely 2 or 4, postoculars; 8, rarely 7 or 9, supralabials with 3rd-5th or 4th-6th in contact with eye; 9 -10, rarely 8, infralabials; 1, rarely 2 or 3, anterior temporals; 2, rarely 1 or 3, posterior temporals.
Distribution within PNG: Altitide records 900-2150m. WEST: upper Fly R. CENT: Owen Stanley Mts. M.BAY: Goodenough Is. E.HIGH: Goroka, Aiyura and Kratke Mts. ENGA: Wabag and Wapenamanda. SIM: Dumun. W.HIGH: Kaironk, Jimi, Waghi and Baiyer R., Mt Hagen. S.HIGH: Hurria R. MOR: Mt Rawlinson and Garaina. E.SPK: Sepik R. and Wewak.
Distribution outside PNG: Central and northern Irian Jaya.

Extralimital Irian Jaya species yet to be reported from PNG:
'MOLUCCAN KEELBACK'
Tropidonophis elongatus
Maximum length: 1.16m.
Description: scalation: DMB 15; V 155 -175; SC 85-107; 2, rarely 1 or 3, preoculars; 3, rarely 4, postoculars; 8, rarely 7 or 9, supralabials with 3rd-5th or 4th-6th in contact with eye; 9-10, rarely 8, infralabials; 2, rarely 1, 3 or 5, anterior temporals; 2-3, rarely 1, 4 or 5, posterior temporals.
Distribution outside PNG: N. and W. Irian Jaya, Biak and Numfor to the north, Salawatti, Seram, Ambon and Halmahera to the west.

'NORTHERN IRIAN MONTANE KEELBACK'
Tropidonophis montanus
Maximum length: 1.14m.
Description: scalation: DMB 15; V 152-171; SC 71-89; 2, rarely 3, preoculars; 3, rarely 2 or 4, postoculars; 8, rarely 7 or 9, supralabials with 3rd- 5th or 4th-5th in contact with eye; 8-9, rarely 7, infralabials; 2, rarely 1 or 3, anterior temporals; 2, rarely 1, 3 or 4, posterior temporals.
Distribution outside PNG: Central and western Irian Jaya.

TREESNAKES
Dendrelaphis spp.

(6 species) Maps: 15a-b

NON VENOMOUS
Average/Maximum length: 0.9 - 1.4m / 2.5m.
Description:

a) physique: Extremely slender with long 'whip-like' tails; head barely distinct from the neck; eye very large with a round pupil. The ventral scales exhibit a sharp ridge running down either side presenting an 'arch-shape' in cross-section which enables traction on vertical tree trunks.
b) colouration: Highly variable interspecifically and intraspecifically depending on geographical location. Dorsally from green to yellow brown or brown to black with lighter underbellies and lacking longitudinal stripes for most or all of the body (*see* key to the species and individual accounts below). Although the colour and patterning of the interstitial skin of the neck, which is exposed during threat, may provide a useful means of identification of live specimens locally, when considered for New Guinea as a whole this is, infraspecifically, a highly variable characteristic and one which also shows considerable interspecific overlap. Interstitial skin colour is not, therefore, included here as a means of differentiating between species of *Dendrelaphis*. Interstitial skin pigmentation is discussed by McDowell (1984).
c) scalation: DMB 13, arranged obliquely, all smooth with apical pits; otherwise fairly variable interspecifically (*see* key to the species and individual accounts below).

Habitat: Inhabiting a wide variety of habitats from rainforest, along trails and in clearings, to banana gardens and coconut plantations.

Habits: Highly agile diurnal arboreal treesnakes of the genus *Dendrelaphis* inflate their necks, exposing brightly coloured interstitial skin, and gape as a defensive response to being threatened. Large specimens of *D. punctulatus* can appear threatening and are often killed, especially all-black Western Province specimens, in mistake for Papuan blacksnakes. Oviparous with clutch sizes from 8 - 16.

Diet: Primarily frogs, geckos and small skinks.

Taxonomic note: The Australo-Papuan-Solomons treesnakes of the genus *Dendrelaphis* are a fairly confusing group. Due to the considerable interspecific overlap of external morphological characteristics present within the genus precise specific identity, especially of dark specimens, may be difficult to determine. Further complicating the situation, wide-ranging species (*D. calligastra* and *D. punctulatus*) often exhibit considerable variation in both colouration and patterning from province to province. Authors generally recognise six Australo-Papuan species which McDowell (1984) placed in three groups based on the structure of their hemipenes; the monotypic *D. lorentzi* and *D. papuensis* groups and the *D. punctulatus* group which contains the remaining four species. The genus *Dendrelaphis* is a complicated taxa containing numerous earlier species described from individual specimens. Parker (1982) considered *Dendrophis katowensis* (Macleay) from Western Province to be a synonym of *Dendrelaphis calligastra* (Günther). McDowell (1984) reported *D. meeki* (Boulenger) as a synonym of *D. gastrostictus* (Boulenger). Several other synonyms are also listed in these papers.

Distribution within PNG: Throughout mainland New Guinea, excluding the highlands, (apart from *D. gastrostictus*) and all the major archipelagoes.

Distribution outside PNG: Several PNG species also occur in Irian Jaya, northern Australia or the Solomon Islands.

Top: Green Treesnake
Dendrelaphis calligastra
(Kar Kar Is., Madang Province)

Left: Treesnake
Dendrelaphis sp.
(Riwo, Madang Province)

Below: Common Treesnake
Dendrelaphis punctulatus
(Bannon Bridge, Central Province)

Key to the treesnakes, genus *Dendrelaphis*, of New Guinea

McDowell's (1984) characters include the architecture of the hemipenes, dentition and structure of the skull, characters of little value to field workers or natural historians. *See* McDowell for a discussion of these characters which are otherwise omitted from this guide.

1a. Dark stripe present along upper edge of lip scales, separating light labial scales from darker pigment of dorsum of head; scales of dorsum of head lacking pits; average length 0.7 - 1.0m, maximum length 1.3m. 2

1b. No dark stripe along upper edge of lips, dark dorsal pigment invading lighter labial scales; large indented pits present on dorsal scales of head; average length 1.2m, maximum length 1.6m. *Dendrelaphis punctulatus*

2a. Fewer than 140 subcaudal scales; small nostril, nasal scale not completely divided. 3

2b. More than 140 subcaudal scales; large nostril, nasal scale completely divided. 5

3a. Mainland provinces. 4

3b. North Solomons and Milne Bay (Misima Is.) *Dendrelaphis salomonis*

4a. Ventral scales 156-173; pale vertebral stripe absent. *Dendrelaphis lorentzi*

4b. Ventral scales 183-203; pale vertebral stripe present or absent. *Dendrelaphis papuensis*

5a. Highland and montane habitats and Milne Bay (Normanby and Fergusson Is.); ventrals 160 -180; subcaudals 147-164. *Dendrelaphis gastrostictus*

5b. Coastal lowland habitats, Bismarck Arch. and Milne Bay (Normanby Is.); ventrals 169-197; subcaudals 142-157. *Dendrelaphis calligastra*

5c. Lowland island habitats, North Solomons and Milne Bay (Misima Is.); ventrals 173-191; subcaudals 124-166. *Dendrelaphis salomonis*

Note: Parker (1982) reports a possible undescribed species from Lake Murray and Kiunga, Western Province. It is a large species (no lengths given) with a black head and neck, body pale brown anteriorly with black transverse bars which increase in frequency and width and become closer together posteriorly resulting in a dark grey to olive brown posterior body and a black tail. Interstitial skin is reported as white or yellow anteriorly and grey-blue posteriorly. Labials and chin are cream, the anterior belly olive -grey and the underside of the tail, black. No scale counts are provided.

COCONUT or GREEN TREESNAKE

Dendrelaphis calligastra

Average/Maximum length: 1.0m / 1.27m.

Description:

b) colouration: Fairly variable dorsally from green to light or bronze brown; underside pale brown; skin between scales often brightly coloured, usually light blue; a broad black stripe from the snout, under and through the eye and running onto the anterior flank of the body separating the light labial scales from the darker dorsal head scales of the head and neck.

c) scalation: DMB 13 arranged obliquely, all smooth; V 169-197; SC 142-157, all paired; anal plate divided; loreal present; 8, rarely 9, supralabials with 4th and 5th or 5th and 6th, in contact with eye; 1 preocular; 2, rarely 3, postoculars.

Distribution within PNG: Altitude records 0-1150m. W.SPK: Aitape and Telefomin; E.SPK: Wewak. MAD: Saidor, Madang, Astrolabe Bay, Adelbert Mts., Kar Kar Is. and Ramu R. delta. MOR: Huon Pen., Lae and Mt Missim. E.HIGH: eastern edge only, no precise locality data. ORO: Popondetta. M.BAY: Fife

Bay and d'Entrecasteaux Is., Normanby Is. NCD: Port Moresby. CENT: Mekeo, Brown R. and Hood Bay. GULF: Omati. WEST: Daru Is., Oriomo Plateau, Trans-Fly, Lake Murray, upper Fly R., Aramia R., Ningerum and Kiunga. ENB: Nakatani Mts. WNB: Garu and Lake Dakatana.
Distribution outside PNG: Irian Jaya and islands to west and south, islands of Torres Strait and Queensland, Australia.

'MONTANE or SPOTTED-BELLIED TREESNAKE'
Dendrelaphis gastrostictus
Average/Maximum length: 1.0m / 1.12m.
Description:
b) colouration: Bronze above on head and body, tail darker; underside, throat and labials yellow with black spotting; a fine black line from the behind the eye, which may continue to the snout, separates the dark dorsal surface of the head from the pale lip scales.
c) scalation: DMB 13 arranged obliquely, all smooth; V 160-180; SC 147-164, all paired; anal plate divided; loreal present; 8 supralabials with 4th and 5th in contact with eye; 1 preocular; 3 postoculars.
Distribution within PNG: Altitude records 0-1475m. WEST: Lake Daviumbo, Wipim, Aramia R. and upper Fly R. GULF: Omati. CENT: Owen Stanley Mts. M.BAY: d'Entrecasteaux Is., Normanby and Fergusson Is. ORO: Kokoda. MOR: Huon Pen. and Mt Missim. E.HIGH: no precise locality data. SIM: no precise locality data. W.HIGH: no precise locality data. *D. gastrostictus* appears to generally occur at altitude although the Western Province collection localities and Gusiko, Huon Pen., Morobe Province are virtually at sea level.
Distribution outside PNG: Northern Irian Jaya.

'LORENTZ RIVER TREESNAKE'
Dendrelaphis lorentzi
Average/Maximum length: 0.7m / 0.82m.
Description:
b) colouration: Brown or olive above on head and body, tail darker, some scales black edged and occasional vertebral scales all black; underside immaculate white or grey, throat yellow; a fine black line passing beneath the eye separating the dorsum of the head from the labials.
c) scalation: DMB 13 arranged obliquely, all smooth; V 156-173; SC 122-136, all paired; anal plate divided; loreal present; 8 (WEST) or 9 (MOR) supralabials with 4th and 5th (WEST) or 5th and 6th (MOR) in contact with eye; 10 infralabials; 1 preocular; 2 or 3 postoculars.
Distribution within PNG: Altitude records 0-655m. WEST: Balimo, Aramia R. and Tabubil, upper Fly R. MOR: Gusiko, Huon Pen.
Distribution outside PNG: Southern Irian Jaya (Lorentz and Mimika Rivers).

PAPUAN TREESNAKE
Dendrelaphis papuensis
Average/Maximum length: 0.7m / 1.05m.
Description:
b) colouration: Brown or red-brown above on head and body; underside patterned with large dark and light blotches, speckled with black; a fine black line passing beneath the eye separating the dorsum of the head from the pale throat and labials and continuing onto neck; pale vertebral stripe may be present.
c) scalation: DMB 13 arranged obliquely, all smooth; V 183-203; SC 119-136, all paired; anal plate divided; loreal present; 8, supralabials with 4th and 5th in contact with eye; 1 preocular; 2 postoculars.
Distribution within PNG: WEST: Oriomo Plateau and upper Fly R., Kiunga. NCD: Port Moresby. CENT: Owen Stanley Mts. M.BAY: Cape Vogel, Fife Bay, Fergusson, Trobriand, Woodlark, Tagula and Rossel Is. ORO: Popondetta and Sangara. MOR: Garaina.

Distribution outside PNG: Not recorded, possibly southeast Irian Jaya.

COMMON or BLACK TREESNAKE

Dendrelaphis punctulatus

Average/Maximum length: 1.2m / 1.64m.

Description:

b) colouration: Black or brown above, slightly lighter laterally, labials and throat pale yellow or white but lacking dark stripe separating labials from the dorsum of the head with the result that the darker dorsal pigment may encroach onto lighter labials; underside white.

c) scalation: DMB 13 arranged obliquely, all smooth; V 185-220; SC 118-144, all paired; anal plate divided; loreal present; 7 or 8, rarely 9, supralabials with 4th and 5th or 5th and 6th in contact with eye; 1 preocular; 2 postoculars; large indented pits present on dorsal scales of head.

Distribution within PNG: WEST: Daru and Bobo Is., Oriomo Plateau, Trans-Fly, Bensbach R., Aramia R., Balimo, Emeti, Lake Murray, Boboa Is., upper Fly R., Kiunga and Olsobip. GULF: Uramu Is. and Baimuru. NCD: Port Moresby. CENT: Mekeo, Brown R., Sogeri, Owen Stanley Mts., Kwikila, Hood Bay. M.BAY: d'Entrecasteaux Is., Fergusson Is., no precise mainland locality data. ORO: Popondetta and Kokoda. MOR: Lae and Huon Pen. MAD: Astrolabe Bay. E.SPK: Kairiru Is. W.SPK: no precise locality data. WNB: no precise locality data. ENB: Matalau Is., Duke of York Is.; no precise New Britain locality data. N.IRE: Danfu.

Distribution outside PNG: Northeastern Australia.

SOLOMONS TREESNAKE

Dendrelaphis salomonis

Average length: 1.0m.

Description:

b) colouration: Green to light or dark brown above, often reddish anteriorly with darker indistinct vertical bars; underside grey, yellowish or cream; a black stripe from the snout, through the eye and onto the temporal region and the anterior flank of the body separating the yellow labial scales from the darker dorsal head scales. Although the Misima Is., Milne Bay population of *Dendrelaphis* is melanistic (all black) McDowell places them in *D. salomonis* on the basis of their scale counts, hemipenes and dentition.

c) scalation: DMB 13 arranged obliquely, all smooth; V 173-189 Bougainville and Buka, 178-191 Misima Is, 171-211 Solomons; SC 124-166, all paired; anal plate divided; loreal present; 8-9 supralabials with 4th and 5th or 5th and 6th in contact with eye; 9-10 infralabials; 1 preocular; 2 postoculars.

Distribution within PNG: N.SOL: Buin, Bougainville and Buka Is. M.BAY: Misima Is.

Distribution outside PNG: Solomon Islands.

GROUND SNAKES
Stegonotus spp.

(7 species) Maps 16a-c

NON VENOMOUS
Average/Maximum length: 0.6 - 1.0m / 1.3m.
Description:

a) physique: Body of moderate build; head slightly distinct from the neck; eye small to moderate, protruding with round pupil.
b) colouration: Unicolour dark grey or brown above with pale underside or either light yellow-grey with black edging to the scales or dark with light scale edging.
c) scalation: Highly variable both interspecifically and intraspecifically, with sex and geographic location (*see* key to the species and individual accounts below).

Habitat: Inhabiting a wide variety of habitats from rainforest to gardens and coconut plantations, on the ground under debris and arboreally in epiphytic plants or, in the case of small specimens, inside honey-combed ant plants. Also found inside buildings and commonly encountered after rain, crossing roads at night.

Habits: Ground snakes are crepuscular or nocturnal and although extremely agile they are also semi-fossorial and can rapidly disappear into leaf litter. Quite 'defensively aggressive', these snakes readily bite if handled. Although *Stegonotus* spp. are considered non-venomous they do possess enlarged teeth in the rear of the mouth with which they 'chew' vigorously. I have received a number of bites which have shown localised effects such as slight swelling, stiffening of the finger, localised redness and pain so the potential presence of mildly toxic saliva should not be discounted. Many technically harmless colubrid snakes around the world are capable of causing similar reactions and bites should be avoided, especially from large specimens of *S. cucullatus*. Oviparous with a clutch size of 6-12.

Diet: Primarily reptile eggs, but also small frogs, lizards, snakes, small mammals and possibly invertebrates.

Distribution within PNG: Throughout the mainland and island provinces.

Distribution outside PNG: Widespread through Irian Jaya, including neighbouring islands, the Moluccas, the Kei Islands, possibly the Aru Islands, and Northern Territory and Queensland, Australia, including the Torres Strait Islands.

Slatey-grey snake – *Stegonotus cucullatus*
(Goldie River, Central Province)

Slatey-grey snake – *Stegonotus cucullatus*
(Wau, Morobe Province) Photo: Mike McCoy

Diehl's little ground snake
Stegonotus diehli
(Northcoast Road,Madang Province)

Common ground snake
Stegonotus parvus
(Northcoast Road, Madang Province)

Juvenile Diehl's little ground snake
Stegonotus diehli
(Riwo, Madang Province)

Milne Bay ground snake
Stegonotus guentheri
(Fergusson Is., Milne Bay Province)
Photo: Daniel Houclin

Northern ground snake
Stegonotus modestus
(Siar, Madang Province)

Key to the ground snake genus *Stegonotus* of New Guinea

1a. 15 dorsal scale rows at midbody. 2
1b. 17 or 19 dorsal scale rows at midbody. 3

2a. A pair of well developed apical pits present on every dorsal scale of the body and anterior tail (fig. 1); juveniles with black blotches on a pale head; Sepik Provinces, and Madang. *Stegonotus diehli*
2b Apical pits absent (fig. 2) or, if present, scattered and vestigial; juveniles patterned as adults; Milne Bay Islands. *Stegonotus guentheri*

3a. A pair of well developed apical pits present on every dorsal scale of the body and anterior tail (fig. 1); juveniles with black blotches on a pale head. *Stegonotus diehli*
3b Apical pits absent (fig. 2) or, if present, scattered and vestigial; juveniles patterned as adults. 4

4a. 17 or 19 dorsal scale rows at midbody; 8 or 9 supralabials; mainland provinces excluding West Sepik and Madang. *Stegonotus cucullatus*
4b. 17 dorsal scale rows at midbody, (19 in Manus); 7 supralabials. 5

5a. Fewer than 194 ventral scales; fewer than 88 subcaudal scales; Bismarck Archipelago. *Stegonotus heterurus*
5b. More than 194 ventral scales; more than 84 subcaudal scales; Momase* and Manus provinces. *Stegonotus modestus*

6a. Ventral scales 165 - 196; subcaudal scales 82 - 99; mainland provinces and satellite islands. *Stegonotus parvus*
6b. Ventral scales 184 - 218; subcaudal scales 73 - 90; Bismarck Archipelago. *Stegonotus* cf. *parvus*

* Momase = Morobe, Madang and both Sepik provinces.

Stegonotus from Madang and Sepik populations, but not those from Morobe, can easily be identified using a combination of the three characteristics listed below:

	Dorsal scales at midbody	Supralabials	Scale colour under tail
S. diehli	15	7	Scales with dark edges
S. parvus	17	8	Scales without dark edges
S. modestus	17	7	Scales with dark edges

Fig 1:
Well developed apical pits present on each dorsal scale tip.

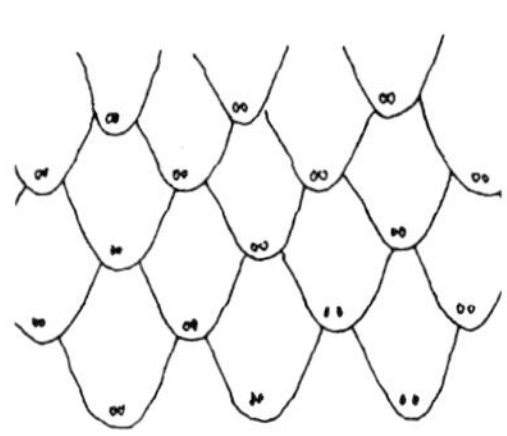

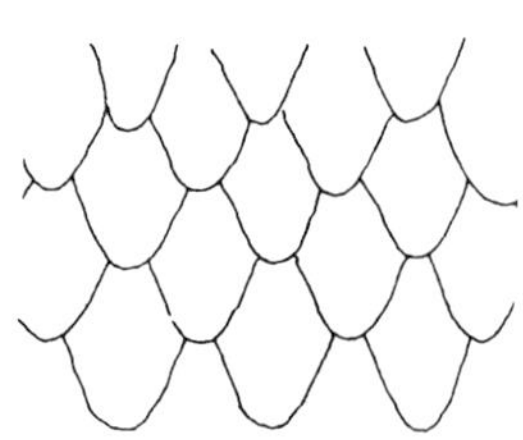

Fig 2:
Apical pits absent

SLATEY-GREY SNAKE

Stegonotus cucullatus

Average/Maximum length: 1.0m / 1.3m.

Description:

b) colouration: Two colour phases appear to exist; either head and body unicolour dark grey or brown above running onto outer edges of otherwise immaculate pale ventral scales (Western Province, and Woodlark and Rossel Islands, Milne Bay Province), or pale yellow-grey with black edging to all dorsal scales presenting a reticulate effect (Central Province and most of the rest of the range); underside of tail often with dark pigment overlaying lighter background.

c) scalation: DMB 17-19, with or without scattered apical pits, decreasing to 15 or 17 rows anterior to cloaca, all smooth; V 187-228; SC 69-101, all paired; anal plate entire; loreal present; 8 or 9 supralabials with 3rd and 4th, rarely 3rd, 4th and 5th, entering eye; 2, rarely 1, preoculars; 2, rarely 1 or 3, postoculars.

Distribution within PNG: Altitude records 0-1700m. WEST: Daru Is., Oriomo Plateau, Morehead, lower and upper Fly R., Kiunga and Ningerum. GULF: Purari R. and Baimuru. NCD: Port Moresby. CENT: Laloki R., Brown R., and Owen Stanley Mts. M.BAY: Moi-Biri Bay, Cape Vogel, Samarai Is., d'Entrecasteaux Is., Fergusson and Normanby Is., Woodlark Is. and Rossel Is. ORO: Popondetta, Kokoda and Mt Victoria. MOR: Garaina, Wau-Bulolo, Huon Gulf, Lae, Huon Pen., Markham R. E.HIGH: Aiyura and Okapa. SIM: Karimui, Kup and Nondugl. W.HIGH: Mt Hagen. MAD: Gogol R. and Kar Kar Is. E.SPK: Wewak. A doubtful record from N.SOL: Bougainville.

Distribution outside PNG: Irian Jaya, neighbouring islands, and northern Australia.

'DIEHL'S LITTLE GROUND SNAKE'

Stegonotus diehli

Average/Maximum length: 0.6m / 0.77m.

Description:

b) colouration: Adults unicolour dark grey or brown above with pale undersides, dark edging of ventral scales in southern populations, dark pigmentation on subcaudals in all populations. Juveniles are universally dark above, with a white head partially obscured by irregular black blotches which may form bars across the snout and nape.

c) scalation: Highly variable geographically and sexually. Sepik and Madang Provinces: DMB 15, each with a pair of apical pits present, also 15 dorsal scale rows anterior to cloaca, all smooth; V 166-181; SC 65-88, all paired; Morobe, central ranges and southern provinces: DMB 17, 15 dorsal scale rows anterior to cloaca; V 180-208; SC 78-104, all paired; anal plate entire; loreal present; 7, rarely 8, supralabials with 3rd and 4th entering eye; 1, rarely 2, preoculars; 1 or 2 postoculars.

Distribution within PNG: WEST: Daru Is., Aramia R. and upper Fly, Kiunga, Tabubil. GULF: Purari R. NCD: Port Moresby. CENT: Sogeri and Owen Stanley Mts. M.BAY: Cape Vogel. ORO: Mt. Victoria. MOR: Garaina, Lae and Huon Gulf. SIM: Karimui. MAD: Ramu R. delta, Bostrom Bay, Astrolabe Bay, Alexishafen and Saidor. E.SPK: Sepik R. and Wewak. W.SPK: Torricelli Mts.

Distribution outside PNG: Irian Jaya to Fak-Fak and Mansinam Island.

'MILNE BAY GROUND SNAKE'

Stegonotus guentheri

Average length: 1.13-1.15m

Description:

b) colouration: Brown dorsally except lowest scale row which has white margins; ventrals and labials off-white.

c) scalation: DMB 15, 15 anterior to cloaca, with or without scattered apical pits, all smooth; V 176-195; SC 64-78, all paired; anal plate entire; loreal present; 8 supralabials with 4th and 5th entering eye, occasionally 6 or 7 with 3rd and 4th entering eye; 1, rarely 2, preoculars; 2 postoculars.

Distribution within PNG: M.BAY: d'Entrecasteaux Is. - Goodenough, Fergusson and Normanby Is., Trobriand Is. - Kiriwina and Kitava Is.

Distribution outside PNG: No records, a PNG endemic.

'BISMARCK GROUND SNAKE'

Stegonotus heterurus

Average/Maximum length: 0.8/1.12m

Description:

b) colouration: Dark above, lighter brown on flanks; ventral and labials off-white.

c) scalation: DMB 17, reducing to 15 rows anterior to the cloaca, with or without scattered apical pits, all smooth; V 178-193; SC 76- 87 all or mostly entire; anal plate entire; loreal present; 7 supralabials with 3rd and 4th entering eye; 1, rarely 2, preoculars; 2 postoculars.

Distribution within PNG: ENB: Gazelle Pen., Rabaul and Duke of York Is. N.IRE: Namatanai and Muliama.

Distribution outside PNG: No records, a PNG endemic.

'NORTHERN NEW GUINEA GROUND SNAKE'

Stegonotus modestus

Average/Maximum length: 1.0/1.3m

Description:

b) colouration: Brown or grey above; labials and underside immaculate white anteriorly, subcaudals with darker edges.

c) scalation: DMB 17, rarely 18 or 19 in Manus Province, reducing to 15 anterior to the cloaca, without apical pits, all smooth; V 195- 216; SC 85-91, all paired; anal plate entire; loreal present; 7 supralabials with 3rd and 4th entering eye ; 1, rarely 2, preoculars; 2 postoculars.

Distribution within PNG: W.SPK: Torricelli Mts. E.SPK: Wewak and Kairiru Is. MAD: Astrolabe Bay and Alexishafen. MAN: Manus Is.

Distribution outside PNG: Irian Jaya and the Moluccas.

COMMON GROUND SNAKE

Stegonotus parvus

Average/Maximum length: 0.7/1.1m

Description:

b) colouration: Brown or grey above; labials and underside immaculate white or yellow-white.

c) scalation: Exhibiting considerable geographical variation; DMB 17, reducing to 15 anterior to the cloaca, without apical pits, all smooth; V 165-196; SC 82-99 all or mostly entire; anal plate entire; loreal present; 7 supralabials with 3rd and 4th entering eye (Morobe), 8 with 4th and 5th entering eye (Sepik Provinces and Madang); 2, occasionally 1, preoculars; 2 postoculars.

Distribution within PNG: WEST: Daru Is., Fly R. and Lake Daviumbo. MOR: Huon Pen., Lae, Watut, Markham and Ramu R. MAD: Ramu R., Astrolabe Bay, Bostrom Bay, Adelbert Mts., Kar Kar and Manam Is. E.SPK: Sepik R. delta and Wewak. W.SPK: Torricelli Mts. W.HIGH: Tigi and Baiyer R.

Distribution outside PNG: Irian Jaya, Jobi Island, and Murray Island, Torres Straits, far northern Queensland.

GROUND SNAKE
Stegonotus cf. *parvus*
Average/Maximum length: not available
Description:

b) colouration: Brown above including labials; underside immaculate white posteriorly grey.
c) scalation: DMB 17, reducing to 15 anterior to the cloaca, without apical pits, all smooth; V 184-218; SC 73-90 all paired; anal plate entire; loreal present; 8 with 4th and 5th entering eye; 2 preoculars; 2 postoculars.

Distribution within PNG: WNB: Whiteman Mts. ENB: Gazelle Pen. and Duke of York Is.
Distribution outside PNG: No records, a PNG endemic.

MANGROVE, MUD AND WATERSNAKES
Subfamily - Homalopsinae

(6 species) Maps: 17a-c

MILDLY VENOMOUS

Average/Maximum length: 0.4m - 0.62m / 1.0 - 1.2m (*Fordonia* & *Cerberus*).

Description:

a) physique: Snakes of moderately slender to stout build with fairly long, non-'paddle-shaped' tails; generally resembling terrestrial snakes but, when handled, exhibit less apparent 'muscle-tone' due to their aquatic existence; head just distinct from the neck, except *Cerberus* which has a broad head; eye small with round pupil; nostrils valvular and positioned on top of the head.

b) colouration: Highly variable interspecifically and in some species (*Fordonia*) intraspecifically (*see* key to the species and individual accounts below).

c) scalation: Highly variable interspecifically (*see* key to the species and individual accounts below).

Habitat: Coastal mangrove swamps, estuary mouths and mudflats. *Enhydris* is more usually associated with freshwater and may be found considerable distances inland inhabiting creeks and swampy grasslands but is also common in coastal areas. *Fordonia, Cantoria* and *Cerberus* have been collected in freshwater although they are primarily estuarine species which occur in fully marine conditions.

Habits: Mud and mangrove snakes are usually found nocturnally in brackish conditions, on mud-flats or in mangrove swamps where some species (*Fordonia, Myron* etc.) inhabit crab holes. All New Guinea homalopsine mud and mangrove snakes are viviparous with litter sizes ranging from 2-18, the largest litters being reported for *Enhydris* and *Fordonia*.

Diet: Crabs, other crustaceans, mudskippers and fish. *Fordonia* is a crab specialist, a carcinophage, although it will occasionally feed on other crustaceans, mud lobsters (thalassimids) or shrimps. Crabs are pinned in their burrows and 'chewed' until they are dismembered. *Fordonia,* the specialist, is not thought to prey on fish whereas *Enhydris*, the generalist, preys on fish, frogs and tadpoles. *Myron* and *Cerberus* feed only on fish.

Distribution within PNG: Western, Gulf and Central Provinces.

Distribution outside PNG: Irian Jaya and northern Australia, several species also occuring in Indo-Malaysia.

Venom: Nothing is recorded concerning the effects of homalopsine venom on man but bites are unlikely as these snakes are fairly inoffensive with small mouths and the fangs, although enlarged, are positioned in the rear of the mouth.

Smooth watersnake
Enhydris polylepis
(Irupi, Western Province)

Grey mangrove snake
Myron richardsoni
(Tapila, Fly River, Western Province)

White-bellied mangrove snakes
Fordonia leucobalia
(Binaturi River, Western Province)

Key to the mangrove, mud and watersnakes, subfamily Homalopsinae of New Guinea
(adapted from Ko Ko Gyi, 1970)

1a. Nasal scales in contact (fig. 1). 2
1b. Nasal scales separated by internasal scale (fig. 2). 3

2a. Dorsal scales smooth, in 21 - 27 (21 - 23 in Trans-Fly) rows at midbody; parietal scales entire (fig. 1). *Enhydris polylepis*
2b. Dorsal scales keeled, in 23 - 25 rows at midbody; parietal scales fragmented (fig. 3). *Cerberus rynchops novaeguineae*
3a. Loreal scale present (fig. 1); 7 - 9 supralabials (fig. 2). 4
3b. Loreal scale absent (fig. 4); 5 supralabials (fig. 4). *Fordonia leucobalia*

4a. Nasal cleft present, dividing nasal scale into two separate halves. 5
4b. Nasal cleft absent, nasal scale entire. *Cantoria annulata*

5a. Dorsal scales weakly keeled, in 19 - 21 rows at midbody; <150 ventrals; <45 subcaudals. *Myron richardsoni*
5b. Dorsal scales smooth, in 27 rows at midbody; >150 ventrals; >60 subcaudals. *Heurnia ventromaculata*

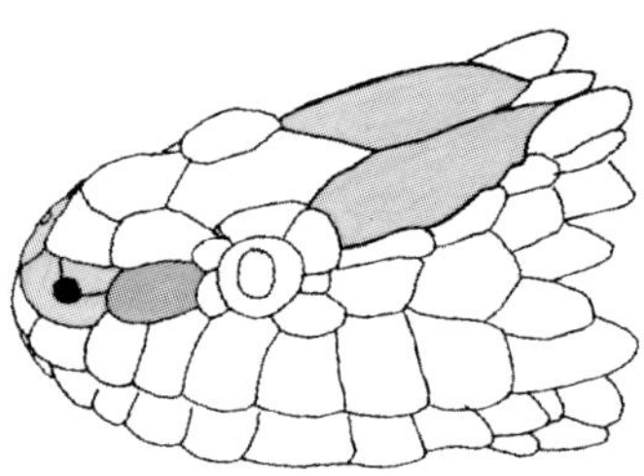

Fig 1: *Enhydris polylepis* - a watersnake.
Nasal scales (anterior, light shaded) in contact; loreal scale (dark shaded) present, separating nasal and preocular scales; parietal scales (posterior, light shaded) entire and unfragmented.

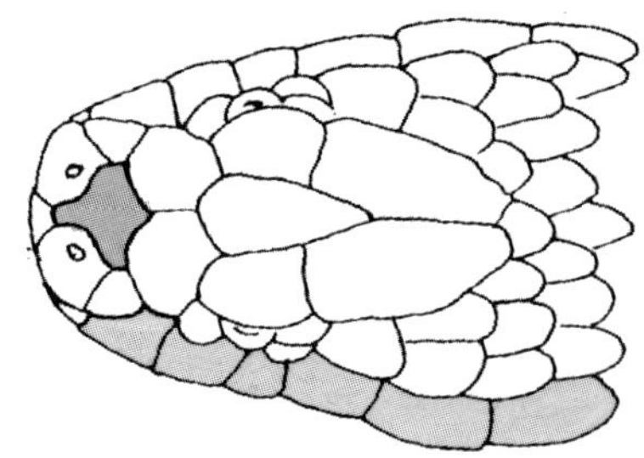

Fig 2: *Myron richardsoni* - a mangrove snake.
Nasal scales separated by a single internasal scale (dark shaded) 7 or more supralabials (light shaded).

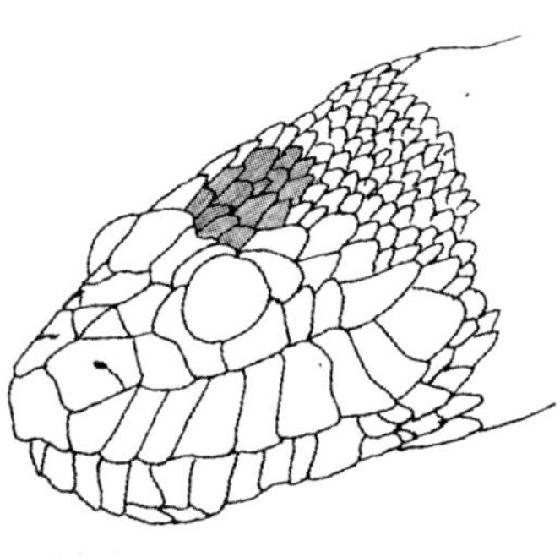

Fig 3: *Cerberus rynchops* - a watersnake.
Parietal scales fragmented (shaded).

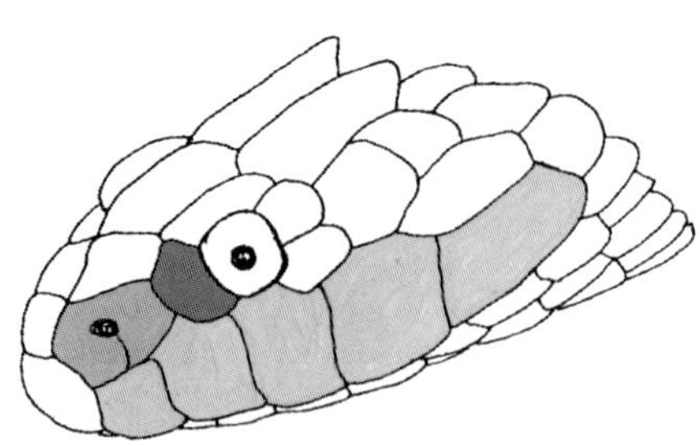

Fig 4: *Fordonia leucobalia* - a mangrove snake.
Loreal scale absent, preocular scale (dark shaded) in contact with nasal scale (light shaded); 5 supralabials (v. light shaded).

BANDED WATERSNAKE

Cantoria annulata

Average/Maximum length: 0.4m / 0.63m.

Description:

b) colouration: Dark brown with numerous pale reddish or yellow-brown blotches which may either continue over the back to form bars or break at the vertebral line; light crossbars also on the snout, supralabials and from the eye to the angle of the jaw; underside of body and tail white or grey with a fine dark longitudinal midline or series of grey blotches and dark edging to the ventral scales.

c) scalation: DMB 21, all smooth; V 165-184; SC 44-52, all paired; anal plate divided; loreal present; 8 supralabials with 4th entering the eye; 9 infralabials.

Distribution within PNG: WEST: Daru and Bobo Is. and Abam, Oriomo R.

Distribution outside PNG: Irian Jaya (Prince Frederik Hendrik Island).

MACLEAY'S SMOOTH WATERSNAKE

Enhydris polylepis

Average: 0.8m.

Description:

b) colouration: Olive green or brown to black dorsally, occasionally faintly striped, with a yellow-orange stripe from the angle of the jaw which may continue as a pale light stripe along the lower flanks; underbelly cream with a dark median longitudinal stripe.

c) scalation: DMB 21-23 (Trans-Fly), 23-27 (upper Fly), all smooth; V 138-147 (Trans-Fly), 151-158 (upper Fly); SC 38-54 (Trans-Fly), 43-51 (upper Fly), all paired; anal plate divided; loreal present; 8, rarely 9, supralabials with 5th or 5th and 6th, rarely 4th and 5th, entering eye; 11-12 infralabials.

Distribution within PNG: Altitude records 0-360m. WEST: Oriomo Plateau and s. Trans-Fly, Sigabaduru, Bensbach, Fly and Aramia R., Balimo, upper Fly R., Kiunga, Ningerum. Formerly two species were recognised for Western Province but McDowell is of the opinion that only one highly variable species is found there.

Distribution outside PNG: Irian Jaya (Lorentz River) and Northern Australia from eastern Queensland to Western Australia.

WHITE-BELLIED MANGROVE SNAKE

Fordonia leucobalia

Average/Maximum length: 0.7m / 0.93m.

Description:

b) colouration: Extremely variable dorsally, dark grey or brown with irregular light spotting, or light grey to yellow or orange with dark spotting; underside and lower flanks off-white lower flanks off-white, immaculate or with small black spots.

c) scalation: DMB 23-29, all smooth; V 130-160; SC 24-45, mostly or all paired; anal plate divided; **loreal absent** (this is the only terrestrial/semi-terrestrial New Guinea **non-elapid** to lack a loreal scale although some pythons have fragmented loreals); 5 supralabials with 3rd or 2nd and 3rd entering the eye.

Distribution within PNG: WEST: Daru and Bobo Is., Pahoturi, Binaturi and Oriomo R., Fly R. to Palmer Junction and Aramia R., Balimo. GULF: Moka, Turama R. NCD: Port Moresby, Napa Napa, Ela Beach. CENT: Porebada. The isolated specimen from Palmer Junction on the Upper Fly River was found in freshwater.

Distribution outside PNG: Irian Jaya, coastal Australia from eastern Queensland to Western Australia and Indo-Malaysia.

RICHARDSON'S GREY MANGROVE SNAKE
Myron richardsoni
Average/Maximum length: 0.3m / 0.4m.
Description:
b) colouration: Light or dark grey to brown with irregular darker barring or spotting and a black stripe behind the eye; ventral surfaces, lower flanks and supralabials light grey to off-white.
c) scalation: DMB 19-23, weakly keeled; V 129-145; SC 30-40, all paired; anal plate divided; loreal present; 8 or 9 supralabials with 4th or 4th and 5th entering the eye; 11 infralabials.
Distribution within PNG: WEST: Oriomo, Mai Kussa and lower Fly R. and Daru Is. NCD: Port Moresby, Napa Napa.
Distribution outside PNG: Irian Jaya and the Aru Islands and northern coastal Australia from Northern Territory to Western Australia.

Extralimital Irian Jayan species yet to be reported from PNG:

BOCKADAM
Cerberus rynchops novaeguineae
Average/Maximum length: 0.8m / 1.2m.
Description:
b) colouration: Dorsally grey to red-brown with numerous transverse black bars; underside yellow with irregular dark crossbands.
c) scalation: DMB 23-25, all keeled; V 140-150; SC 43-48, all paired; anal plate divided; loreal present; 8-9 supralabials with 5th or 6th entering the eye; 12-13 infralabials.
Distribution outside PNG: Irian Jaya (Merauke and southern coast). The nominate subspecies, *Cerberus rynchops rynchops,* occurs throughout India, Sri Lanka, Indo-Malaysia and the Philippines. Australian populations, from northern coastal Australia, are sometimes referred to as a separate subspecies, *Cerberus rynchops australis.*

'MAMBERAMO RIVER WATERSNAKE'
Heurnia ventromaculata
Average length: 0.73m.
Description:
b) colouration: Dorsally grey black with lowest two rows of body scales yellow; underside pale yellow with darker flecking; posterior supralabials white and all infralabials white with black spots.
c) scalation: DMB 27, smooth; V 160; SC 62, all paired; anal plate divided; loreal present; 7 supralabials with 4th or 5th entering the eye; 13 infralabials. Separated from similar *Enhydris* by the presence of a large internasal scale between the smaller nasal scales.
Distribution outside PNG: Irian Jayan endemic (Mamberamo River).

BROWN CAT SNAKE or BROWN TREESNAKE

Boiga irregularis

Map: 18

MILDLY VENOMOUS

Average/Maximum length: 1.5-2.0m / 2.3m

Description:

a) physique: Generally slender bodied, (excepting the very large 'python-like' specimens on Guam), with elongate tails; the head is broad and extremely distinct from the narrow neck; the eye is very large with a vertically elliptical pupil (hence 'cat snake'). The ventral scales of the cat snake exhibit a sharp ridge running down either side presenting an 'arch-shape' in cross-section which enhances the climbing and gap-bridging abilities of these snakes.

b) colouration: Usually light brown to yellow or red-brown dorsally with darker irregular barring or spotting present or absent, lighter yellow below. Juveniles may be almost pink with translucent undersides. Eye yellow or brown with a dark pupil.

c) scalation: DMB 19-23 in oblique rows with vertebral row enlarged, all smooth; V 217-272; SC 100-125, mostly paired; anal plate entire; loreal present; 9 supralabials with 4th to 6th in contact with eye.

Habitat: Inhabiting a wide variety of habitats from coastal mangrove swamps and montane rainforest to cultivated gardens and villages.

Habits: Commonly encountered around buildings after dark, the nocturnal or crepuscular brown cat snake is aggressive and quick to bite if disturbed or molested. Oviparous with clutch sizes of 3-11.

Diet: Frogs, sleeping lizards and birds, bird and reptile eggs, bats, small mammals and other snakes.

Distribution within PNG: Altitude records 0-1400m (Mt Hagen, WHP).

WEST: Trans-Fly, Oriomo Plateau, Daru and Bobo Is., Bensbach, Aramia and Nomad R., Lake Murray and Boboa Is. and lower and upper Fly, Gigabip. GULF: Baimuru. NCD: Port Moresby, Kanudi, Kaugere. CENT: Mekeo, Owen Stanley Mts., Brown, Laloki, Goldie and Vanapa R., Sogeri, 14 Mile, Sogeri, Subitana and Hood Bay. M.BAY: Trobriand Is., Kiriwina Is., and d'Entrecasteaux Is., Goodenough, Fergusson and Normanby Is., no precise mainland locality data. ORO: Popondetta. MOR: Wau-Bulolo, Huon Gulf, Lae and Huon Pen. MAD: Ramu R. and delta, Astrolabe Bay, Adelbert Mts., Schrader Mts, Simbai and Kar Kar Is. E.SPK: Wewak. W.SPK: Aitape and Seleo Is. E.HIGH: no precise localities. W.HIGH: Tigi and Mt Hagen. WNB: Moramora and Garu. ENB: Rabaul and Duke of York Is. N.IRE: Kavieng. MAN: Manus and Baluan Is. N.SOL: Bougainville.

Boiga irregularis is the only 'venomous' land snake reported from Manus Province.

Distribution outside PNG: Irian Jaya and its satellite islands, the Moluccas and Sulawesi, the Solomon Islands and northern and eastern Australia. Accidentally introduced onto Guam where it has had a devastating effect on the native flightless-bird population.

Venom: Within New Guinea the bite of the brown cat snake has been described as like a bee sting with symptoms restricted to localised swelling and itching, and it is generally not considered dangerous, although large specimens are capable of delivering a painful, fairly bloody, bite. However, sleeping infants bitten by this species on Guam developed lethargy, localised swelling, discolouration and blood blistering of the skin, drooping eyelids (ptosis), slow pupillary responses, both visual and potentially serious breathing difficulties and, in two instances, depressed heart rates. However, these bites were probably caused by un-naturally out-sized specimens which are reputed to 'grow to the size of pythons' on a diet of native flightless birds. It is unlikely that bites from this species pose any serious threat to people in PNG where the species is in equilibrium with its environment, rarely exceeds 2.0m and has natural predators and competitors.

Antiserum: None available or recommended. Bites to children, from large specimens, should be treated symptomatically by medically qualified persons, with particular attention to potential neurological or respiratory problems. Although defensive bites usually entail a rapid single strike and release, or multiple series of strikes and releases, resulting largely in 'cosmetic' injuries, feeding bites, such as those recorded as occurring in houses at night on Guam (*see* '*Boiga*' in further reading p.230) may be more serious and potentially life-threatening for infants. Following the initial bite a cat snake will maintain its grip and begin to chew hard to bring the enlarged grooved fangs, positioned beneath the eyes on the rear of the upper jaw, into play. To minimise the painful lacerating injuries caused by the bite of this snake the mouth of the snake should initially be held closed on the skin until a stick, pencil or similar object can be located with which to prize the jaws away. Holding the stick at right angles to the snake's head, begin by inserting it across the front of the upper jaw so that it protrudes from both sides of the mouth above the bitten member. Then begin to release the pressure on the snake's mouth, roll the stick backwards into the mouth carefully to disengage the teeth from the skin of the victim. Concentrate on the upper jaw first and continue until it is totally disengaged and the mouth is gaping. The process may be assisted by the victim easing the bitten limb back into the snake's mouth slightly to unhook the recurved teeth. Do not be tempted to pull the bitten limb from the snake's mouth as this may cause deep lacerations if any teeth remain in contact with the skin. Finally unhook the flexible lower jaw from the underside of the bitten member by again sliding the bitten limb backwards to disengage the teeth. Whilst the cat snake is biting it will also coil its body around its victim and apply constriction. Lacking the strength of a python this represents no serious threat to adults or older children but unattended infants may be at risk if a coil were to be passed around the neck. However, these coils will provide the snake with an anchorage point so the coils should be unwound and held by another person to prevent them recoiling and exerting pressure. The same procedure may be used to remove non-venomous pythons.

Brown Cat snake (juvenile)
– *Boiga irregularis*
(Kar Kar Is., Madang Province)

Brown Cat snake
– *Boiga irregularis*
(Kar Kar Is., Madang Province)

PARKER'S SNAKE - AN ENIGMA !

In addition to the known non-venomous keelbacks (*Tropidonophis*) and file snakes (*Acrochordus*), mildly venomous mangrove, mud and watersnakes (*Fordonia, Myron, Cantoria* and *Enhydris*) and highly venomous sea kraits (*Laticauda*) and true seasnakes (*Hydrophis* etc.) occurring in freshwater or saline habitats in southern Western Province, Parker (1982) reported an unknown species which apparently caused rapid deaths in three children bitten near the village of Wipim in the early 1970's. A similar snake is reported by villagers as far afield as Morehead and Bensbach although no deaths are documented.

The snake is said to be aquatic and very rare. Parker reports that villagers say it is 'normal' looking with smooth scales, enlarged ventrals and a short cylindrical tail. Colouration is yellowish-brown to brown dorsally, pale yellow to white ventrally. It is said to occur in freshwater habitats such as small inland streams and sago swamps, rather than larger rivers or open swampy grassland. It apparently hides on the muddy bottom, although it has also been observed to bask on dry land. Maximum length is reported to be 1.5-2.0m.

This snake is reportedly very highly venomous with deaths occurring within a few minutes of the bite (much faster than those caused by the taipan or even seasnakes - unless venom was injected intravenously). The three deaths at Wipim, in 1972 and 1973, involved young girls who were apparently bitten whilst bathing in the Ouwe Creek. All three reportedly died after walking only a few hundred metres from the creek.

Parker considered all the known Western Province snake species and also those Irian Jayan species which have yet to be recorded from PNG i.e. *Cerberus rynchops* and *Pseudechis australis*, but none of them fitted the descriptions provided by villagers. The rapid deaths of the three children would initially suggest a seasnake and I did collect a seasnake, believed to be *Enhydrina schistosa*, some 60km inland in the Oriomo River at Old Zim, into which the Ouwe Creek feeds, in 1986, but the description of the unknown snake's tail coupled with the sea-going Kiwai people's familiarity with true seasnakes would seem to also rule out that possibility. Another possibility might be the elapid *Micropechis ikaheka*, a snake frequently associated with semi-aquatic habitats, but the speed with which the venom appears to have acted on the young girls would seem to discount this species also. Despite field trips to Wipim by Parker, myself and others to search for this species it still evades science. Therefore, until a specimen of this unusual, and obviously extremely important, snake species is collected, preserved and forwarded to the National Museum in Port Moresby we will remain ignorant of its true identity.

Parker made such a plea in 1982 and a specimen has still not been forthcoming.

Ouwe Creek Pool
Wipim, Western Province
Site of three fatal snakebites from an unidentified snake in the 1970's

FRONT-FANGED VENOMOUS LANDSNAKES
Family Elapidae

The family Elapidae is quite diverse in PNG containing terrestrial and amphibious species (the sea kraits) and both secretive burrowing species with small mouths and some of the largest and most dangerously venomous snakes in the world. Some species show close relationships with the elapids of Australia but others, especially the more secretive montane or insular species, are endemic to New Guinea and its satellite islands. 36% of the truly terrestrial snakes of PNG are elapids, 59% of those are New Guinea / Solomons endemics.

Key to the Venomous Terrestrial Snakes of Mainland New Guinea: Family Elapidae
Separate keys follow for the North Solomons and the Milne Bay island elapids.
The sole terrestrial elapid recorded from the Bismarck Archipelago is *Aspidomorphus muelleri*.
Highly venomous genera and species are indicated in **bold**.

1a. Tail flattened and 'paddle-shaped' (fig. 1); body patterned with blue and black rings; saltwater aquatic and coastal. ***Laticauda* spp.**
1b. Tail not flattened or 'paddle-shaped'; body not ringed or, if ringed, not blue and black; primarily terrestrial or associated with freshwater. 2

2a. Supraocular scale, above eye, may be slightly or considerably raised as a 'horn-like' projection (fig. 4); subocular scales present; head 'viperine' in appearance; body short, squat and cryptically patterned; tail slender with a short terminal 'spike' (fig. 2). ***Acanthophis* spp.**
2b. Supraocular scales not raised as 'horns'; subocular scales absent; head rounded, not 'viperine'; body not unduly squat though may be short; tail without terminal 'spike'. 3

3a. Dorsal scales arranged in 19 or more rows at midbody. 4
3b. Dorsal scales arranged in 17 or fewer rows at midbody. 5

4a. Dorsal scales arranged in 21 or 23 rows at midbody; subcaudal scales all paired; anal scale entire (fig. 3a); head elongate with shelved supraocular scale shielding large eye from above (fig. 5); body colouration usually brown, grey or black with an orange vertebral stripe present. ***Oxyuranus scutellatus***
4b. Dorsal scales arranged in 19, rarely 21, rows at midbody; subcaudal scales mostly single, a few paired posteriorly; anal scale usually divided (fig. 3b); head rounded with small eye not shielded from above by supraocular scale (fig. 6); body colouration usually black, glossy or matt, occasionally brown but lacking vertebral orange stripe. ***Pseudechis papuanus***

5a. Dorsal scales arranged in 17 rows at midbody. 6
5b. Dorsal scales arranged in 15 or fewer rows at midbody. 8

6a. 6 supralabial scales with temporolabial scale present between 5th and 6th (fig. 7). 7
6b. 6 supralabial scales, temporolabial scale absent (fig. 8). ***Pseudonaja textilis***

7a. Preocular scale and nasal scales in broad contact, prefrontal scale separated from 2nd supralabial (fig. 9); maximum length in excess of 2.0m; dorsal colouration generally brown without bar across nape. ***Pseudechis australis***

7b. Preocular and nasal scales separated by downward projection of prefrontal scale which contacts 2nd supralabial (fig. 7); maximum length less than 1.0m; dorsal body colour grey brown with or without lighter bar across nape of neck. *Furina tristis*

8a. Eye extremely large; snake active by day, fast moving. *Demansia* spp.

8b. Eye very small; snake secretive or active by night. 9

9a. Dorsal body generally white to yellow with a series of darker, red to brown, bands beginning at midbody and continuing in frequency and width posteriorly to a dark tail; head grey; maximum length 2.0m. ***Micropechis ikaheka***

9b. Dorsal body pattern either unicolour grey to brown without multiple bands, markings if present confined to a single nape band, irregular 'ocelli' blotches centred on the head or longitudinal stripes on the head and body; maximum length less than 1.0m. 10

10a. Anal scale divided (fig. 3b). 11

10b. Anal scale entire (fig. 3a). 12

11a. Body pattern either unicolour grey to brown with head markings confined to light nape or head bars, or body and head dark brown with centre of each head and anterior body scale with a cream spot; eye very small. *Toxicocalamus* spp.

11b. Body pattern brown or red with striking and contrastingly darker head with multiple dark and light stripes and/or 'ocelli' passing onto the neck and anterior body; eye moderately small. *Aspidomorphus* spp.

12a. 4 or 5 supralabial scales (fig. 10); body colouration unicolour with patterning, if present, confined to a light bar across the head or nape of the neck. *Toxicocalamus* spp.

12b. 6 supralabial scales (fig. 11); body colouration either unicolour or pink with a black head running into a black vertebral stripe. *Rhinoplocephalus* spp.

Keys to the elapid genera are to be found on the following pages:

Genus	**Page**
Toxicocalamus	129
Rhinoplocephalus	138
Aspidomoprhus	148
Demansia	154
Acanthophis	158
Laticauda	171

ELAPID TAILS

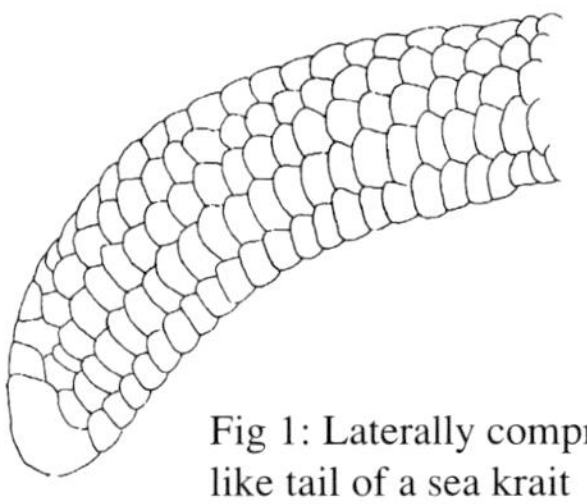

Fig 1: Laterally compressed paddle-like tail of a sea krait

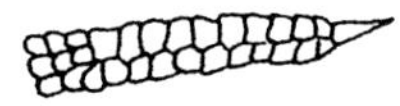

Fig 2: Tail tip of death adder exhibiting a short terminal spine.

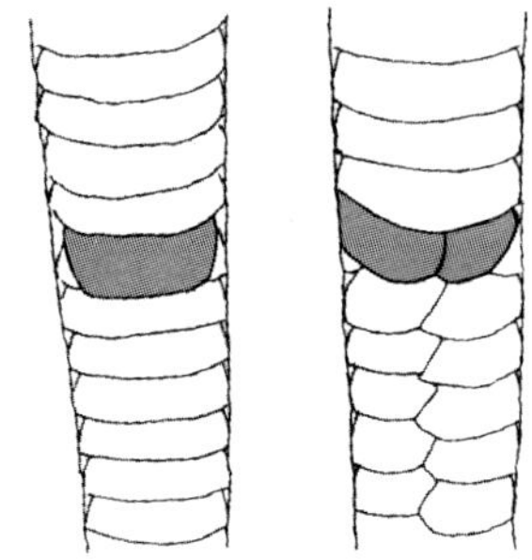

Fig 3a (left): *Oxyuranus* anal plate entire (shaded); subcaudals paired.
Fig 3b (right): *Pseudechis* anal plate usually divided (shaded); subcaudals single anteriorly, paired posteriorly.

ELAPID HEADS

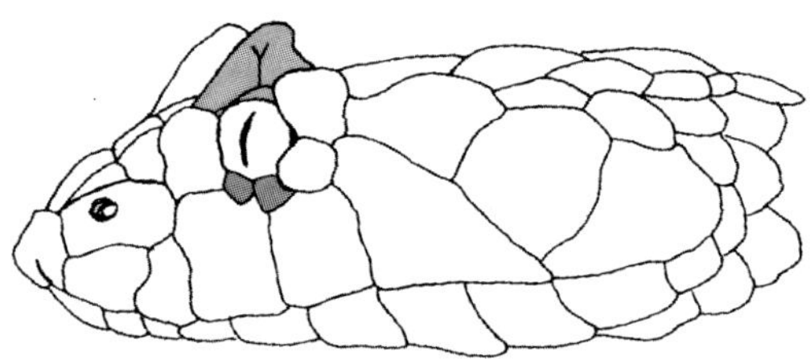

Fig 4: *Acanthophis* sp. – a death adder.
Raised 'horn-like' supraocular scale often present over the eyes (light shaded); subocular scales present between the eye and the labial scales (dark shaded).

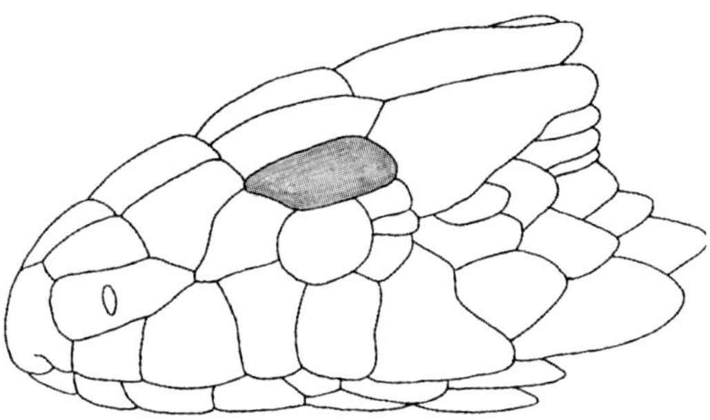

Fig 5: *Oyuranus scutellatus* – a taipan.
Supraocular scale (shaded) large and 'shelved' over a large eye presenting a 'scowling expression'.

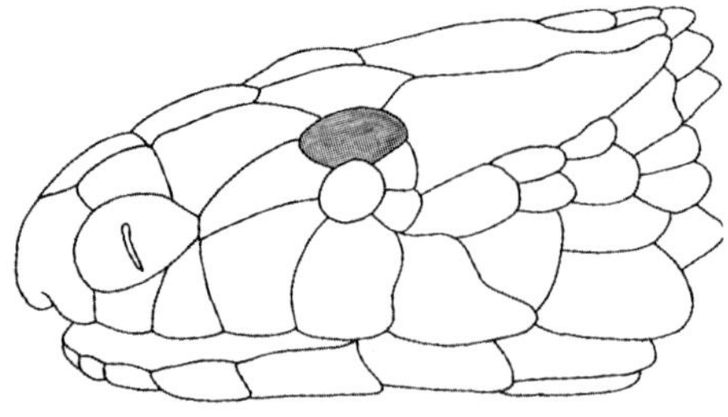

Fig 6: *Pseudechis papuanus* – a blacksnake.
Supraocular scale (shaded) not especially large, or shelved over a small eye.

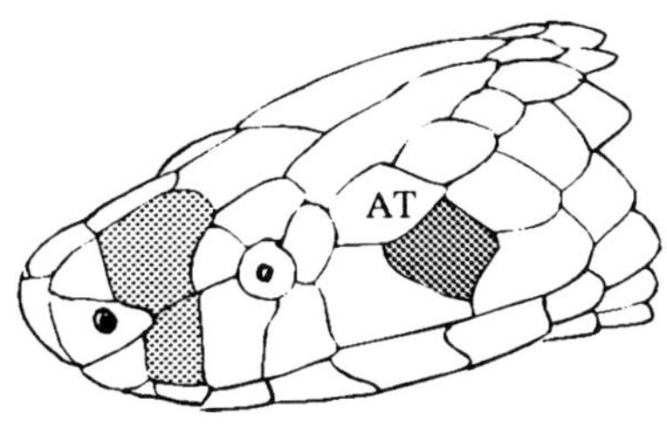

Fig 7: *Furina tristis* – an elapid.
Temporolabial scale present (dark shaded) betwen 5th and 6th supralabials, (AT = anterior temporal); downward process of prefrontal scale contacts 2nd supralabial (both light shaded) preventing contact between preocular and nasal scales.

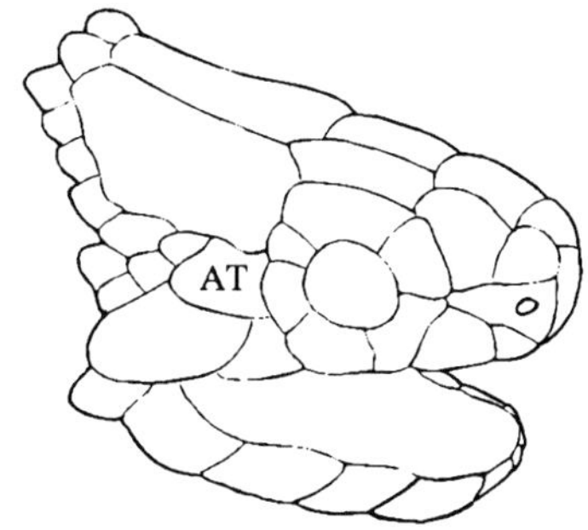

Fig 8: *Pseudonaja textilis* – a brownsnake.
Temporolabial scale absent, (AT = anterior temporal).

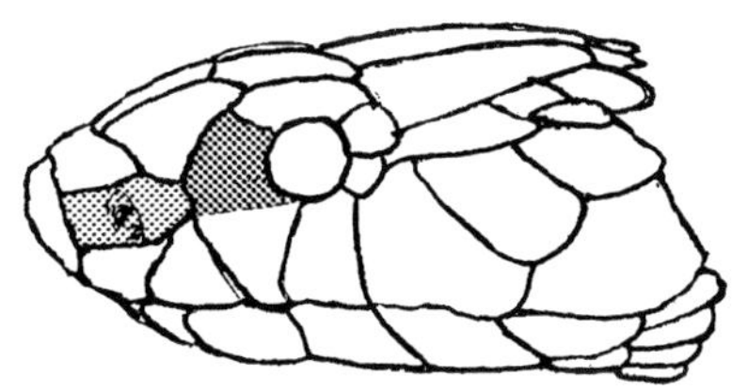

Fig 9: Pseudechis australis – a king brownsnake.
Preoculars (dark shaded) and nasal scales (light shaded) in contact, not separated by downward process of prefrontal.

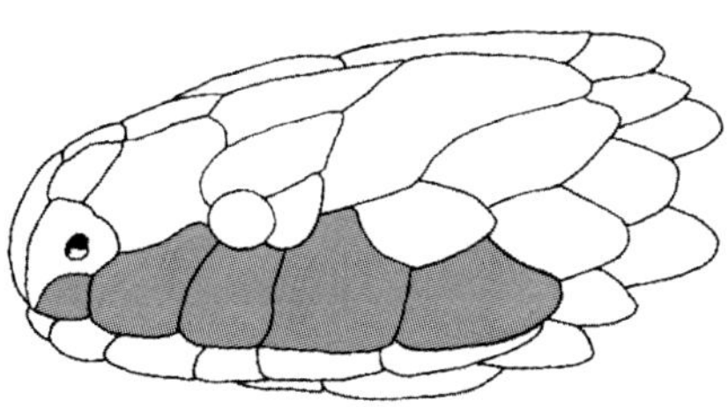

Fig 10: *Toxicocalamus stanleyanus* – an elapid.
Four or five supralabials (shaded).

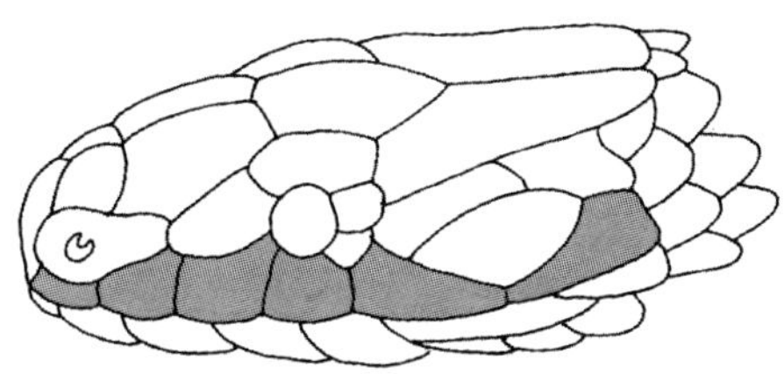

Fig 11: *Rhinoplocephalus nigrostriatus* – an elapid.
Six supralabials (shaded).

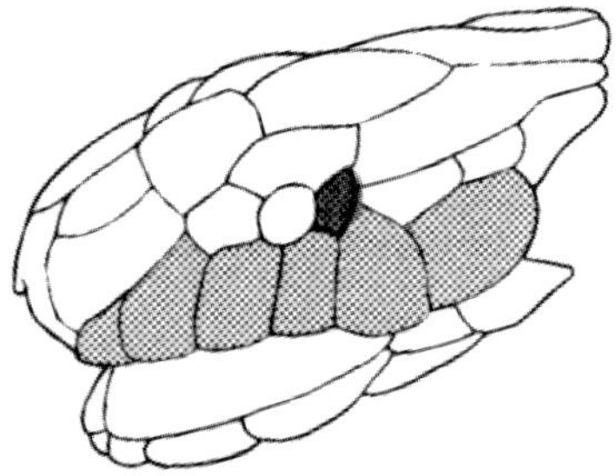

Fig 12: *Parapistocalamus hedigeri* – an elapid.
Six supralabials (light shaded); one postocular (dark shaded).

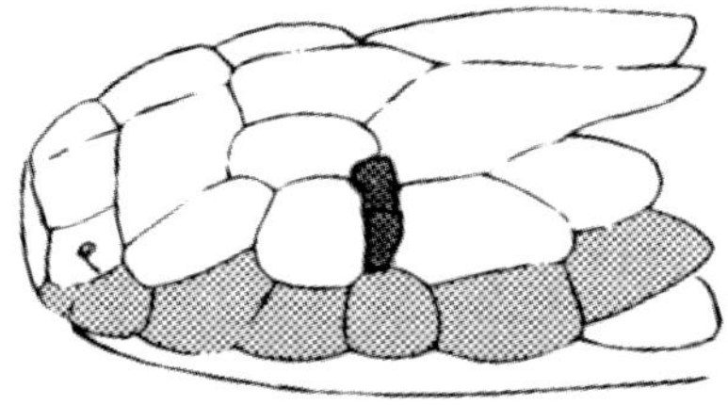

Fig 13: *Salomonelaps par* – an elapid.
Seven supralabials (light shaded); two postoculars (dark shaded).

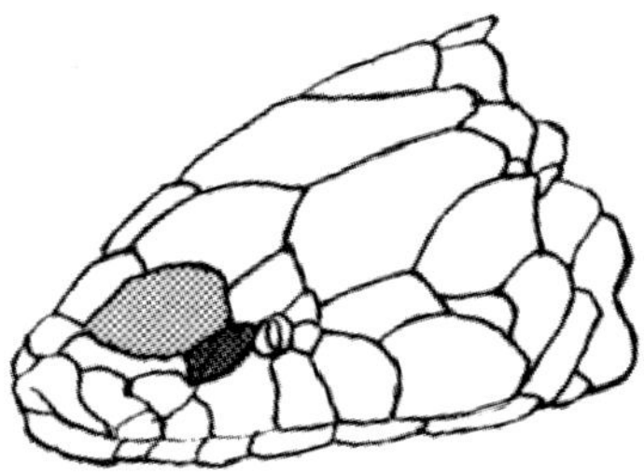

Fig 14: *Toxicocalamus holopelturus* - an elapid.
Preocular present (dark shaded) distinct from prefrontal (light shaded).

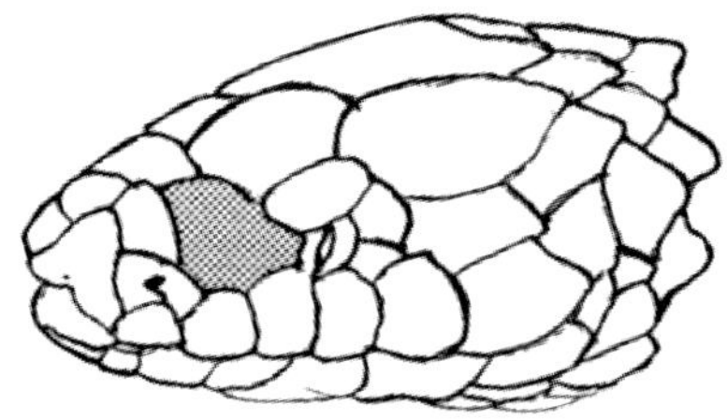

Fig 15: *Toxicocalamus misimae* – an elapid.
Preocular absent, fused with prefrontal (light shaded).

Key to the Venomous Terrestrial Snakes of the North Solomons Province, PNG
including *Loveridgelaps elapoides* from Shortland Islands, Solomon Islands (Fred Parker *pers. comm.*).

1a. Dorsal scales arranged in 15 rows at midbody. 2
1b. Dorsal scales arranged in 17 rows at midbody. 3

2a. 6 supralabial scales; one postocular scale present (fig. 12); single preocular scale present or absent; 35 or fewer subcaudal scales, all paired; anal scale entire or divided; body colouration fairly unicolour except for a light crossband over the rear of the head (Bougainville). *Parapistocalamus hedigeri*

2b. 7 supralabial scales; two postocular scales present (fig. 13); single preocular scale present; 38 or more subcaudal scales, both single and paired; anal scale divided; body colouration may comprise a series of darker vertebral blotches (Buka, & Fauro & Shortlands - Solomon Islands). *Salomonelaps par*

3a. Fewer than 181 ventral scales; subcaudal scales 38 or more, both single and paired; anal scale divided (fig. 3b); body colouration may comprise a series of darker vertebral blotches. *Salomonelaps par*
3b. In excess of 190 ventral scales; subcaudal scales 38 or fewer, all paired; anal scale entire (fig. 3a); body patterning consists of a white head followed by orange and black vertebral crossbars which may or may not ring the body (Shortlands Is., Solomon Islands) *Loveridgelaps elapoides*

Key to the Venomous Terrestrial Snakes of the islands of Milne Bay Province, PNG

1a. Dorsal scales arranged in 17 rows at midbody (Fergusson & Woodlark). *Toxicocalamus longissimus*
1b. Dorsal scales arranged in 15 rows at midbody. 2

2a. Subcaudals fewer than 42 (Goodenough, Fergusson, Normanby, Misima, Tagula, Trobriand & Woodlark). *Aspidomorphus lineaticollis*
2b. Subcaudals in excess of 42. 3

3a. Preocular scale present (fig. 14), distinct from prefrontal scale (Rossel). *Toxicocalamus holopelturus*
3b. Preocular scale absent (fig. 15), fused with prefrontal scale (Misima). *Toxicocalamus misimae*

Note: the single specimen of '*Toxicocalamus loriae*' from Normanby Island is not included in this key as details of its scalation were unavailable. This specimen may represent an undescribed species (Ken Aplin, W. Aust. Mus. *pers. comm.*)

NEW GUINEA FOREST SNAKES
Toxicocalamus spp.

(9 species, *see* Taxonomic note) Maps: 19a-c

MILDLY VENOMOUS

Average/Maximum length: 470-770mm / 960mm (*T. grandis* - Irian Jaya).

Description:

a) physique: These are slender cylindrical snakes with narrow flattened heads which are barely distinct from the neck; the tail is fairly short, blunted and with a terminal scale at the tip; the eye is small with a vertically elliptical pupil.

b) colouration: Head and upper body fairly uniform dark or light brown although head markings may be present in juveniles and some adults consisting of i) a pale 'prefrontal band' across the snout anterior to the eyes; ii) a pale 'parietal band' across the head posterior to the eyes and iii) a pale 'nuchal band' across the neck behind the head. This last mark may be followed by an area of darker pigment referred to as a 'nuchal blotch'. Regardless of other colouration or markings the vertebral scale row is generally darker than the adjacent scale rows, resulting in a dark 'vertebral stripe'. The remaining dorsal scale rows, excluding the lowest 2-3 rows, are also dark. Fine dark median lines may be present on each dorsal scale either forming a series of longitudinal stripes which are more apparent on the lower dorsal scales, or a broken series of dark spots along the body. Ventral surfaces, throat, supralabials and lowest 2-3 dorsal scale rows often immaculate yellow or white, frequently a dark median longitudinal stripe is visible running through each lower dorsal scale row, and a pair of similar stripes may be present on the ventrals continuing onto the subcaudals, which are frequently darker than the anterior ventral scales. For fuller species descriptions refer to McDowell (1969).

c) scalation: Variable and species specific (*see* key to the species and individual accounts below). Dorsal scales are all smooth and loreal and subocular scales are always absent.

Habitat: Lowland and montane rainforest, kunai grass and gardens.

Habits: Very secretive almost totally fossorial (burrowing) snakes which inhabit soil, leaf litter and fallen logs. Usually inoffensive and of such small size that most specimens are probably incapable of biting a human. Active by day, and probably also by night. Oviparous but generally little known.

Diet: Soft-bodied invertebrates eg. earthworms. Fly pupae and small snails are also recorded and large specimens may possibly feed on small frogs or reptile eggs.

Taxonomic note: The genus *Toxicocalamus* in its currently recognised form comprises four original genera: *Apistocalamus, Pseudapistocalamus, Toxicocalamus* and *Ultrocalamus.* McDowell (1969) reduced *Apistocalamus* and *Ultrocalamus* to subgenera of *Toxicocalamus* and synonymised *Pseudapistocalamus nymani* with *Toxicocalamus loriae.* The formerly recognised species *Apistocalamus pratti, A. loennbergi* and *A. lamingtoni* are also synonymised within this widely distributed species. (*T. loriae* of McDowell may constitute a species-complex, Ken Aplin, Western Australian Museum, *pers. comm.*). In addition McDowell resurrected *Toxicocalamus (Ultrocalamus) buergersi* from synonymy within *T. (U.) preussi* and he named three new species: *T. (Apistocalamus) spilolepidotus*; *T. (A.) holopelturus* and *T. (Toxicocalamus) misimae.* At least one new species of *Toxicocalamus* is *in prep* (Ken Aplin, *pers. comm.*). The name forest snakes was coined in collaboration with Patrick David and Ivan Ineich.

Distribution within PNG: Mainland New Guinea and islands of Milne Bay Province. Several species are extremely poorly known. Of the nine currently recognised species, two are only known from single specimens, another from two specimens and a third from three specimens (*see* individual accounts below).

Distribution outside PNG: Several species have been collected in Irian Jaya and *T. grandis* is only reported from there (*see* individual accounts below).

Confusing species: The small burrowing elapid snakes of genus *Toxicocalamus* may be confused with the blindsnakes, *Ramphotyphlops* spp. and *Typhlops* spp; ground snakes, *Stegonotus* spp; brown-headed snake, *Furina tristis*; and Australian small-eyed snakes, *Rhinoplocephalus* spp.

Venom: Nothing is known of the venom and although small specimens are probably harmless to man, due to small mouth size, larger specimens should be treated with caution. *T. loriae* is a relatively large species and adult specimens should be considered potentially dangerous.
Antiserum/initial dose: Not available.

Key to the forest snakes, genus *Toxicocalamus*, of New Guinea

1a. Preocular scale present, distinct from prefrontal scale (fig. 1); ventral scales broad, at least 4x wider than scales of first lateral row. 2
1b. Preocular scale absent, fused with prefrontal scale as single scale (fig. 2); ventral scales narrow, 2 - 3x wider than scales of first lateral row. 5

2a. Internasal scale in contact with preocular scale; third supralabial scale in broad contact with nasal scale, preventing contact of second supralabial scale with preocular scale (fig. 3); first subcaudal scale paired, remainder single. *Toxicocalamus holopelturus*
2b. Internasal scale usually not in contact with preocular scale, allowing prefrontal scale to contact nasal scale; third supralabial scale not in contact, or only point contact, with nasal, allowing second supralabial scale to contact preocular scale (fig. 4); all subcaudal scales paired. 3

3a. Nasal scale not divided; maximum size greater than 900mm. *Toxicocalamus grandis*
3b. Nasal scale divided; maximum size less than 800mm. 4

4a. Four infralabial scales in contact with anterior chinshields (fig. 5a); preocular scale in contact with nasal scale (fig. 6); dorsal colouration chocolate brown with each scale (except vertebral scale row) possessing a cream central spot which increases in size moving away from the vertebral row until ventral scales are cream with chocolate brown edges; scales of head also spotted with cream; cream ring around terminal spine of tail. *Toxicocalamus spilolepidotus*
4b. Three infralabial scales in contact with anterior chinshields (fig. 5.b); preocular scale in contact with nasal scale (fig. 7) or separated from nasal by prefrontal (fig. 8); dorsal colouration light brown with light pigment confined to a pair of small bars on the nape. *Toxicocalamus loriae*

5a. Internasal scale present, distinct from prefrontal scale; temporal scale present, distinct from rear-most supralabial scale, preventing supralabial contact with parietal scale; 5 - 7 supralabials (fig. 9); tail tip pointed or rounded with conical terminal scale. 6
5b. Internasal absent, fused with prefrontal as single scale; temporal absent, fused with rear-most supralabial, allowing supralabial contact with parietal; 4-5 supralabials (fig. 10); tail tip blunt with flattened, keeled terminal scale. 8

6a. Anal plate divided; 6, rarely 7, supralabials with 3rd and 4th in contact with eye (fig. 9). 7
6b. Anal plate entire; 5 supralabials with 2nd and 3rd in contact with eye (fig. 11). *Toxicocalamus stanleyanus*

7a. 17 dorsal scale rows at midbody; 1-2 postoculars; (figs. 12 & 13). *Toxicocalamus longissimus*
7b. 15 dorsal scale rows at midbody; 2 postoculars; (fig. 14). *Toxicocalamus misimae*

8a. 13, rarely 15, dorsal scale rows at midbody; postocular scale usually present and distinct from supraocular scale preventing contact of supraocular scale with 3rd and 4th supralabial scales; 5 supralabials (fig. 15). 9

8b. 15 dorsal scale rows at midbody; postocular scale absent, fused with supraocular scale, allowing contact of supraocular scale with 3rd and 4th supralabial scales; 4 supralabials (fig. 16). *Toxicocalamus buergersi*

9a. 5th supralabial separated from postocular by 4th supralabial (fig. 17), rarely making point contact only; nuchal band narrow, <60% length of parietals. *Toxicocalamus preussi preussi*

9b. 5th supralabial in broad contact with postocular (fig. 18); nuchal band broad, >80% length of parietals. *Toxicocalamus preussi angusticintus*

NB The Garaina, Morobe Province population of *Toxicocalamus* does not fit any key or description for known species. McDowell (1969) considered them to be representatives of introgressive hybridisation between *T. loriae* and *T. stanleyanus*.

(Species accounts continue on p. 134)

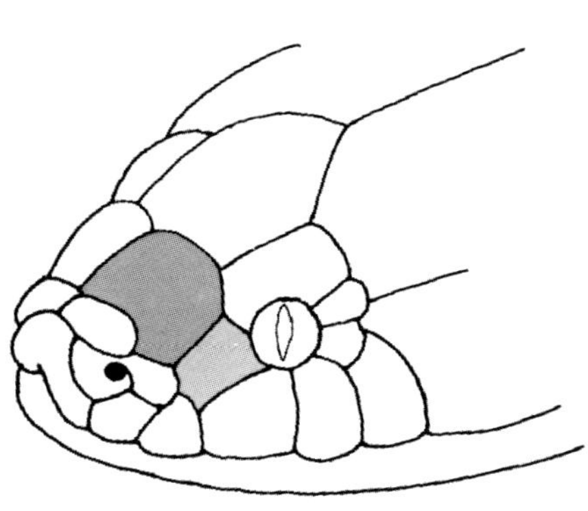

Fig 1: Preocular scale present, distinct from prefrontal scale.

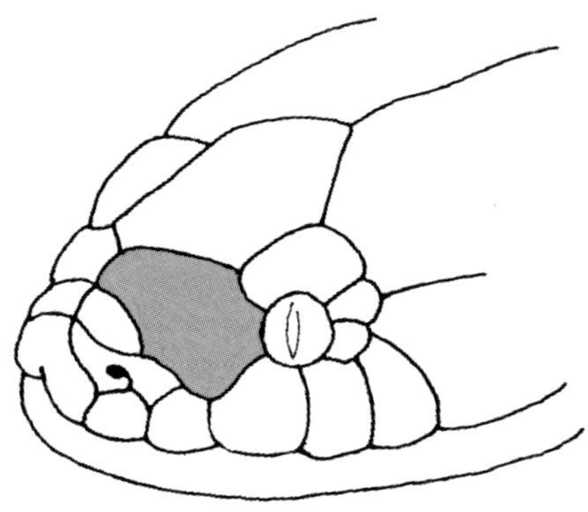

Fig 2: Preocular scale absent, fused with prefrontal scale.

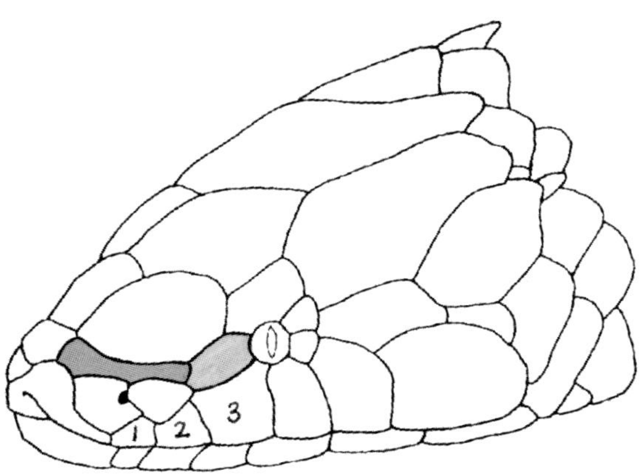

Toxicocalamus holopelturus AMNH 76660
Mt Rossel, Rossel Is., Louisiade Archipelago,
Milne Bay Province
(after McDowell 1969)
Fig 3: Internasal scale in contact with preocular; 3rd supralabial in contact with nasal scale, 2nd supralabial not in contact with preocular.

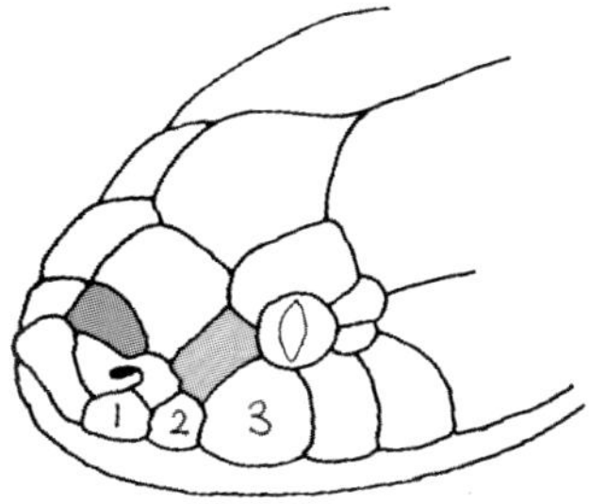

Fig 4: Internasal scale not in contact with preocular; 3rd supralabial not in contact with nasal scale, 2nd supralabial in contact with preocular.

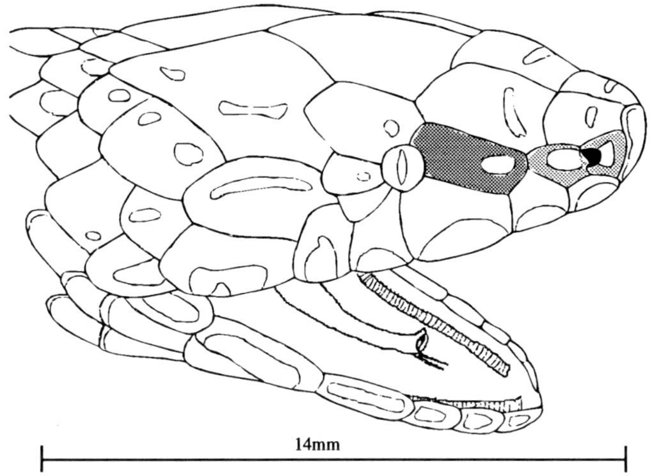

Toxicocalamus spilolepidotus PNGM 22132
Yaiya, Kratke Mts., Eastern Highlands Province
Fig 6: Preocular scale in contact with nasal scale.

Fig 5a: Four infralabials in contact with anterior chin shield

Fig 5b: Three infralabials in contact with anterior chin shield

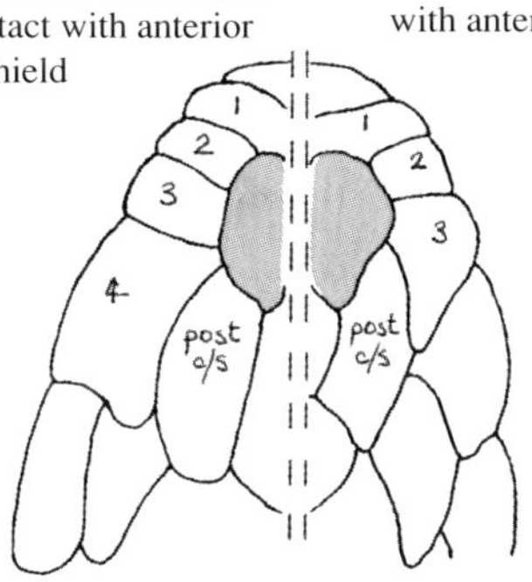

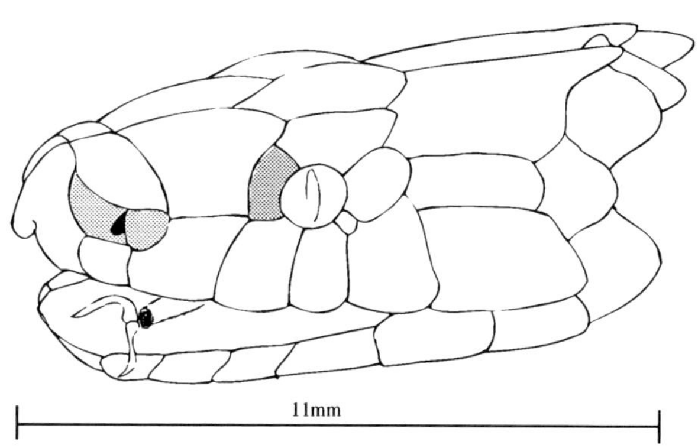

Toxicocalamus loriae PNGM 22160
Dobel, Mt Hagen, Western Highlands Province
Fig 8: Preocular scale separated from nasal scale by prefrontal scale.

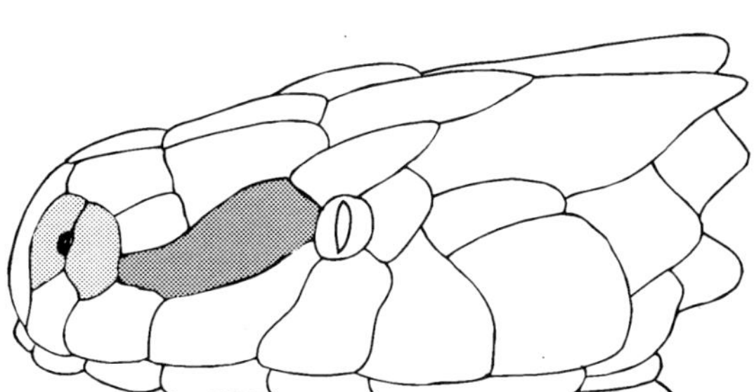

Toxicocalamus loriae BMNH 1935.5.10.174
Mafulu, Owen Stanley Range, Central Province
Fig 7: Preocular scale in contact with nasal scale.

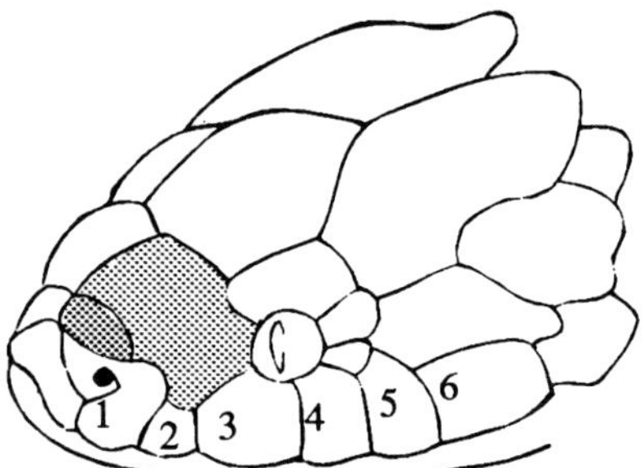

Fig 9: Internasal scale present, distinct from prefrontal; temporal scale present between rear-most supralabial and parietal scale; 5-7 supralabials.

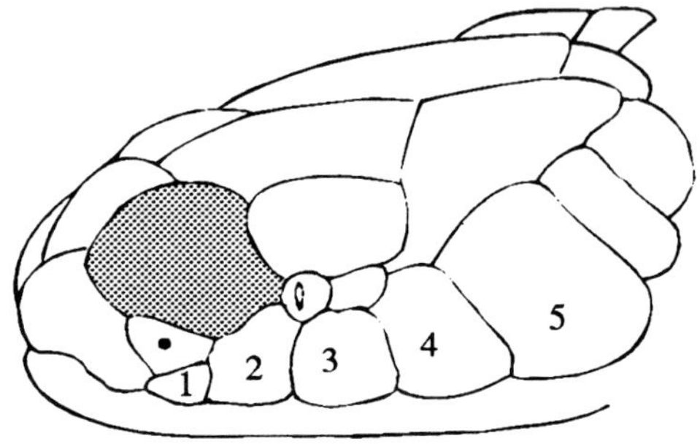

Fig 10: Internasal scale absent, fused with prefrontal; temporal scale absent, fused with rear-most supralabial; 4-5 supralabials.

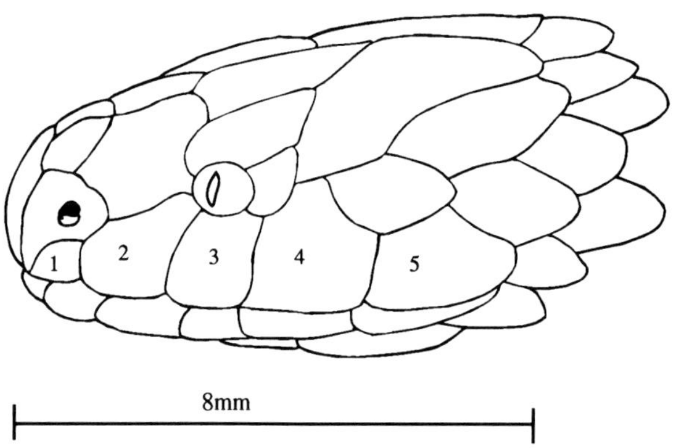

Toxicocalamus stanleyanus BMNH 1904.3.17.13
Dinawa, Owen Stanley Range, Central Province.
Fig 11: 5 supralabial scales with 2nd and 3rd in contact with eye.

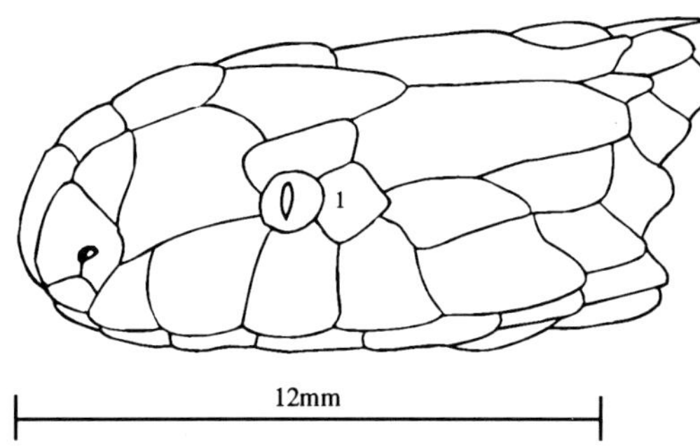

Toxicocalamus longissimus BMNH 1904.11.1.60
Fergusson Is., d'Entrecasteaux Arch.,
Milne Bay Province
Fig 12: 17 dorsal scale rows at midbody;
1 postocular scale.

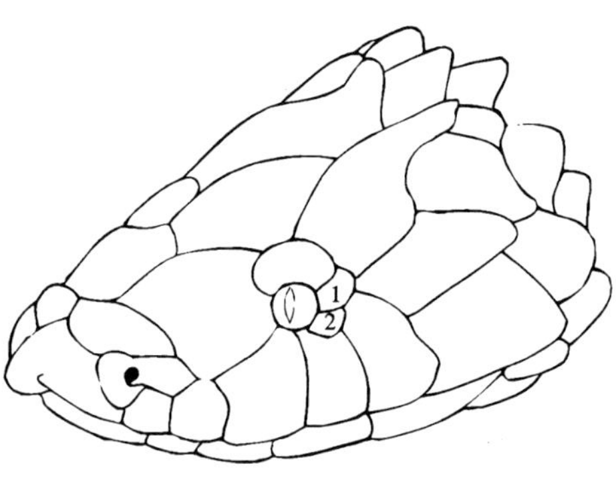

Toxicocalamus longissimus AMNH 76629
Woodlark Is., Milne Bay Province
Fig 13: 17 dorsal scale rows at midbody; 2 postocular scales.

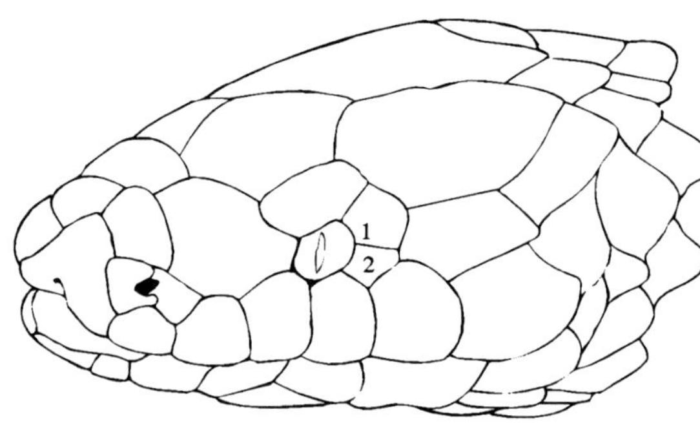

Toxicocalamus misimae AMNH 76684
Misima Is., Louisiade Arch., Milne Bay Province
(after McDowell 1969)
Fig 14: 15 dorsal scale rows at midbody;
2 postocular scales.

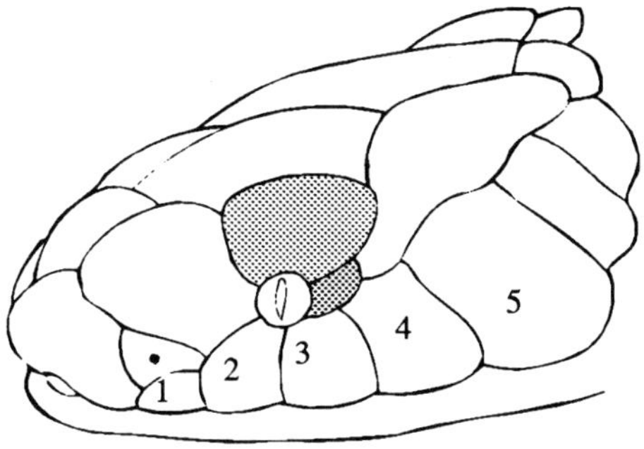

Fig 15: Postocular scale usually present and distinct from supraocular scale; 5 supralabial scales.

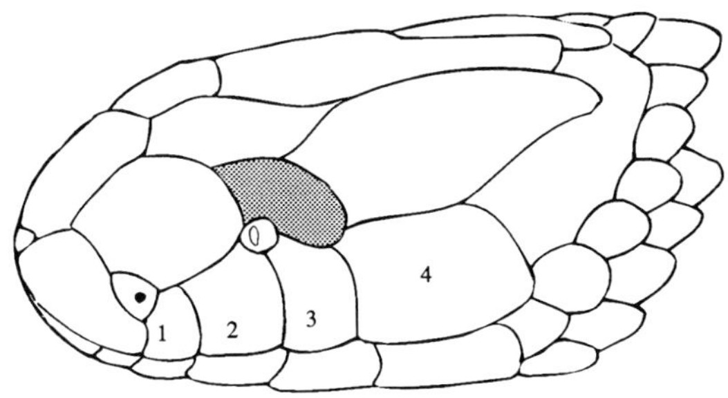

Toxicocalamus buergersi AMNH 75001
Wewak, East Sepik Province
(After McDowell 1969)
Fig 16: Postocular scale absent, fused with supraocular scale, in contact with 3rd and 4th supralabials; 4 supralabial scales.

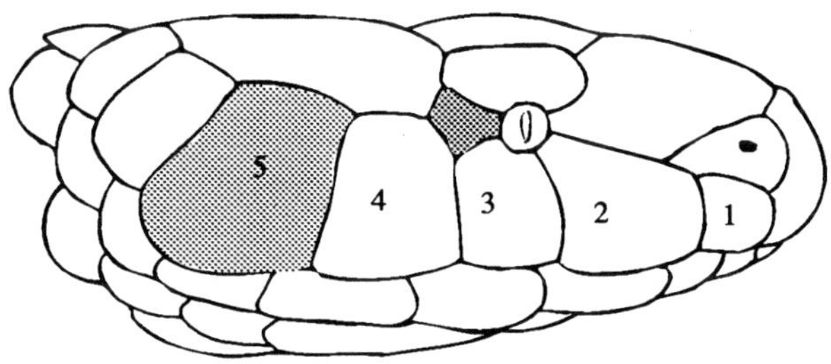

Toxicocalamus preussi preussi AMNH 62470
Idenburg R., West Irian
(after Bogert & Matalas 1945)
Fig 17: 5th supralabial separated from postocular by 4th supralabial.

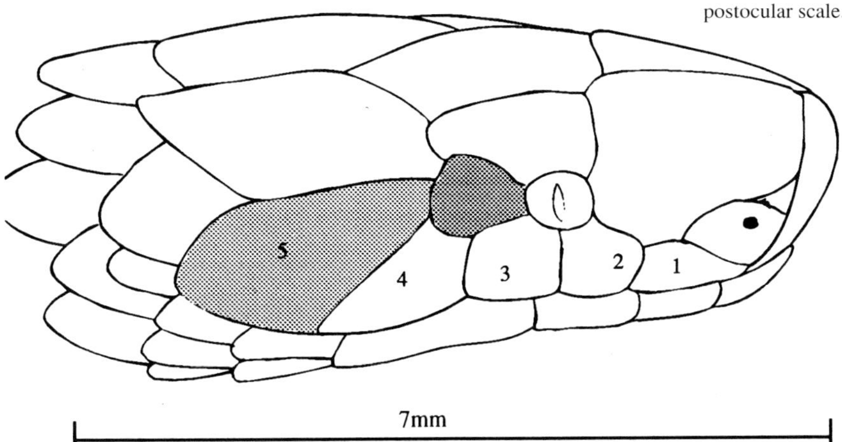

Toxicocalamus preussi angusticintus
PNGM 24584
Olsobip, Star Mts., Western Province
Fig 18: 5th supralabial in contact with postocular scale.

'BUERGERS'S FOREST SNAKE'

Toxicocalamus (Ultrocalamus) buergersi (fig. 16)

Maximum length: 365mm

Description:

b) colouration: Dorsally brown, scales of outer row light; upper lips and venter yellow.

c) scalation: DMB 15; V 293-319; SC 24-40, mostly paired, terminal scale elliptical; anal plate entire; 4 supralabials with 2nd and 3rd in contact with eye; preocular fused with internasal and prefrontal as a single scale; postoculars fused with supraocular; 4th supralabial in contact with parietal.

Distribution within PNG: W.SPK: Torricelli Mts. E.SPK: Prince Alexander Mts and Wewak. Known only from five specimens.

Distribution outside PNG: No records, a possible PNG endemic.

'MT. ROSSEL FOREST SNAKE'

Toxicocalamus (Apistocalamus) holopelturus (fig. 3)

Maximum length: 620mm.

Description:

b) colouration: Dorsum dark grey-brown except lowest two rows which are white with black spots; undersides yellow to white anteriorly with dark edging, subcaudals grey with pale edging; lips, throat and chin white.

c) scalation: DMB 15; V 226; SC 59, first paired, remainder single, terminal scale conical; anal plate divided; 6 supralabials with 3rd and 4th in contact with eye; single preocular, in contact with internasal; 2 postoculars.

Distribution within PNG: Altitude records 700m. M. BAY: Louisiade Arch., Rossel Is., Mt. Rossel. Known only from a single specimen.

Distribution outside PNG: No records, a PNG endemic.

'WOODLARK or FERGUSSON ISLAND FOREST SNAKE'

Toxicocalamus (Toxicocalamus) longissimus (figs. 12 & 13)

Maximum length: 650mm.

Description:

b) colouration: Dorsum grey-brown; sides white with a grey brown streak to each scale; red bar on snout, red spots behind eye and on rear of head; upper lips yellow with dark vertical bar from mouth through eye and from angle of jaw over neck; venter white with brown longitudinal stripes.

c) scalation: DMB 17; V 262 - 305; SC 26 - 48, paired, terminal scale conical; anal plate divided; 6, rarely 7, supralabials with 3rd and 4th in contact with eye; preocular fused with prefrontal; 1 - 2 postoculars.

Distribution within PNG: M.BAY: d'Entrecasteaux Arch. - Fergusson Is., and Woodlark Is. Known only from eight specimens.

Distribution outside PNG: No records, a PNG endemic.

Toxicocalamus longissimus (Fergusson Is., Milne Bay Province) (BMNH 1904.11.1.60)

'LORIA FOREST SNAKE'

Toxicocalamus (Apistocalamus) loriae (figs. 7 & 8)

Maximum length: 690mm.

Description:

b) colouration: Dorsum olive brown; upper lips yellow; yellow nape bars may be present on the neck; ventral surfaces yellow, the latter with several rows of small dark spots, subcaudals dark with light edges.

c) scalation: DMB 15; V 173 - 220; SC 23 - 49, mostly paired, terminal scale conical; anal plate usually divided (entire in Mt Lamington population); 6 or 7 supralabials with 3rd and 4th in contact with eye; 1, rarely 2, preoculars, in contact or not in contact with nasal; 1 - 2 postoculars.

Distribution within PNG: Altitude records 60-1500m. WEST: Ningerum, upper Fly R. CENT: Owen Stanley Mts. ORO: Popondetta, Kokoda and Mt Lamington. M.BAY: d'Entrecasteaux Arch. - Normanby Is. MOR: Huon Pen., Markham R. and Garaina. MAD: Kaironk R., Adelbert and Finisterre Mts. and Kar Kar Is. W.SPK: Torricelli Mts. WHP: Baiyer and Waghi R and Mt Hagen. S. HIGH: Mt Sisa. SIM: Kundiawa and Waghi R. E.HIGH: Yaiya R. *T. loriae* may be a species-complex containing several species. A hybrid race between *T. loriae* and *T. stanleyanus* occurs at Garaina, Morobe Province. The Normanby Island specimen may represent an undescribed species (K.Aplin *pers. comm*).

Toxicocalamus loriae (Nemadaso, Southern Highlands Province)
Photo: Steve Donnellan, S. Aust. Mus.

Distribution outside PNG: Irian Jaya as far west as Fak Fak on the Onin Peninsula.

'MISIMA ISLAND FOREST SNAKE'

Toxicocalamus (Toxicocalamus) misimae (fig. 14)

Maximum length: 468mm.

Description:

b) colouration: Dorsum dusky brown, venter yellowish anteriorly, dusky posteriorly; lips, throat and chin yellow; head dark with faint bands and spots (remnants of juvenile patterning).

c) scalation: DMB 15; V 231; SC 48, paired, terminal scale conical; anal plate divided; 6 supralabials with 3rd and 4th in contact with eye; preocular fused with prefrontal; 2 postoculars.

Distribution within PNG: M.BAY: Louisiade Archipelago - Misima Is. CENT: Mekeo - 100km nw. Port Moresby. (K. Aplin *pers. comm*). Known only from three specimens, although the Central specimen may represent a new species due to the 700km gap between the Mekeo and the type locality.

Distribution outside PNG: No records - a PNG endemic.

'PREUSS'S FOREST SNAKE'

Toxicocalamus (Ultrocalamus) preussi (figs.17 & 18)

Maximum length: 765mm

Description:

b) colouration: Dorsally dark brown, lower two rows yellow; head as body with broad yellow band across rear of head between eyes and neck (fading with age); lips, throat and underside of body and tail yellow, latter with brown bars across each scale.

c) scalation: DMB 13 (rarely 15); V 280-330; SC 16-54, mostly paired, terminal scale conical; anal plate entire; 5 supralabials with 2nd and 3rd in contact with eye; preocular fused with internasal and prefrontal as a single scale; single postocular, occasionally absent.

Distribution within PNG: Altitude records 50-850m. WEST: Upper Fly R. GULF: Purari R., Omati and Baimaru. S.HIGH: Mt Sisa. E.SPK: Wewak. W.SPK: Seleo Is., Bewani Mts. and Torricelli Mts. Bogert & Matalas (1945) referred the Sepik/Irian Jaya population to *T. preussi preussi* and the Western Province population to *T. p. angusticintus*. A new species from Mt Boobiani is currently *in prep.* (K. Aplin, *pers. comm.*).

Distribution outside PNG: Northern Irian Jaya.

Toxicocalamus preussi angusticintus (Waro, Southern Highlands Province) (Photo: S. Donnellan, S. Aust. Mus.)

'SPOTTED FOREST SNAKE'

Toxicocalamus (Apistocalamus) spilolepidotus (fig. 6)

Maximum length: 780mm.

Description:

b) colouration: Dorsum of body and head dark chocolate brown with a distinctive cream spot in the centre of each scale, excluding the vertebral row, which increases in size away from the vertebral scale row. Ventral surfaces yellow to cream posteriorly, darker with pale centres anteriorly and under head.

c) scalation: DMB 15; V 202-205; SC 33, mostly paired, terminal scale elliptical; anal plate divided; 6 supralabials with 3rd and 4th in contact with eye; single preocular in contact with nasal; 2 postoculars.

Distribution within PNG: E.HIGH: Kratke Mts. Known only from two specimens.

Distribution outside PNG: No records.

Toxicocalamus spilolepidotus (Yaiya, Kratke Mts., Eastern Highlands Province) (PNGM22132)

'OWEN STANLEY RANGE FOREST SNAKE'

Toxicocalamus (Toxicocalamus) stanleyanus (fig. 11)

Maximum length: 610mm

Description:

b) colouration: Dorsally black-brown with lower scale rows white with black centres; head as body, upper lips yellow, with dark bar from mouth, through eye to dorsum of head and angle of jaw over rear of head followed by yellow collar; underside white with two rows of black spots.

c) scalation: DMB 15; V 218 - 262; SC 25 - 51, mostly paired, terminal scale conical; anal plate entire; 4 or 5 supralabials with 2nd and 3rd in contact with eye; preocular fused with prefrontal; single postocular, occasionally absent.

Toxicocalamus stanleyanus (Waro, Southern Highlands Province) Photo: Steve Donnellan, S. Aust. Mus.

Distribution within PNG: Altitude records 100-1300m. WEST: Upper Fly R. GULF: Baimuru and Omati. CENT: Owen Stanley Mts. S.HIGH: Mt. Sisa. W.SPK: Torricelli Mts. A hybrid race between *T. stanleyanus* and *T. loriae* occurs at Garaina, Morobe Province.
Distribution outside PNG: Southern Irian Jaya to Lorentz River and Onin Peninsula at Fak Fak.

Extralimital Irian Jayan species yet to be reported from PNG:

'SETEKWA RIVER FOREST SNAKE'
Toxicocalamus (Apistocalamus) grandis
Maximum length: 960mm.
Description:
b) colouration: Dorsally dark brown with irregular white spots; lips and venter white.
c) scalation: DMB 15; V 207; SC 27, mostly paired, terminal scale conical; anal plate divided; 6 supralabials with 3rd and 4th in contact with eye; single preocular; single postocular.

Distribution outside PNG: Setekwa River, south central Irian Jaya. Known only from a single specimen.

AUSTRALIAN SMALL-EYED SNAKES

Rhinoplocephalus spp.

(2 species) Map: 20

MILDLY VENOMOUS

Average/Maximum length: 500mm / 615mm

Description:

a) physique: *R. nigrostriatus* is an extemely slender cylindrical snake with a narrow flattened head only slightly distinct from the neck; tail is long and whip-like (21-33 % SVL [snout to vent length] in Australian specimens, 27-38% SVL New Guinea specimens); the eyes are small and slightly protruding. *R. boschmai* is similar though shorter and stouter; tail shorter (14-22% SVL).

b) colouration: Species specific (*see* key to the species and individual accounts below).

c) scalation: 15 scale rows at midbody, all smooth; ventral and subcaudal scale count vary with species (*see* key to species and individual accounts below); anal plate single; loreal and subocular scales absent.

Habitat: Savanna woodlands and gardens.

Habits: Nocturnal, secretive, almost totally fossorial (burrowing) snakes which inhabit soil to a depth of 10cm, leaf litter, fallen logs, termite mound debris, cut grass piles. Inoffensive and of such small size that they are believed incapable of biting a human. *Rhinoplocephalus* is a viviparous genus and litter sizes for New Guinea species are *R. nigrostriatus* 4 - 9, *R. boschmai* 5 - 11.

Diet: Known to feed primarily on small diurnal skinks such as *Carlia*. *R. boschmai* has also been recorded taking small unidentified snakes. Two Australian species, *R. monachus* and *R. nigriceps*, are occasionally cannibalistic.

Taxonomic note: Snakes of the genus *Rhinoplocephalus* have variously been placed in the genera *Unechis, Denisonia* and *Suta* by earlier workers, and *Cryptophis* by some recent authors (Ehmann 1992).

Distribution within PNG: Southern Western Province.

Distribution outside PNG: Northeastern Queensland. The related species, *R. boschmai*, has been collected in Irian Jaya and may also eventually be recorded in Western Province, west of Morehead.

Confusing species: The small whip snakes, genus *Rhinoplocephalus,* may be confused with the small burrowing elapids of genus *Toxicocalamus.*

Venom: Nothing is known but probably harmless to man due to the small size of the mouth.

Antiserum/initial dose: Not available.

Key to the Australian small-eyed snakes genus, *Rhinoplocephalus,* of New Guinea

1a. Body colour pink to brown with black head and vertebral stripe 1 to 3 scale rows wide; ventrals more than 170; subcaudals more than 60; nasal scale in contact with preocular, preventing contact between prefrontal and supralabials (fig. 1); tail equal to 21-38% of the snout to vent body length. *Rhinoplocephalus nigrostriatus*

1b. Body colour uniform dark brown to grey-brown, dark vertebral stripe absent but lower-most row of dorsal scales marked with dark spots; ventrals fewer than 170; subcaudals fewer than 50; nasal scale not in contact with preocular, allowing prefrontal to contact 2nd supralabial (fig. 2); tail equal to 14-22% of the snout to vent body length. *Rhinoplocephalus boschmai*

BLACK-STRIPED SNAKE

Rhinoplocephalus nigrostriatus (fig. 1)

Maximum length: 615mm.

Description:

b) colouration: Reddish-brown to coral pink with a grey or brown patch on the head (black in juveniles), bordered behind by a black 'V' which continues to form a black vertebral stripe 1-3 scales wide down the midline of the back; head same colour as stripe; lowest lateral scales of dorsum pink or red with scattered cream spots; underside white or cream.

c) scalation: DMB 15; V 170-186; SC 64-79; nasal and preocular scales in contact.

Distribution within PNG: WEST: s. Trans-Fly, Oriomo R. to Bensbach R.

Distribution outside PNG: Queensland, Australia.

Extralimital Irian Jayan species yet to be reported from PNG:

CARPENTARIA WHIP SNAKE

Rhinoplocephalus boschmai (fig.2)

Maximum length: 560mm.

Description:

b) colouration: Body colour uniform dark brown to grey-brown or black, excepting dorsal scale row adjacent to ventrals which is light anteriorly and marked with dark spots posteriorly; head pale grey-brown; underside cream with dark spots and a dark stripe under the tail.

c) scalation: DMB 15; V 150-168 (150 in Merauke specimen); SC 32-48 (46 in Merauke specimen); prefrontal and 2nd supralabial scales in contact.

Taxonomic note: This species was formerly referred to as *Suta* or *Unechis carpentariae.*

Distribution outside PNG: Queensland, Northern Territory and Western Australia, and southeast Irian Jaya (Merauke).

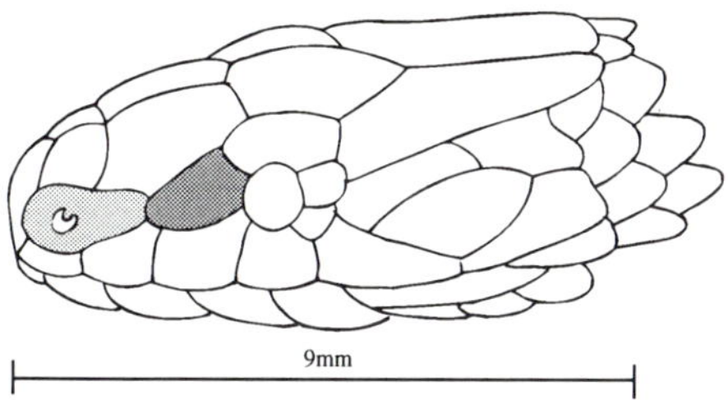

Rhinoplocephalus nigrostriatus PNGM 22130
Balamuk, Bensbach R., Western Province
Fig 1: nasal scale in contact with preocular, preventing contact between prefrontal and 2nd supralabial.

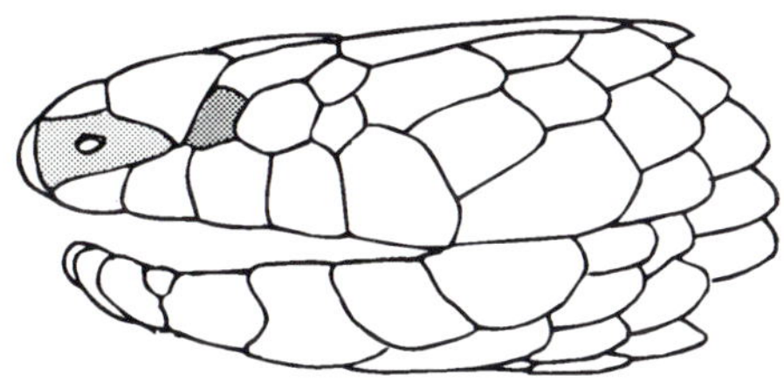

Rhinoplocephalus boschmai RMNH 10874
Merauke, West Irian (after Brongersma & Knaap-van Meeuwen 1964)
Fig 2: nasal scale separated from preocular by contact between prefrontal and 2nd supralabial.

Black-striped snake
– *Rhinoplocephalus nigrostriatus*
(Queensland, Australia)
Photo: J. Weigel, Aust. Reptile Park

BROWN-HEADED or GREY-NAPED SNAKE

Furina tristis

Map: 21

MILDLY VENOMOUS

Average/Maximum length: 0.7m / 1.0m.

Description:

a) physique: A slender snake with a narrow head only slightly distinct from the neck; tail moderately long; eye very small with round pupil.

b) colouration: Head and upper body glossy dark grey-brown to grey with lateral scales light edged; nape of neck with a yellow bar present or absent in adults, a white nape bar usually present in juveniles; underside immaculate white. The Australian common name 'brown-headed snake' does not describe New Guinea populations.

c) scalation: DMB 17 all smooth; V 160-190; SC 30-60 usually all paired; anal plate divided; loreal absent; nasal and preocular scales separated by downward projection of the prefrontal scale which contacts 2nd supralabial (fig. 1); subocular scales absent; 6 supralabials with 3rd and 4th in contact with eye and temporolabial scale between 5th and 6th; single preocular and 2 postoculars.

Habitat: Damp habitats, monsoon forests or rainforests, coconut plantations and gardens.

Habits: A fairly secretive, nocturnal and semi-fossorial snake which inhabits coconut husk piles and piles of decaying vegetation. Although specimens react vigorously to being handled they do not readily bite However, since it is venomous and of moderately large size this species should be treated with caution. Oviparous with a clutch size of six.

Diet: Primarily skinks, especially the genus *Sphenomorphus*.

Distribution within PNG: WEST: S.Trans-Fly, Fly R., Bamu R., Lake Murray and Daru Is. NCD. CENT: Mekeo, Owen Stanley Mts., Brown and Laloki Rivers., Loloata Is.

Distribution outside PNG: Cape York Peninsula, far-northern Queensland, and the islands of the Torres Strait. In common with the carpet python, the taipan, the king brownsnake, whip snakes (*Demansia*) and the black-headed snakes (*Rhinoplocephalus*), this species forms part of a radiation of Australian species to New Guinea. Also recorded from the Aru Islands, Indonesia, and probably present in southern Irian Jaya around Merauke.

Confusing species: The brown-headed snake *Furina tristis* may be confused with the non-venomous ground snakes, *Stegonotus* spp., or, due to its dark head and banded neck, with the highly venomous small-eyed snake, *Micropechis ikaheka*.

Venom: Nothing is known and caution is suggested.

Antiserum/initial dose: Not available.

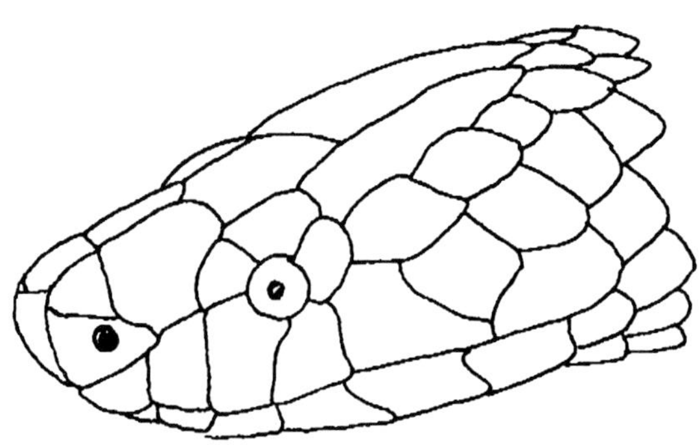

Furina tristis
(Kunini, Binaturi R., Western Province)

Brown-headed snake – *Furina tristis*
(Loloata Island, Central Province)

Brown-headed snake
– *Furina tristis*
(Kunini, Western Province)

Brown-headed snake – *Furina tristis*
(Vaifa'a, Mekeo, Central Province)

SOLOMONS CORAL SNAKE
Salomonelaps par

Map: 22

Average/Maximum length: 0.75m / 1.17m.

Description:

a) physique: A moderately stout snake with a moderately broad head slightly distinct from the neck and a fairly long tapered tail with a rounded terminal tip. The eyes are quite large.

b) colouration: Variable, body reddish-brown to dark brown often with a series of broad reddish vertebral blotches or darker brown transverse bands, occasionally scales may be edged with black giving a reticulate pattern similar to that in *Stegonotus cucullatus*; head either lighter or darker than body; underside white.

c) scalation: DMB 15-17 (15 in 2 Buka males, 17 in 1 Buka female), all smooth; V 158-180 (159-166 Buka); SC 38-59, (42-52 Buka) both single and paired (42-52 Buka); anal plate divided; subocular scales absent; loreal usually absent, occasionally a small loreal present on one side; usually 7 supralabials with 3rd and 4th in contact with eye; single preocular and 2 postoculars.

Habitat: Rainforests, usually in close proximity to forest streams.

Habits: An active and fairly common snake which will flatten its neck and hiss when provoked but does not readily bite. Diurnal, the natural history of this species is little known although it is believed to be mostly active above the ground surface. *Salomonelaps* is thought to be an oviparous species.

Diet: Primarily skinks but also small frogs and typhlopid snakes. Rarer prey items reported include geckoes, a small agamid lizard tail and a beetle.

Taxonomic note: Formerly known as *Hoplocephalus par*, McDowell (1970) considered the species *H. woodfordi* and *H. melanurus*, from the Solomon Islands to be synonyms.

Distribution within PNG: N.SOL: Buka, north of Bougainville, no records from Bougainville.

Distribution outside PNG: *Salomonelaps par* is also recorded from Shortlands and Fauro islands in the Solomon Islands, directly to the south of Bougainville, and many of the major islands of the archipelago, excluding Choiseul and San Cristobal.

Confusing species: The Salomons coral snake, *Solomonelaps par*, may be confused with the Bougainvillian or Hediger's coral snake, *Parapistocalamus hedigeri*, or the Solomons small-eyed snake, *Loveridgelaps elapoides*, which also occurs in the northern Solomons islands .

Venom: Nothing is known but should be treated with caution. Rare bites have occurred but no fatalities are recorded.

Antiserum/initial dose: Not available.

Solomons snake – *Salomonelaps par* (Boromole, Nggela Is., Solomon Islands)
Photo: Allen Allison, Bishop Museum, Hawaii.

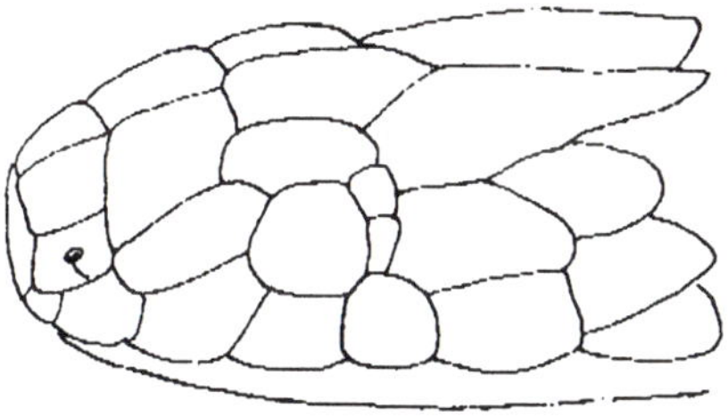

Salomonelaps par
Solomon Islands

Solomons snake – *Salomonelaps par* (Choiseul Is., Solomon Islands)
Photo: Mike McCoy, Honiara.

SOLOMONS SMALL-EYED SNAKE

Loveridgelaps elapoides

Map: 22

MILDLY VENOMOUS

Average/Maximum length: 0.80m / 1.05m.

Description:

a) physique: A snake of average build with a small, slightly compressed head, barely distinct from the neck, and a fairly long tapered tail. The eyes are very small.

b) colouration: Dorsally white, the head being either immaculate apart from black spotting on the snout and lips or white posteriorly and black anteriorly; followed on the body by alternating orange (or yellow) blotches and black bands, the blotches halting halfway down the flanks with the black bands either halting on the lower dorsal scale rows or continuing under the belly to form complete black rings or spreading laterally to join up below the orange blotches as a longitudinal black stripe; ventral surfaces either immaculate white or with heavy black markings, especially below the black bands.

c) scalation: DMB 17, all smooth; V 193-218; SC 31-38, all paired; anal plate entire; loreal and subocular scales absent; nasal scale may be divided or entire; usually 7 supralabials with 3rd and 4th in contact with eye; single preocular and 1 or 2 postoculars.

Habitat: Rainforests, usually in close proximity to forest streams.

Habits: A rare nocturnal or crepuscular species which inhabits forest floor debris, leaf litter and rotten logs. The natural history and reproductive biology of this secretive snake is unknown. Generally *Loveridgelaps* is thought to be an inoffensive snake but caution is required when handling large specimens, especially as this species was once considered a close relative of the highly venomous small-eyed snake *Micropechis ikaheka* (*see* Taxonomic note).

Diet: Small lizards, typhlopid blindsnakes and possibily small frogs.

Taxonomic note: *Loveridgelaps elapoides* was long thought to represent a Solomons subspecies of the New Guinea small-eyed snake, *Micropechis ikaheka*, (Loveridge, 1948). McDowell (1970) considered it to be more closely related to its neighbouring genus *Salomonelaps*. The local Solomons name for this snake is "shark of the jungle", (Shine 1991). The generic name "*Loveridgelaps*" is in honour of the eminent herpetologist Arthur Loveridge (1891-1980).

Distribution outside PNG: Recorded from several islands in the Solomons Group: Isabel, Gizo, Nggela, Malaita, Florida and Guadalcanal, where *Salomonelaps par* also occurs, and recently also collected from Shortlands Island which is adjacent to Bougainville (Fred Parker, *pers. comm.*) so it may eventually be collected from North Solomons Province. No records from PNG.

Confusing species: The Solomons small-eyed snake may be confused with the Bougainvillian or Hediger's coral snake, *Parapistocalamus hedigeri*, the Solomons coral snake, *Salomonelaps par*, which also occurs within North Solomons Province but it most resembles the New Guinea small-eyed snake, *Micropechis ikaheka,* from New Guinea and the semi-aquatic sea kraits, *Laticauda* spp.

Venom: Nothing is known but should be treated with caution. Rare bites have occurred but no fatalities are recorded.

Antiserum/initial dose: Not available.

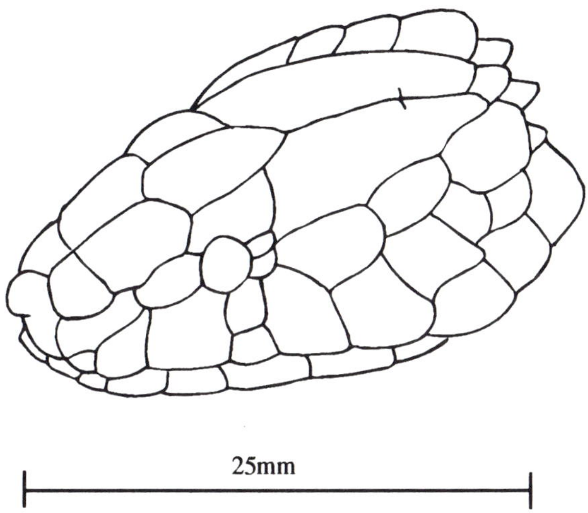

Loveridgelaps elapoides BMNH 193.3.4.2
Bilua, Gizo Is., Solomon Islands

Solomons small-eyed snake – *Loveridgelaps elapoides*
Gizo, Solomon Islands. BMNH 1933.3.4.2

Solomons small-eyed snake - *Loveridgelaps elapoides* (Nggela Is. [Florida Is.], Solomon Islands)
Photo: Mike McCoy, Honiara.

BOUGAINVILLIAN or HEDIGER'S CORAL SNAKE

Parapistocalamus hedigeri

Map: 22

MILDLY VENOMOUS

Average/Maximum length: 300mm / 500mm.

Description:

a) physique: A small, slender species with a narrow head only slightly distinct from the neck; eye small with round pupil.

b) colouration: Head and body unicolour brown although a lighter crossband may be present on the neck and the rear of the head, and both the lip scales and lowest dorsal scales are lighter; underside yellow or light brown and underside of tail similar or suffused with grey.

c) scalation: DMB 15, all smooth; V 159-169; SC 32-35, paired; anal plate entire or divided; loreal and subocular scales absent; usually 6 supralabials with 3rd and 4th in contact with eye; single postocular; preocular absent (fig. 3) or present, either in contact with nasal scale (fig. 4) or prevented from contact by downward process of prefrontal (fig. 2).

Habitat: Forests.

Habits: A very rare and secretive semi-fossorial snake which inhabits rotten logs and leaf-litter. Nocturnal or crepuscular. The natural history of this species and its mode of reproduction is poorly known but it is probably oviparous.

Diet: No stomach contents recorded.

Taxonomic note: The Bougainville population of *P. hedigeri* was formerly recognised as a subspecies of *Denisonia (Salomonelaps) par*, as *D. par furva* (Klemmer 1963). *Parapistocalamus* is now believed more closely related to the Asian / American coral snakes of subfamily Elapinae or the sea kraits of the Laticaudinae than the other Australo-Papuan elapids which comprise the subfamily Oxyuraninae (McDowell 1967; 1969; Mengden 1983). The primary difference between this species and other Australo-Papuan elapids is the presence of a maxillary diastema (toothless gap) behind the fangs.

Distribution within PNG: Altitude record 660m. N.SOL: s. and sw.Bougainville, Buin district, Empress Angusta Bay and Torokina.

Distribution outside PNG: A Bougainville endemic.

Confusing species: Hediger's snake *Parapistocalamus hedigeri* may be confused with the typhlopid blindsnakes, the Solomons coral snake, *Salomonelaps par*, or the Solomons small-eyed snake, *Loveridgelaps elapoides*, which also occur in northern Solomons islands.

Venom: Nothing is known but probably harmless to man due to the small size of its mouth.

Antiserum/initial dose: Not available.

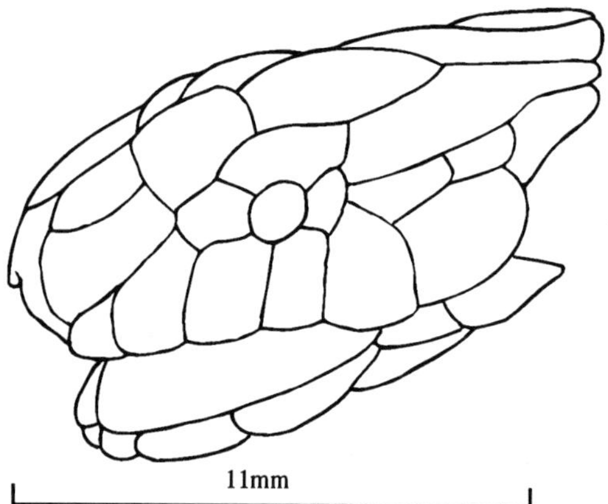

Parapistocalamus hedigeri
BMNH 1965.355
(Empress Augusta Bay, Bougainville, North Solomons Province)

Prefrontal scale in contact with 2nd supralabial, preventing contact between preocular and nasal. CNHM 44840 Empress Augusta Bay, Bougainville, North Solomons Province

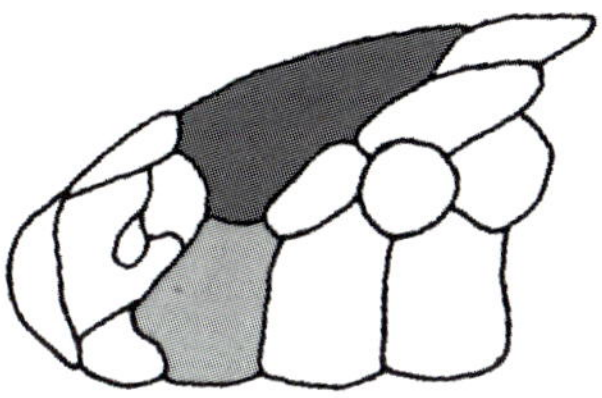

Preocular scale in contact with nasal, preventing contact between prefrontal and 2nd supralabial. CNHM 44842 Empress Augusta Bay, Bougainville, North Solomons Province

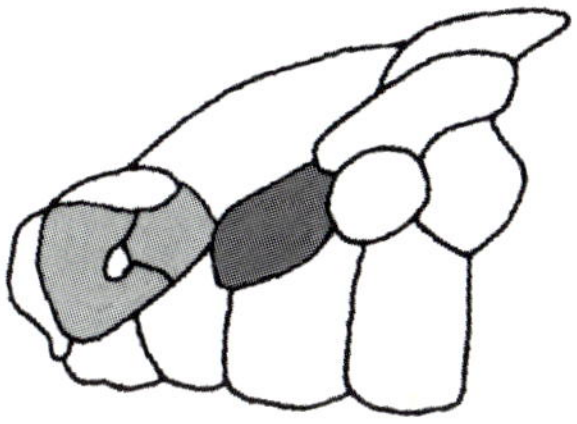

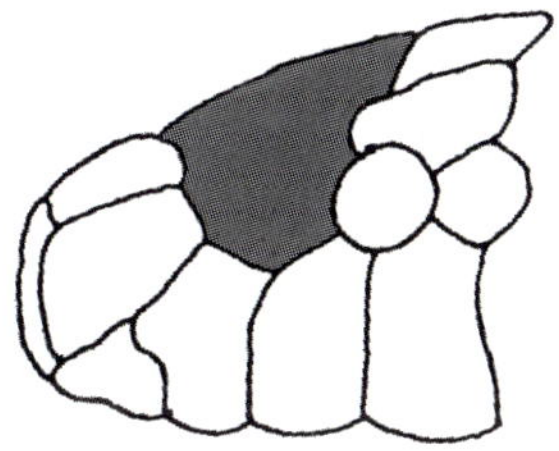

Preocular scale fused as single scale with prefrontal. MCZ67350 Buruburunna, Bougainville, North Solomons Province (after Williams & Parker 1964)

Parapistocalamus hedigeri (Bougainville, North Solomons Province)
Photo: Mike McCoy, Honiara.

NEW GUINEA CROWNED SNAKES

Aspidomorphus spp.

(3 species) Map: 24

MILDLY VENOMOUS

Average/Maximum length: 400mm / 645mm (*A. muelleri*).

Description:

a) physique: Moderately slender snakes with a flattened head, slightly distinct from the neck; tail fairly short; eye small with pupil vertically elliptical, barely apparent in *A. muelleri*.
b) colouration: Highly variable (*see* key and individual species accounts below).
c) scalation: DMB 15, all smooth; V 138-182; anal plate divided; SC 19-41; loreal absent; 6 supralabials with 3rd and 4th in contact with the eye; single preocular, in contact with nasal and 2nd supralabial; 1-2 postoculars; 7 infralabials. (*See* individual accounts below).

Habitat: Lowland rainforests and monsoon forests up to 1500m.

Habits: Secretive terrestrial snakes which inhabit decaying logs, debris and leaf litter. Probably oviparous but the natural history of this genus is not well documented. Probably nocturnal.

Diet: Not known. Possibly food items may include invertebrates small reptile eggs or small leaf-litter lizards and frogs. *A. muelleri* attains a sufficient size to prey on small mammals.

Distribution within PNG: Throughout mainland PNG including the islands of Milne Bay Province and the Bismarck Archipelago.

Distribution outside PNG: Irian Jaya and the Moluccas.

Confusing species: The crowned snakes of genus *Aspidomorphus* may be confused with the harmless ground snakes, *Stegonotus* spp., or the highly venomous small-eyed snake, *Micropechis ikaheka*.

Venom: Effects of bites by *A. muelleri* include localised pain similar to that experienced following a wasp sting, swelling and loss of sensation in the bitten area and a drop in pulse rate, drowsiness and feelings of cold and nausea. Larger specimens must be considered potentially dangerous, especially to infants and children. Nothing is known of the venoms of the other two species.

Antivenom/initial dose: Not available.

Key to the crowned snakes genus *Aspidomorphus* of New Guinea*

1a. Ventrals fewer than 161; subcaudals fewer than 30; distribution: Sepik Provinces and Irian Jaya. *Aspidomorphus schlegeli*

1b. Ventrals usually in excess of 160; subcaudals usually in excess of 30 (both characters for mainland populations, excluding Owen Stanley Range specimens); distribution: mainland, Bismarck Archipelago and Milne Bay Islands. 2

2a. Dark-edged, pale vertebral stripe beginning on the posterior edge of the parietals and running along the neck above a lighter stripe from the eye and a further light stripe from the rear of the eye to the 6th supralabial where a separate light stripe continues a short distance along the lower neck (fig. 1); distribution: mainland and Milne Bay islands. *Aspidomorphus lineaticollis*

2b. Vertebral stripe absent; scales of head with light edged, dark centred 'ocelli' (fig. 2); a short dark diagonal stripe or blotch present, passing from the underside of the eye through lighter supralabials to the jaw; distribution: mainland and Bismarck Archipelago. *Aspidomorphus muelleri*

* Note: Species within the genus *Aspidomorphus* exhibit considerable interspecific overlap in scale counts and geographically intraspecific variation in both scale counts and patterning with the result that the construction of a key to encompass the three species was difficult. Therefore it is suggested the serious field worker refer to the brief descriptions below and the detailed descriptions in Brongersma, 1934, and McDowell, 1967.

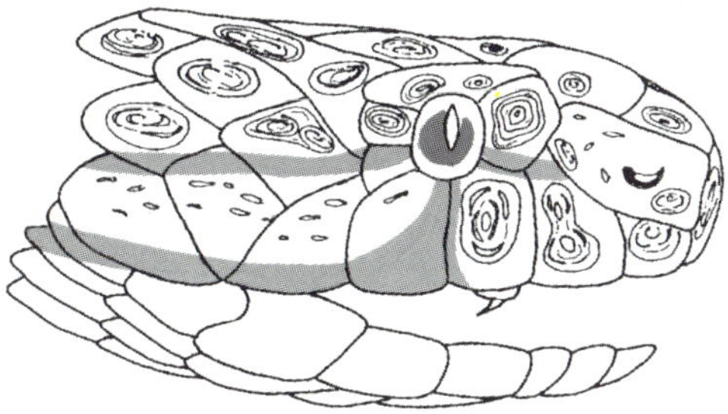

Aspidomorphus muelleri – Muller's crowned snake. Stripe through eye, 'brindle' below eye and ocelli on dorsum of head present.

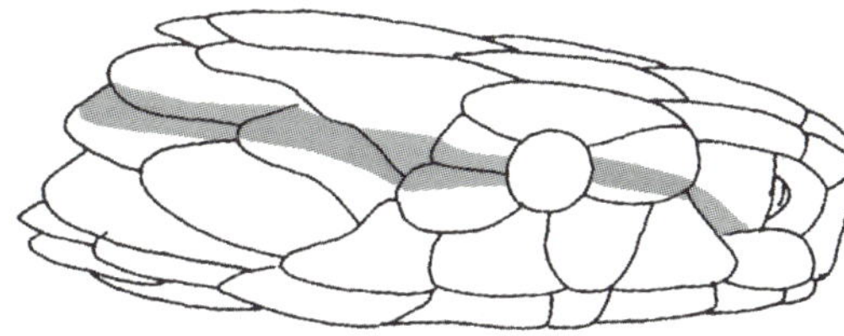

Aspidomorphus schlegeli – Schlegel's crowned snake. Stripe through eye present but 'brindle' and ocelli absent.

Striped crowned snake
– *Aspidomorphus lineaticollis*
(Fergusson Is., Milne Bay Province)
Photo: Daniel Heuclin

Striped crowned snake
– *Aspidomorphus lineaticollis*
(Fergusson Is., Milne Bay Province)
Photo: Daniel Heuclin

'STRIPED CROWNED SNAKE'

Aspidomorphus lineaticollis

Maximum length: 400mm.

Description:

b) colouration: Body and head light or dark brown; a pale vertebral stripe, edged by darker stripes, beginning on the posterior edge of the parietals and running along the back; a second light stripe exiting the rear of the eye across the lower edge of the parietal and parallel to the dark stripe on the neck; a third light stripe runs from the underside of the eye to the 6th supralabial where a further, separate, light stripe continues a short distance along the lower neck; scales of snout, prefrontals and internasals, may possess light edged, dark centred 'ocelli'. Underside of chin, neck and anterior body light to dark grey with mottled speckling, posterior ventral surfaces and subcaudals generally pale.

c) scalation: V 139-174; SC 24-41, usually all paired; other scalation as generic account.

Taxonomic note: Brongersma (1934) referred the Morobe and Madang population to the subspecies *A. mülleri lineaticollis* and the Milne Bay island population to *A. mülleri lineatus*. He also considered the single Misima specimen to be distinct from other Milne Bay populations.

Distribution within PNG: Altitude record 400m. M.BAY: d'Entrecasteaux Arch. - Goodenough, Fergusson and Normanby Is., Louisiade Arch. - Misima and Tagula Is., Trobriand and Woodlark Islands. ORO: Lego. MOR: Huon Pen., upper Watut R., Bulolo. MAD: Adelbert Mts. and Astrolabe Bay.

Distribution outside PNG: No records. A PNG endemic.

MÜLLER'S CROWNED SNAKE

Aspidomorphus muelleri

Maximum length: 460mm (Bismarck Arch.); 660mm (mainland New Guinea).

Description:

b) colouration: Highly variable. Body and head dorsally brown, pale vertebral stripe absent; scales of head and anterior neck with symmetrical grey and black 'ocelli'; short dark diagonal stripe or blotch ('brindle' of McDowell) from the underside of the eye to the jaw, breaking the pale supralabial stripe which is again interrupted by dark pigment on the side of the neck (*A. mülleri interruptus* of Brongersma - *see below*) or continues uninterrupted along the side of the neck and anterior body (*A. mülleri mülleri* of Brongersma). Underside of chin, neck and anterior body dark grey to black, posterior ventral surfaces including subcaudals pale cream to white.

c) scalation: V 160-182; SC 29-40, usually all paired; other scalation as generic account.

Taxonomic note: Brongersma (1934) referred the western New Guinea population to the subspecies *A. mülleri mülleri* and the Bismarck Arch. population to *A. mülleri interruptus.*

Distribution within PNG: Altitude records 500-1500m. W.SPK: Aitape and Telefomin. E.SPK: Wewak and Kairuru Is. W.HIGH: Baiyer R. E.HIGH: Kratke Mts. WEST: S.Trans-Fly and upper Fly R. GULF: Baimuru and Purari R., Uramu Is. S.HIGH: Mt. Sisa. SIM: Karamui. CENT: Owen Stanley Mts. and Brown R. MOR: Garaina. ORO: Popondetta and Owen Stanley Mts. M.BAY: Kwagira R. WNB: Whiteman Mts. and Garu. ENB: Jacquinot Bay, Gazelle Pen., Rabaul, Duke of York Is. N.IRE: Muliama.

Distribution outside PNG: Irian Jaya and the islands of Misool, Salawatti, and Seram.

SCHLEGEL'S CROWNED SNAKE

Aspidomorphus schlegeli

Maximum length: 425mm.

Description:

b) colouration: Generally brown dorsally, with dark-edged, pale vertebral stripe usually present but dark blotch below eye ('brindle') absent and 'ocelli' markings missing from head and neck scales. Underside dark anteriorly, lighter but finely spotted with darker pigment posteriorly and white under the tail.

c) scalation: V 137-161; SC 19-29, usually all paired; other scalation as generic account.

Distribution within PNG: W.SPK: Aitape. Single collection locality for Papua New Guinea.

Distribution outside PNG: Irian Jaya as far as Fak Fak and the islands of Misool, Salawatti, Batanta and Waigeu.

Müller's crowned snake – *Aspidomorphus muelleri* (Wewak, East Sepik Province)
Photo: Mike McCoy, Honiara.

Schlegel's crowned snake – *Aspidomorphus schlegeli* (Iffar, Lake Sentani, Irian Jaya)
(BMNH 1938.6.9.49)

PAPUAN WHIPSNAKES

Demansia spp.

(2 species) Map: 23

MILDLY VENOMOUS

Average/Maximum length: 0.6m / 1.5m.

Description:

a) physique: Slender, rapidly moving snakes with a narrow head slightly distinct from the neck; tail long and whip-like; eye very large with a round pupil.
b) colouration: Body dark brown to black, lightening towards tail which may be red-brown; interstitial skin yellow to white; underside blue-grey, darkening posteriorly; head pale or copper-brown, spotted with dark pigment above but lips lighter, white or yellowish, spotted with brown; chin and throat white; lower neck much lighter than rest of body; eye with brown iris surrounded by ring of yellow on preocular and postocular scales; belly light grey to black.
c) scalation: DMB 15, all smooth; V 160-225; SC 69-105, paired; 6 supralabials with 3rd and 4th in contact with the eye, temporolabial present between 5th and 6th; anal plate divided; loreal and subocular scales absent.

Habitat: Southern savannas and savanna woodlands but also common in cultivated gardens and around villages.

Habits: One of the fastest moving snakes in New Guinea, capable of travelling across the ground so rapidly that the eye has difficulty following. Whip snakes are diurnal and may be active during the hottest part of the day when other species are not in evidence. This is a fairly inoffensive species although specimens which are molested will bite rapidly and large specimens have been observed to rear up in front of walkers or motor vehicles. Most commonly encountered in the drier season. Occasionally pairs of males may be observed in combat over a female, chasing one another, rising up and attempting to force the other to the ground. Oviparous with a clutch size of up to eight.

Diet: Lizards, skinks and small agamid lizards, and frogs.

Taxonomic note: *Demansia* can be a complicated genus and differentiating between the New Guinea species presents difficulties for the taxonomist with Parker (1982) reporting that what may constitute good characters for the identification of Australian species rapidly break down when applied to PNG populations, an opinion shared by myself. It is not at all clear which species of *Demansia* occur in southern New Guinea. Wilson and Knowles (1988), Cogger (1992) and Ehmann (1992) refer the New Guinea whip snakes to *D. papuensis papuensis* yet Storr (1978) recognised both this species and *D. atra* as occurring in southern New Guinea. All specimens I have examined resembled *D. atra* since they possess black sutures on the anterior ventral and temporal scales and combined ventral and subcaudal counts lower than 275, yet several also exhibited spotting on the head, a reported *D. papuensis* characteristic. The whip snakes of southern PNG clearly require further study in order to determine their identity and in the meantime descriptions of both possible species are provided here.

Distribution within PNG: WEST: s.Trans-Fly, Daru Is., Bensbach, Oriomo, Aramia and Fly R. and Balimo. NCD. CENT: Mekeo, Loloata Is., Brown R., Goldie R., Laloki R. Whether *Demansia* occurs on the isolated savannas of Gulf Province has yet to be determined.

Distribution outside PNG: *D. atra* occurs in Western Australia, Northern Territory and Queensland, and a subspecies of *D. papuensis, D. papuensis melaena,* has been described for Western Australia and Northern Territory. *Demansia* spp. also inhabit the southern savannas of Irian Jaya.

Confusing species: The Papuan whip snakes of genus *Demansia* are frequently confused with the highly venomous Papuan blacksnake, *Pseudechis papuanus*, and the Papuan taipan, *Oxyuranus scutellatus canni,* and they may also be confused with the non-venomous black treesnake, *Dendrelaphis punctulatus*, or the keelbacks, *Tropidonophis* spp.

Venom: Whip snakes may be responsible for some of the less serious snakebites exhibiting sharp localised pain, slight swelling and mild haemostatic disturbances. Gow (1976) reports a bite from a 1.5m specimen which caused severe local pain and swelling of the entire arm which took a week to subside and I have also received a bite from a smaller specimen which resulted in sharp pain which lasted 15-20 minutes. Worrell (1963) also reports localised pain and swelling, following a bite, which recurred over a period of two weeks. Urticaria and other anaphylactic features may be seen in patients sensitised by a history of previous snakebites. Serious symptoms are considered unlikely.
Antivenom/initial dose: Antivenom is not generally used to treat whip snake bites but might be required following a serious bite from a large specimen. The CSL Austrao-Papuan polyvalent antivenom would be the most appropriate.

Black whipsnake – *Demansia atra*
(Goldie River, Central Province)

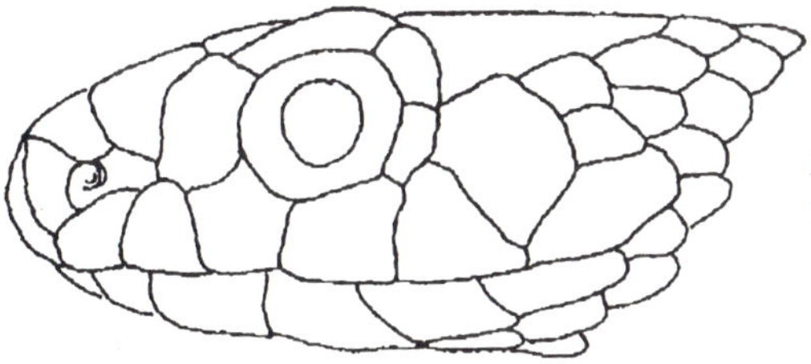

Demansia atra
(Wipim, Western Province)

Black whipsnake – *Demansia atra*
Goldie River, Central Province)

Key to the whipsnakes genus, *Demansia*, of New Guinea

1a. Ventrals plus subcaudals fewer than 275; head dark brown, not spotted; posterior edge of anterior ventrals and temporal scales with black margin. *D. atra*

1b. Ventrals plus subcaudals in excess of 275; head brown, spotted with darker pigment; posterior edge of anterior ventrals and temporal scales without black margin. *D. papuensis papuensis*

LESSER BLACK WHIPSNAKE

Demansia atra

Average/Maximum length: 0.66m / 1.15m.

Description: scalation: DMB 15, all smooth; V 176-197; SC 70-88, paired; V +SC 249-270.

Distribution outside PNG: Southeastern Irian Jaya around Merauke and widely in northern Australia, Northern Territory, Queensland and Western Australia.

PAPUAN or GREATER BLACK WHIPSNAKE

Demansia papuensis papuensis

Average/Maximum length: 0.63m / 1.52m.

Description: scalation: DMB 15, all smooth; V 192-220; SC 78-105, paired; V+SC 281-322.

Distribution outside PNG: Southeastern Irian Jaya around Merauke. A second subspecies, *Demansia papuensis melaena*, occurs in Northern Territory and Western Australia.

NEW GUINEA DEATH ADDERS

Acanthophis spp.

(*see* Taxonomic note) Map: 25

HIGHLY VENOMOUS

Average/Maximum length: 0.3-0.5m/1.0m.

Description:

a) physique: An elapid which has evolved to occupy an Indo-Australian niche, which is occupied elsewhere by the vipers. Characterised by a short, stumpy, viper-like body covered by smooth or slightly keeled scales; tail fairly elongate with a unique laterally flattened tip and a short terminal spine; head broad and viper-like, usually with raised horn-like supraocular scales over the eye, head also with either keeled or smooth scales; fangs relatively long and mobile and pupil of eye vertically elliptical.

b) colouration: Highly variable, red, brown or greyish above, either uniform or with alternating pale and dark transverse bands spotted with black; underside either immaculate white or spotted with black; tail black with white or yellow tip; head as body, brown or red with darker transverse streaks above and white labials and chin spotted with black.

c) scalation: DMB 21-23, weakly keeled or smooth; V 110-134; SC 36-57, anteriorly single, posteriorly paired; anal plate single; loreal scale absent; subocular scales present.

Habitat: Widespread in monsoon and rainforest habitats, both lowland and mid-altitude uplands areas. Also encountered in savanna woodland, upland grassland valleys, coffee plantations and cultivated gardens. Frequently found under rubbish or vegetation debris.

Habits: A nocturnal snake, sluggish during the day, the death adder is commonly encountered on paths and trails where it represents a serious snakebite threat. Although inactive it will respond by biting if trodden upon. Following an initial bite a death adder will often hang on, not attempting to flee in the manner of the other elapids, and it is frequently killed by its victim or his/her relatives. Usually bites occur below the ankle. Death adders are generally considered commonest at the end of the wet season. They are viviparous with a litter size of up to eight.

Diet: Terrestrial skinks and other lizards, birds or small mammals and possibly frogs, using its wiggling tail tip to attract prey, a form of behaviour known as 'caudal luring'.

Taxonomical note: Various scientific names have been proposed for geographical races of New Guinea death adders in the past - *Acanthophis praelongus, A. rugosus* and *A. laevis*. Storr (1981) recognised three species, *Acanthophis antarcticus, A. praelongus* and *A. pyrrhus* for Australia, and McDowell (1984) considered a possible similarity between a) *A. antarcticus* of Australia and "highland" New Guinea specimens and b) *A. praelongus* of Australia and "lowland" New Guinea *Acanthophis*. However, he also noted that whereas Western Australian *A. praelongus* is considered to possess the more raised 'horn-like' supraocular scales amongst Australian death adders, in New Guinea this is more a characteristic of the highland form, or *'A. antarcticus'*, rather than the flat-supraocular-scaled lowland form, *'A. praelongus'*. Storr (1981) listed the following characteristics for *A. antarcticus* and *A. praelongus* in his key and descriptions to the Western Australian death adders (see page 158):

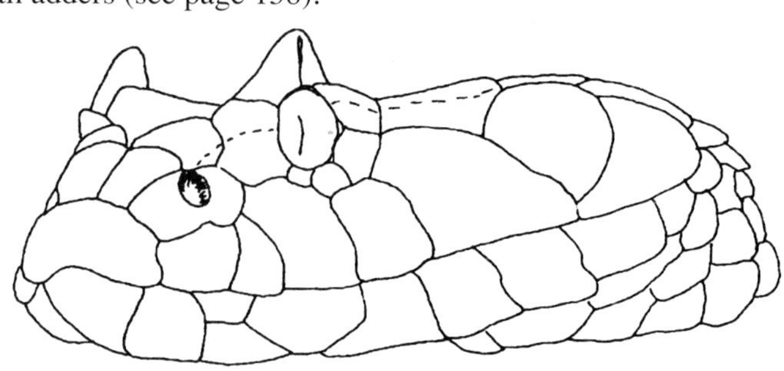

Acanthophis sp.
Tapini, Central Province

New Guinea death adder – *Acanthophis* sp
(Goldie River, Central Province)

New Guinea death adder – *Acanthophis* sp.
(Tapini, Central Province)

New Guinea death adder – *Acanthophis* sp.
(Eboa, Mekeo, Central Province)

New Guinea death adder – *Acanthophis* sp.
(Kar Kar Is., Madang Province)

New Guinea death adder – *Acanthophis* sp
(Riwo, Madang Province)

Death adder (Storr 1981)	W. Australian *A. antarcticus*	W. Australian *A. praelongus*
Head scales	smooth or weakly rugose (rough or keeled).	moderately rugose (rough or keeled).
Anterior dorsals	smooth or weakly keeled	strongly keeled (except juveniles)
Supraocular scale	not raised	raised
Supralabials	boldly patterned white and black (or dark brown)	not boldly patterned, usually as ground colour or darker
Dorsal scale rows at midbody	usually 21 (rarely 23)	usually 23 (rarely 21 or 19)
Dorsal scale rows on neck	not usually reduced	usually reduced by 2 or 4

A detailed taxonomic study of the Indonesian/New Guinea death adders may be required before the taxonomy of this widespread and complicated genus can be fully understood and documented.

Distribution within PNG: Altitude records 0-1800m (SIM).
Specimens: WEST: Oriomo Plateau, Balamuk, Balimo and Emeti, Lake Murray, Boboa Is., lower and upper Fly, Aramia and Nomad R. GULF: Purari and Vailala R., Omati and Ihu. NCD: Port Moresby. CENT: Mekeo, Goldie R., Laloki R., Brown R., Vanapa R., Veimauri R., Owen Stanley Mts. ORO: Buna. MOR: Watut R., Wau, Bulolo, Huon Gulf, Lae, Huon Pen. and Markham R. E.HIGH: Henganofi and Dunantina. SIM: Kratke Mts., Chuave and Karimui. W.HIGH: Baiyer R. S.HIGH: Waefe and Wasu, Mt Bosavi. MAD: Ramu R., Astrolabe Bay and Kar Kar Is. E.SPK: Sepik R. W.SPK: Sepik R. In error reports from ENB: Gazelle Pen. and Rabaul.
Snakebites: WEST: Ningerum and Balimo. NCD: East Boroko, Gerehu, Bomana. CENT: Woitape, Kabadi, Boera, Veimauri, Vanapa and Brown R., 14 Mile, Sabusa. ORO: Kokoda.
A small montane race (to 300mm) is known from Henganofi, Eastern Highlands, and large specimens (up to 1.0m) are reported from the Sepik, Ramu and Markham River valleys. Reports of specimens from New Britain are considered doubtful.
Distribution outside PNG: Indonesia from Irian Jaya to Obi to the west and the Aru Islands. Three species occur in Australia, (not Tasmania).
Confusing species: The death adders of genus *Acanthophis* are often confused with the New Guinea ground boa, *Candoia aspera,* and in the Mekeo region, with the keelbacks, *Tropidonophis* spp.

Venom: Death adder venom is NEUROTOXIC. Wounds do not bleed, no coagulapathy, and tests show no rhabdomyolysis. The neurotoxic action consists of progressive post-synaptic paralysis, beginning with ptosis and external ophthalmoplegia and ending in fatal bulba and respiratory paralysis. A post-synaptic neurotoxin isolated from Australian *Acanthophis antarcticus* is known as acanthophin A. Average adult death adder venom yields vary from 42-85mgs.
Antivenom/initial dose: CSL Monovalent Death Adder/6,000 units. Death adder antivenom has been used very effectively to reverse symptoms and save patients and recent work also suggests that neuromuscular transmission may be improved by the administration of anticholinesterases.

NEW GUINEA SMALL-EYED or IKAHEKA SNAKE

Micropechis ikaheka

(*see* Taxonomic note) Map: 26

HIGHLY VENOMOUS

Average/Maximum length: 1.0-1.5m / 2.0m (Kar Kar and Sepik).

Description:

a) physique: A fairly stocky species with relatively short tail; head narrow and only slightly broader and distinct from the neck; eyes very small with round pupil.

b) colouration: Variable even locally. The head is light to dark grey and distinct from the yellow or cream neck, which may be either immaculate, common in northern populations, or with the scales tipped with darker pigment, especially southern specimens. Further down the body the dark scale tipping becomes more prevalent until by midbody the dark pigment appears as red or brown, often darker edged, crossbands which increase in width and frequency posteriorly until the tail is entirely dark. Underside creamish yellow edged with black or brown. Labials, throat and chin may be yellow. The light colouration of this snake has lead to the commonly applied local name of 'white snake'.

c) scalation: DMB 15 at midbody, all smooth; V 178-223; SC 37-55, all paired; anal divided; 6 supralabials with 3rd and 4th contacting eye and temporolabial between 5th and 6th; loreal and subocular scales absent.

Habitat: Confined to monsoon and rainforest areas, swamps and similar habitats from sea level to 1500m. Small-eyed snakes also inhabit discarded husk piles in worked coconut plantations. The name "ikaheka" is derived from an Irian Jayan dialect name for "land eel".

Habits: Generally a secretive semi-fossorial species which inhabits leaf litter, loose soil or piles of decaying vegetation, coconut husks etc. There is also recent evidence, from Irian Jaya, that *Micropechis* may at times become arboreal. A snake responsible for a human snakebite was captured in a bird trap. Small-eyed snakes are usually only encountered in the open at night. They react with aggression if handled or molested, striking rapidly to the side and delivering a tenacious chewing bite. Small-eyed snakes appear to be more commonly encountered in husk piles during the drier months but may be encountered crossing roads in the wet or dry seasons. They are usually nocturnal but may occasionally be active by day. Oviparous.

Diet: Little known but frogs, lizards and small mammals probably constitute the main prey items. Snakes, including the New Guinea ground boa, *Candoia aspera,* and conspecifics (cannibalism), are reported as occasional prey and invertebrates such as earthworms may also be taken.

Taxonomic note: Two subspecies, *M. ikaheka ikaheka* and *M. i. fasciatus*, have been recognised but their distributional ranges may be a source of some dispute. While more recent authors recognising *M. i. fasciatus* record it as an Aru Islands endemic (Klemmer 1963), Loveridge (1948) originally extended its range from the Aru Is. to Aitape (West Sepik) and Gusiko (Morobe) in 'Australian New Guinea' (now PNG) and apparently confined *M. i. ikaheka* to 'Dutch New Guinea' (Irian Jaya) based on a Fak Fak specimen. Loveridge considered the two Aitape and Gusiko specimens to be 'more or less banded' and provided the following scale counts: ventrals 180-185, subcaudals 37-39, but these counts overlap greatly with those obtained from the Fak Fak specimen. Since these scale counts, obtained from a very small sample, are quite similar, expecially when compared with the Doré type specimen scale counts (*see* Scalation above), coupled with the variable characteristic of degree of banding, they would appear to provide insufficient data to differentiate between these two subspecies on mainland New Guinea so the small-eyed snake is referred to here without any subspecial designation i.e. *Micropechis ikaheka.*

Loveridge's third 'race', *M. i. elapoides* from the Solomon Islands is now recognised as a separate species (*see Loveridgelaps elapoides*).

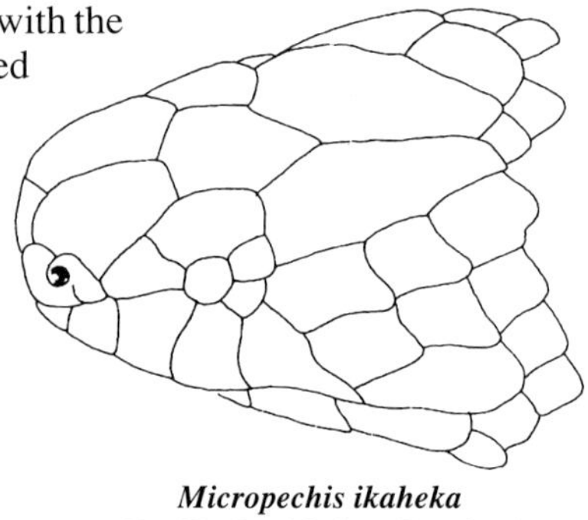

Micropechis ikaheka
(Kar Kar Is., Madang Province)

Distribution within PNG: Altitude records 0-1500m (Telefomin, W.SPK).
Specimens: WEST: Kiunga, Ningerum, Tikam and Star Mts. GULF: Purari R. delta, Kikori, Kerema and Omati. CENT: Dinawa, Owen Stanley Mts., Sogeri, Musgrave R. and Koroko. M.BAY: Loani on mainland opposite Samarai Is. and Kwagira R. ORO: Owen Stanley Mts., Kokoda, Ilimo, Sirorata and Popondetta. MOR: Wau, Garaina, Ivau, Markham R., Fransisco R., Lae and Huon Pen. MAD: Kaironk R., Rai Coast, Astrolabe Bay, Adelbert Mts., Kar Kar Is., Hansa Bay and Bogia. E.SPK: Wewak, Dreikikir, Pagwi, Hunstein R., Torricelli Mts., Mt. Sumbau and Walis Is. W.SPK: Amanab, Aitape, Idam R. and Telefomin. SIM: Karimui and Tive Plateau. E.HIGH: Aiyura. W.HIGH: Jimi R. S.HIGH: Mt. Sisa and Mt. Bosavi.
Snakebites: ORO: Kokoda. MAD: Kar Kar Is., Rai Coast, Saidor and Bogia. E.SPK: Wewak.
Distribution outside PNG: A New Guinea/Aru Is. endemic distributed throughout Irian Jaya as far as Arfak, the Vogelkop and including some of the neighbouring islands to the north, west and southeast. The population from the Aru Kslands, south of Irian Jaya, is often referred to as a separate subspecies, *Micropechis ikaheka fasciatus*.
Confusing species: Within its range the small-eyed snake *Micropechis ikaheka* may be confused with the non-venomous ground snakes, *Stegonotus* spp., or the mildly venomous elapid *Furina tristis*. The ringed python, *Bothrochilus boa*, of the the Bismarck Archipelago may be mistaken for the small-eyed snake even though the latter has not been recorded from New Britain or New Ireland.

Venom: Small-eyed snake venom contains a post-synaptic NEUROTOXIN, a calcium channel blocker and possibly also a pre-synaptic NEUROTOXIN. It is ANTICOAGULANT, and also HAEMOLYTIC *in vitro*, and may be MYOTOXIC with the result that patients may produce dark urine, haemoglobinuria and/or myoglobinuria. Minor symptoms include nausea, severe headache and prolonged weakness, whilst serious envenomings may result in loss of consiousness, severe neuromuscular paralysis and respiratory arrest in a few hours. To date only two human fatalities, from Bulolo, Morobe Province and Kar Kar Island, have been positively confirmed for this species by ELISA testing of human blood serum with *Micropechis* venom although published reports suggest that this species may be responsible for several other fatalities in East Sepik, Madang, Morobe and Oro Provinces. It now appears that the threat posed by this species, especially to plantation workers, was previously underestimated. Venom yields from 1.2m Kar Kar Is. specimens exceeded 120mgs (dry wt.).
Antivenom/initial dose: CSL Polyvalent Australo-Papuan. Monovalent Tiger Snake in large doses was also reportedly used to treat bites from this species on Kar Kar Island. **Death adder antivenom is believed ineffective.**

Adult small-eyed snake – *Micropechis ikaheka* (Kar Kar Is., Madang Province)

Juv. small-eyed snake – *Micropechis ikaheka* (Kar Kar Is., Madang Province)

PAPUAN TAIPAN

Oxyuranus scutellatus canni

Maps: 27a-b

HIGHLY VENOMOUS

Average/Maximum length: 1.83-2.44m / 3.36m.

Description:

a) physique: A large slender species with an elongated head, which is distinct from the narrow neck; tail long and whip-like; eye moderate-sized with round pupil; supraocular scale over eye sharply shelved giving the taipan a scowling expression and partially obscuring the eye from above.

b) colouration: Body olive or dark brown (Western Province) or grey to black, or light blue (Central Province), usually with a broad orange stripe running down the centre of the back, at least from midbody to above the cloaca, although this may be absent or less obvious in young specimens or in brown adults. Underside cream or white, with or without orange speckling; head usually coloured as body although may be paler or darker, snout, labials and chin usually light grey; iris of eye brown.

c) scalation: DMB 21, or 23, anteriorly may be weakly keeled; V 220 - 250; SC 45 - 80, all paired; anal plate single; 6 supralabials with 3rd and 4th in contact with eye and temporolabial between 5th and 6th; 8 infralabials loreal and subocular scales absent.

Habitat: Lowland savanna and dry savanna woodland but common in gardens and relatively built-up areas, even major towns. Taipan do not occur in rainforest or wet monsoon forest.

Habits: Nervous and retiring but will strike with speed and aggression if startled, threatened or molested, arching the body back, flattening the head and striking forward and upwards. Moves rapidly over the ground, often with head and forepart of body raised, and known to pursue fleeing animals or humans over a short distance. Possesses very acute senses which usually prevent human encounters, but when aroused the taipan is known for its characteristic rapid and multiple "strike and release" bites which can lead to enormous quantities of venom being injected. Commonest during the beginning of the wet season but may be encountered at any time. Diurnal and crepuscular, but nocturnal in very hot weather. Oviparous with a clutch size of 13-18 eggs.

Diet: Primarily, if not entirely, mammals i.e. rats and bandicoots, and possibly ground nesting birds.

Distribution within PNG: Altitude records 0-360m although there is a dubious snakebite record from Fane, Owen Stanley Mts. at 1220m.

Specimens: WEST: Fly R. mouth, Oriomo Plateau, Daru Is., Morehead, Bamustu, Bensbach, Balimo, Lake Daviumbu and Lake Murray. GULF: Kukipi and Vailala R. NCD. CENT: Mekeo, Goldie, Brown and Vanapa R., Sogeri, Kupiano, Kivori, Amazon Bay and Abavi Is. Questionable record – M.BAY: mainland opposite Samarai Is. Whether the taipan occurs on the isolated savannas of Gulf Province between Kikori and Kukipi has yet to be determined.

Snakebites: Throughout urban NCD. CENT: entire lowlands from Mekeo, Bereina and Vaifa'a, Rigo, Kwikila and Kupiano to Amazon Bay. The altitudinal record from Fane (1220m) was based on a low ELISA reading and therefore may be suspect.

Distribution outside PNG: The Papuan taipan is an endemic New Guinea subspecies confined to southern Papua and southeastern Irian Jaya at least as far west as Senggo, Kepi and the Wildoman River, north of Kolepom, Prinz Frederik Hendrik Island. Its close relative, the Australian coastal taipan *Oxyuranus scutellatus scutellatus*, occurs in Queensland and Northern Territory, Australia.

Confusing species: Due to its geographical variation in body colour the Papuan taipan *Oxyuranus scutellatus canni* may be confused with a number of other species occuring within its range, most notably the Papuan blacksnake, *Pseudechis papuanus;* whip snakes, *Demansia* spp.; brown cat snake, *Boiga irregularis*; black treesnake, *Dendrelaphis punctulatus*; common keelback, *Tropidonophis mairii*; Papuan olive python, *Apodora papuana*, and possibly also the eastern brownsnake, *Pseudonaja textilis,* if the presence of that species in Central Province is eventually proven (*see* Eastern brownsnake account).

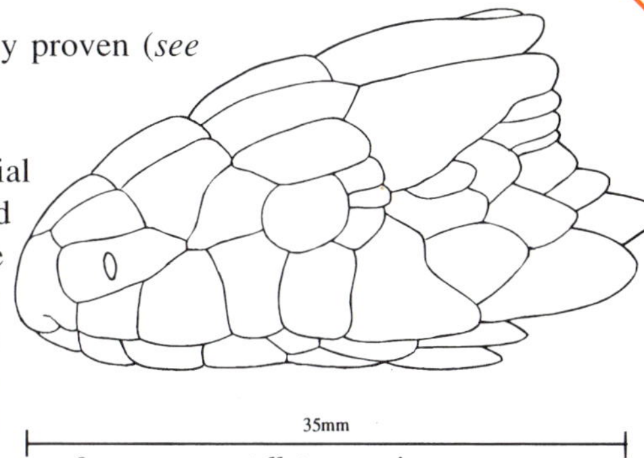

Oxyuranus scutellatus canni BMNH 1992.542
Note enlarged, 'shelved' supraocular scale above large eye.

Venom: The Papuan taipan is the most venomous terrestrial snake in New Guinea. The taipans as a group are considered by many experts to be the most venomous land snakes in the world and the inoffensive inland taipan *Oxyuranus microlepidotus* of central Australia is thought to possess the most toxic snake venom known. The venom of the Papuan taipan is strongly NEUROTOXIC, capable of causing rapid respiratory paralysis and death, coupled with a powerful PROCOAGULANT. Common early signs include sudden faints, prolonged bleeding both from the site of the bite and from'first aid' razor-cuts and ptosis of the eyelids. The venom is weakly HAEMOLYTIC and also possibly MYOTOXIC and CARDIOTOXIC (*see* Glossary). Taipan venom contains both pre-synaptic and post-synaptic neurotoxins and a multifunctional complex calcium channel blocker (taicatoxin). Pre-synaptic neurotoxins cannot easily be reversed with antivenom and one such neurotoxin isolated from taipan venom, known as taipoxin, is considered the second most toxic neurotoxin yet isolated from a snake venom. A 2.0m taipan can inject 100-200 mgs of venom in one bite and it may bite several times in rapid succession. The lethal human dose for a 70Kg adult male has been estimated at 7mg. In view of the high toxicity, considerable yield and pre-synaptic nature of taipan venom **urgent medical treatment is vital for survival.**

Antivenom/initial dose: CSL Monovalent Taipan/12 000 units.

Papuan Taipan – *Oxyuranus scutellatus canni* (juv.)
(Eboa, Mekeo, Central Province)

Papuan Taipan – *Oxyuranus scutellatus canni*
(Goldie River, Central Province)

Papuan Taipan – *Oxyuranus scutellatus canni*
(Kuru Two, Western Province)

Papuan Taipan – *Oxyuranus scutellatus canni*
(Mekeo, Central Province)

PAPUAN BLACKSNAKE

Pseudechis papuanus

Map: 28

HIGHLY VENOMOUS

Average/Maximum length: 2.14m / 2.44m.

Description:

a) physique: A strong, stout bodied snake with a broad, flattened and rounded head which is only slightly distinct from the neck; tail fairly long; eyes small with round pupil; supraocular not distinctly shelved (*see* Papuan taipan).

b) colouration: Body uniformly glossy or matt black above, occasionally brown; blue-grey or gunmetal-grey below; head as body but may be lighter on labials; neck yellow to off-white with black speckling.

c) scalation: DMB 19, or rarely 21, all smooth; V 221-230; SC 49-63 with first 25-45 single and remainder paired; anal scale divided; 6 supralabials with 3rd and 4th in contact with eye and temporolabial between 5th and 6th; loreal and subocular scales absent.

Habitat: Lowland savanna and savanna woodland but showing a greater preference for damper, swampy ground and also extending further into forests than the taipan.

Habits: Nervous and inclined to flee when approached, but when cornered the blacksnake is reported to attack with power and tenacity unrivalled by any other Australo-Papuan species and one which has earned this species the Mekeo language name of "Auguma" - to bite again. However, this may equally be true of the Papuan taipan, *Oxyuranus scutellatus.* Most commonly encountered in late dry season. Generally diurnal but crepuscular, avoiding hottest part of day, in extremely dry weather. Oviparous.

Diet: Primarily frogs, but also small mammals, lizards and possibly ground nesting birds.

Distribution within PNG: Specimens: WEST: Oriomo Plateau, Morehead Bamustu, Aramia R. and Boset, Fly R. NCD. CENT: Mekeo, Bereina and Vaifa'a, Yule Is., Kupiano, Marshall Lagoon and Amazon Bay. M.BAY: southern coast, no precise locality data.

Snakebites: CENT: Inawabui, Vanapa, Koira, Sogeri/Kerea, Bootless Bay and Rigo.

It is interesting that *P. papuanus* has been reported from Yule Island but not Daru Island whilst the reverse is true for *Oxyuranus scutellatus.* Formerly, blacksnakes were also probably much more common than they are today. Their decline may be due to a combination of habitat destruction and the introduction of the highly toxic-skinned South American cane toad, *Bufo marinus* as a biological crop pest control in the 1930's. Native amphibian-eating snakes such as death adders and blacksnakes would soon succumb if they attempted to prey on this toad.

Distribution outside PNG: The Papuan blacksnake is a New Guinea endemic possibly only occurring outside PNG along the southern coast of Irian Jaya as far west as Prince Frederik Hendrik Island or Kolepom (Parker 1982).

Confusing species: The Papuan blacksnake, *Pseudechis papuanus*, is a rare snake yet people continually report sightings. This is usually caused by the mis-identification of one of the following common species: the Papuan taipan, *Oxyuranus scutellatus canni*; the whip snakes, *Demansia* spp.; the common black treesnake, *Dendrelaphis punctulatus*, or the keelbacks, *Tropidonophis* spp.

Venom: The high venom toxicity and considerable venom yield combine to make the blacksnake the second most venomous land snake in PNG, after the taipan. As with the taipan, the venom of the blacksnake contains an extremely powerful NEUROTOXIN (*see* Glossary) causing death through respiratory paralysis. Blacksnake venom is also weakly COAGULANT and both HAEMORRHAGIC and anti-platelet in effect causing circulatory damage and leakage, and continued bleeding from the site of the bite and any lacerations present. The neurotoxic paralysis caused by blacksnake venom is of the pre-synaptic type which is difficult to reverse with antivenom.

Antivenom/initial dose: CSL Monovalent Papuan Blacksnake/18,000 units or Polyvalent Papuan/ 40,000 units.

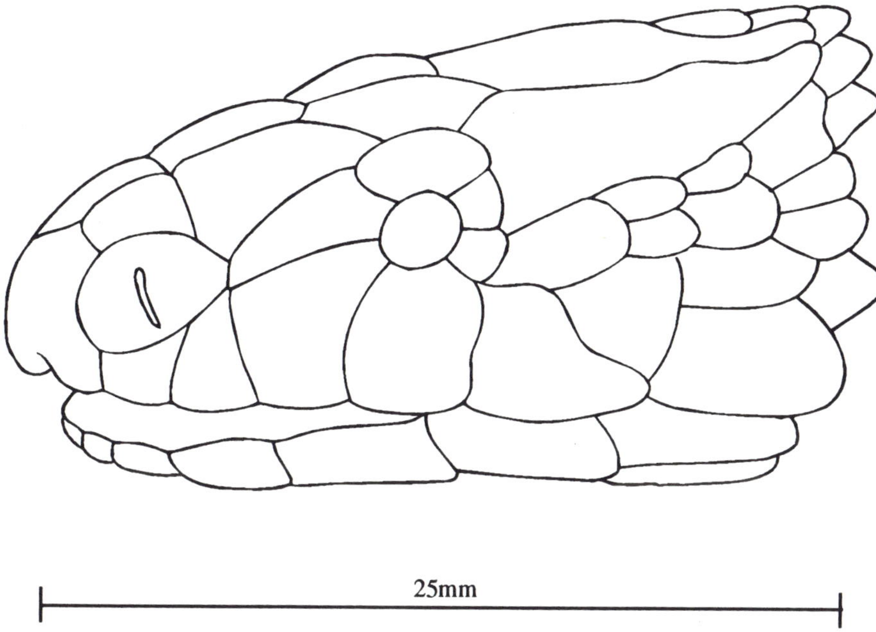

25mm

Pseudechis papuanus
BMNH 97.12.10.127 (Rigo, Central Province)
Note 'normal' supraocular over small eye.

Papuan blacksnake – *Pseudechis papuanus* (Bamustu, Aramia R., Western Province)

MULGA or KING BROWNSNAKE

Pseudechis australis

Map: 29

HIGHLY VENOMOUS

Average/Maximum length: 2.44m / 2.75m.

Description:

a) physique: A heavy bodied species with a broad flattened head which is distinct from the neck, especially in large specimens which may have bulbous cheeks; eye small with round pupil.

b) colouration: Tan brown dorsally, both head and body; underside yellowish cream.

c) scalation: DMB 17, all smooth; V 185-225; SC 50-75, all single except extreme posterior few which may be paired; anal plate usually divided; 6 supralabials with 3rd and 4th in contact with eye and temporolabial between 5th and 6th; loreal and subocular scales absent.

Habitat: Savanna and savanna woodlands in New Guinea but also tropical forests and deserts in Australia.

Habits: A relatively slow-moving diurnal species which may become crepuscular or even nocturnal to avoid excessively hot weather. King brownsnakes are capable of injecting huge quantities of venom although the venom is not considered as toxic as that of the Papuan blacksnake. Reputedly unpredictable, this species will flatten its head and strike with rapidity and aggression, often holding on and chewing after the initial bite. Oviparous.

Diet: Lizards, small mammals and frogs, but also occasionally ground-nesting birds, reptile eggs or other snakes.

Distribution within PNG: Not yet recorded for PNG but possibly present in Western Province west of the Fly River, probably in the Morehead/Bensbach region near the frontier with Irian Jaya. Parker (1982) reports a possible sighting from the Kwari R. between Weam and Obo, near the Irian Jaya border.

Distribution outside PNG: Throughout most of Australia, except extreme southwest, southeast and Tasmania, in a wide variety of habitats. Also from southeastern Irian Jaya in the vicinities of Etna Bay and Merauke near the PNG frontier.

Confusing species: The king brownsnake, *Pseudechis australis* may be confused with brown specimens of the Papuan blacksnake, *Pseudechis papuanus*, the Papuan taipan, *Oxyuranus scutellatus canni*, and the Papuan olive python, *Apodora papuana*.

Venom: Although the venom of the king brownsnake has been shown to be HAEMOLYTIC both *in vitro* and *in vivo* in laboratory tests, haemolytic effects are rarely reported in humans. King brownsnake venom is also CYTOTOXIC, weakly NEUROTOXIC and MYOTOXIC, affecting the heart muscle. A potent myotoxin known as mulgotoxin has been isolated from the venom which is also reported to be ANTICOAGULANT, or containing anticoagulant properties, and possibly also HAEMORRHAGIC, causing continual wound bleeding.

Antivenom/initial dose: CSL Monovalent Blacksnake/18,000 units.

King brownsnake – *Pseudechis australis* (Merauke, Irian Jaya)

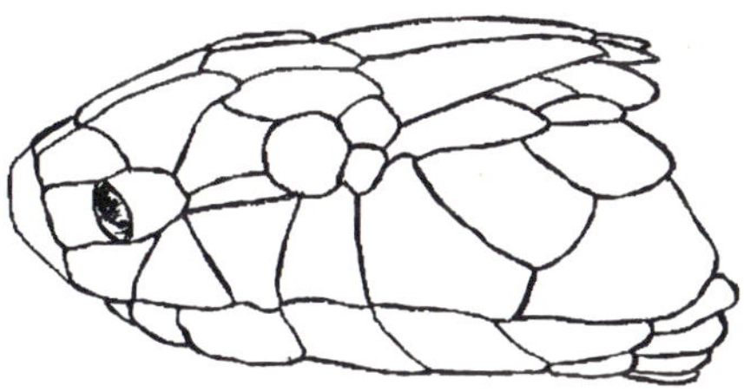

Pseudechis australis
(Australia)

King brownsnake – *Pseudechis australis* (NT Australia captive)

EASTERN or COMMON BROWNSNAKE

Pseudonaja cf. *textilis*

(*see* Taxonomic note) Map: 30

HIGHLY VENOMOUS
Average/Maximum length: 1.83m / 2.14m.
Description:

a) physique: A slender snake with the head only slightly distinct from the neck; tail moderate length; eye medium-sized with round pupil.
b) colouration: Brown or black above, often with darker cross-bands which may be particularly evident in juveniles; underside off-white to creamy-yellow, speckled with pink, brown or grey; head coloured as body with lighter throat.
c) scalation: DMB 17, all smooth; V 185-235 (200-205 Milne Bay specimens); SC 45-75, usually all paired but occasionally with a few single anterior scales; anal plate divided; 6 supralabials with 3rd and 4th in contact with eye; loreal, temporolabial and subocular scales absent. The absence of the temporolabial scale serves to differentiate this species from other large PNG elapids; taipan, Papuan blacksnake etc.

Habitat: Grassy woodlands, open country and sandy or rocky heathlands in coastal areas.
Habits: Very fast moving and inclined to flee from human approach but if threatened the brownsnake is prepared to defend itself vigorously by taking up a raised S-shaped stance with coiled neck and mouth open wide in readiness for the strike. Known to rush towards potential threats and press home an attack often striking several times in quick succession from the S-stance. Particularly common in very warm weather and usually diurnal. Oviparous with clutches of up to sixteen.
Diet: Primarily small lizards and mammals but also frogs.
Taxonomic note: McDowell (1967) reported the first two New Guinea specimens of *Pseudonaja textilis*, collected from localities in Milne Bay Province by the 1953 Archbold Expedition, as resembling Queensland and Northern Territory specimens in all characters and scale counts except the number of maxillary teeth. The Milne Bay specimens exhibited twelve solid teeth behind the fangs in the left maxilla compared to the usual Australian count of nine, ten or eleven. Oro Province brownsnakes are reportedly also much darker than Australian specimens.
Distribution within PNG: First recorded from PNG in the late 1950s, and then believed to be confined to a fairly localised area in Milne Bay and Oro Province. The brownsnake was thought to have been introduced accidentally from Australia in the form of eggs transported with military or agricultural equipment and supplies during the previous 10-15 years. This theory was strengthened by the apparent absence of the species from the southern Papuan savannas which lay between this apparently isolated New Guinea population and that of the widespread Australian eastern brownsnake complex. However, although no eastern brownsnakes have actually been collected south of the mountains, human serum samples obtained from several snakebite patients treated at Port Moresby General Hospital, and analysed at the Liverpool School of Tropical Medicine in the United Kingdom, have shown positive ELISA test results for brownsnake venom. These patients originated from locations scattered through Central Province. The snakebite data suggests that this species may occur south of the mountain chain but this cannot be confirmed until a specimen of *Pseudonaja* is obtained from Central Province.
Specimens: ORO: Embogo and Popondetta. M.BAY: Menapi, Cape Vogel, Dogura and Baiawa, Moi Biri Bay. Snakebites: NCD: 9 mile. CENT: Goldie R., Tapini 850m, also possibly Kapari and Imairu.
Distribution outside PNG: Eastern Queensland, New South Wales, Victoria and southern South Australia. Also isolated localities in Northern Territory.
Confusing species: The eastern brownsnake, *Pseudonaja* cf. *textilis,* in New Guinea, may be confused with the Papuan taipan, *Oxyuranus scutellatus canni*; the brown cat snake, *Boiga irregularis*; or the Papuan olive python, *Apodora papuana.*

Venom: The eastern brownsnake possesses the most toxic terrestrial snake NEUROTOXIN yet isolated. Known as textilon, it is considered of higher toxicity than the taipoxin of the taipan. The venom is also strongly COAGULANT, causing haemorrhage from mucous membranes of the mouth etc. and both MYOTOXIC and NEPHROTOXIC. Venom yields vary from 2-68mgs.
Antivenom/initial dose: CSL Monovalent Brownsnake / 1000 units; Monovalent Taipan.

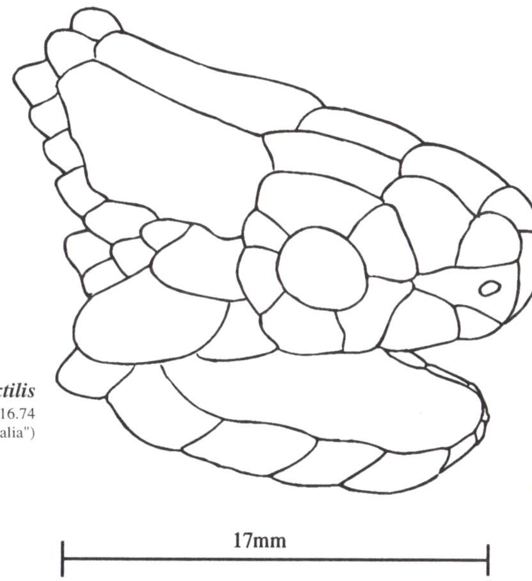

Pseudonaja textilis
BMNH 63.6.16.74
("Australia")

Eastern Brownsnake – *Pseudonaja textilis* (Armidale, NSW, Australia) Photo: Hank Jenkins

BANDED SEA KRAITS

Laticauda spp.

(2 species) (*see* Taxonomic note) Map: 31

HIGHLY VENOMOUS

Average/Maximum length: 0.75-1.0m / 1.5m.

Description:

a) physique: Body elongate and rounded with the head barely distinct from the neck; tail strongly laterally compressed posteriorly and paddle-like; nostrils positioned laterally; eyes small with round pupils.
b) colouration: Body with 20-70 regular gunmetal-blue and black bands, the blue-grey bands being lighter below; underside cream or white; head black with yellow snout and lips.
c) scalation: DMB 19-25, all smooth; V 210-250, well developed; SC 27-47; anal divided; 7-8 supralabials with 3rd and 4th entering eye; single preocular; 2 postoculars; large azygous shield between the prefrontal scales present or absent; internasal scales present between nasal scales.

Habitat: Coastal marine with a particular preference for coral reefs, rocky shorelines and mangrove entanglements.

Habits: Graceful swimmers but also extremely capable on land, the amphibious sea kraits are known to venture onto land, usually at night, to scale cliffs and travel overland for considerable distances. They may often be found asleep ashore during the day under boulders or in caves. Sea kraits are generally inoffensive snakes, disinclined to bite, but they are highly venomous and as the consequences of a bite are extremely serious, even apparently docile sea kraits should be considered potentially dangerous. Oviparous with a clutch size of three (*L. laticaudata*) to six eggs (*L. colubrina*) which are laid on land, usually in a cave or rocky depression.

Diet: Fish and eels including a species of snake eel (*Myrichthys colubrina*) which mimics the sea krait in its patterning.

Taxonomic note: Two species, *Laticauda colubrina* and *L. laticaudata*, have long been recorded from the coastlines of PNG but a third species, *L. schistorhynchus*, has also been listed on a number of checklists as occurring in "New Guinea, Melanesia and Polynesia". Cogger (1975) comments that this species is generally considered an endemic of the Tongan and Samoan Island groups and a report of a specimen from Tendanye or Bertrand Is. near Wewak, E.Sepik Prov., cannot be confirmed. Heatwole and Guinea (1993) report this species as a subspecies of the Far East Asian *L. semifasciata* and confine it to Niue Is., Tonga and Samoa. According to Cogger *L. schistorhynchus* can be distinguished from *L. colubrina* and *L. laticaudata* by the presence of a horizontally divided rostral scale (undivided in both New Guinea species) and a count of fewer than 200 ventral scales (210-250 in *L. colubrina*, 225-245 in *L. laticaudata*) and these characteristics should serve to identify the species should another specimen be found. A fifth species, *L. crockeri*, is endemic to Lake Te-Nggano on Rennel Island in the Solomon Islands to the east.

Seas: Indian and Pacific oceans; South China, Celebes, Java, Flores, Banda, Timor, Arafura, Coral and Bismarck (both species).

Distribution within PNG: Sea kraits are common along the coastlines and on the numerous small offshore islands along the Central and Madang coastlines.

Distribution outside PNG: Both species are found from India through Indo-Malaysia to the Torres Strait islands, the Solomon Islands and the Pacific. A third species occurs in the Solomon Islands and two others are known from Indonesia and the Fijian Islands.

Confusing species: The sea kraits of genus *Laticauda* are only likely to be confused with the entirely marine seasnakes, especially those belonging to banded genera such as *Hydrophis*.

Venom: The venom of these species is considered to be highly toxic but no snakebites are on record. A post-synaptic NEUROTOXIN has been isolated from the venom of *L. colubrina* and two NEUROTOXINS, Laticotoxins a and b, from the venom of *L. laticaudata*.
Antiserum/initial dose: One ampoule of CSL Seasnake antivenom is required to neutralise 5.8mg of *L. laticaudata* venom.

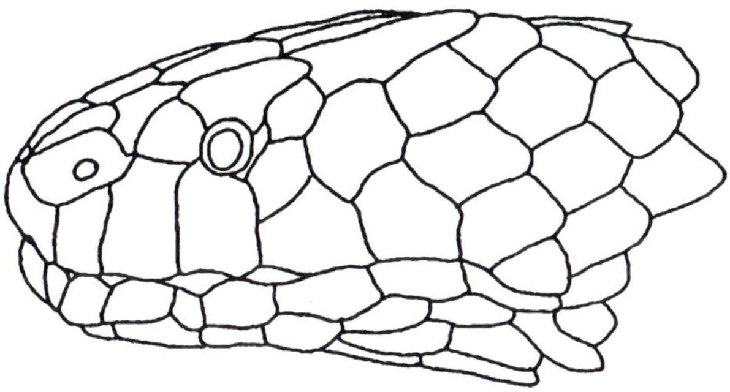

Laticauda colubrina
(Jais Aben, Madang Province)

Common sea krait - *Laticauda laticaudata*
(Makira Is. [San Cristobal Is.] Solomon Is.)
Photo: Mike McCoy

Yellow-lipped sea krait – *Laticauda colubrina*
(Lion Is., Central Province)

Key to the sea kraits, genus *Laticauda,* of New Guinea

1a. Dorsal scales in 21-25 rows at midbody; large azygous scale present between prefrontals (fig. 1) although it may be incompletely separated from either prefrontal (fig. 2) or absent (fig.3). *Laticauda colubrina*

1b. Dorsal scales in 19 rows at midbody; azygous scale absent (fig. 3). *Laticauda laticaudata*

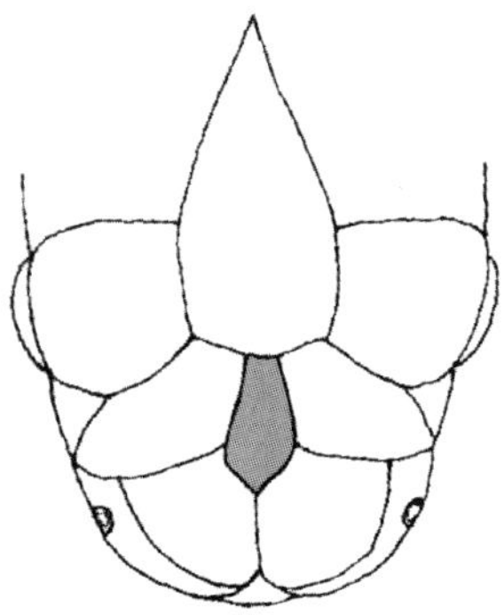

Fig 1: ***Laticauda colubrina***
Azygous scale (shaded) present between prefrontals, prefrontals not in contact

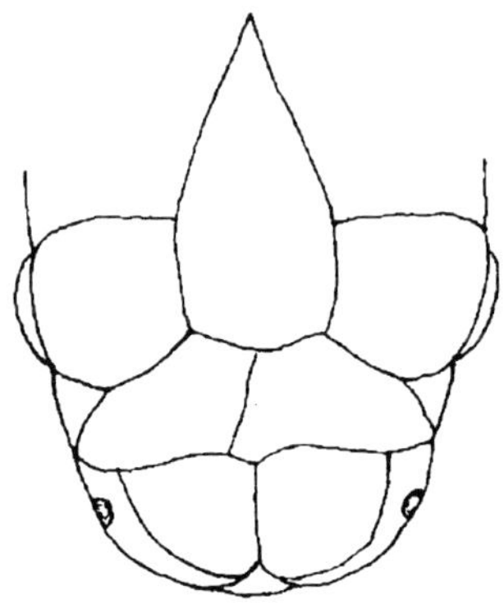

Fig 2: ***Laticauda colubrina***
Azygous scale fused with or incompletely separated from prefrontals, prefrontals in contact.

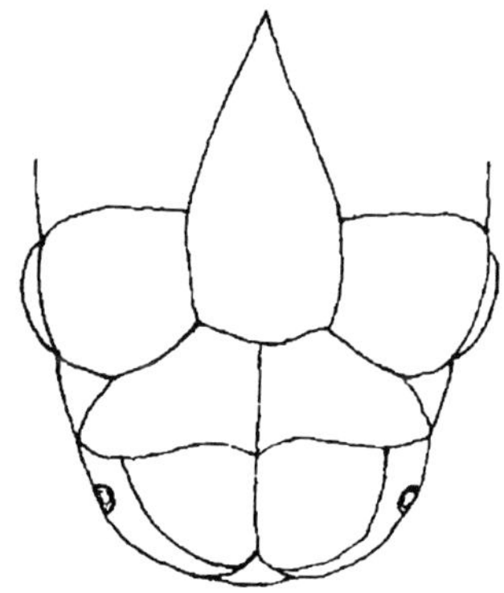

Fig 3: ***Laticauda colubrina*** or ***L. laticaudata***
Azygous scale absent, prefrontals in contact on midline.

YELLOW-LIPPED, WIDE-LIPPED or COLUBRINE SEA KRAIT

Laticauda colubrina

Average/Maximum length: 0.8 /1.5m.

Description: scalation: DMB 21-25; all smooth; V 210-250; SC 27-47 all paired; anal plate divided; 7-8 supralabials with 3rd and 4th in contact with eye; rostral undivided; single preocular; 2 postoculars; a large azygous scale present between prefrontal scales although it may be incompletely separated from prefrontals.

Distribution within PNG: Coastlines and inshore islands of Central, Milne Bay, Madang, New Ireland and North Solomons Provinces. NCD: Port Moresby, Manubada Is., Lion Is., Kila Kila. M.BAY: Orangerie Bay, Samarai Is. and Louisiade Arch. (without precise locality data). MOR: Huon Pen. Finschhafen. MAD: Saidor, Astrolabe Bay mainland and offshore islands, Kar Kar Is. W.SPK: Aitape. ENB: Gazelle Pen. and Mioko Is. WNB: no precise locality data. MAN: Manus and Pigeon Is. N.SOL: Bougainville.

Distribution outside PNG: India and Indo-Malaysia to Torres Strait Islands, the Solomon Islands and the western Pacific.

BLACK-BANDED, LARGE-SCALED or COMMON SEA KRAIT

Laticauda laticaudata

Average/Maximum length: 0.7 / 1.0m (outsized specimens over 3.0m are reported).

Description: scalation: DMB 19; all smooth; V 225-245; SC 30-47, all paired; anal plate divided; 7 supralabials with 3rd and 4th in contact with eye; rostral undivided; single preocular; 2 postoculars; azygous scale absent from between prefrontal scales.

Distribution within PNG: Coastlines and inshore islands of Central and Milne Bay Provinces. NCD: Port Moresby and Bavo Is. M.BAY: Samarai Is. MAD: no precise locality data. ENB: Duke of York Is. N.IRE: no precise locality data.

Distribution outside PNG: India and Indo-Malaysia to Torres Strait Islands, the Solomon islands and the western Pacific.

Mark O'Shea with sea krait – *Laticauda colubrina*
(CRI, Madang Province)

Yellow-lipped sea krait – *Laticauda colubrina* (Nagada harbour, Madang Province)

TRUE SEASNAKES
Hydrophiidae

Maps: 32a-c

True seasnakes differ from the more familiar sea kraits (genus *Laticauda*) in that they are almost entirely marine and virtually helpless on land (although at least two monotypic species, *Parahydrophis mertoni* and the Australian *Ephalophis greyi*, are known to be semi-aquatic, moving into the intertidal zone of mangrove creeks, when juvenile) while the sea kraits are amphibious, travelling over land with ease and even scaling cliffs. True seasnakes are also viviparous, giving birth at sea, while sea kraits are oviparous, coming ashore to lay their eggs. The fully aquatic seasnakes have their nostrils positioned dorsally whereas sea kraits possess the laterally positioned nostrils typical of terrestrial snakes. McDowell (1967 & 1972) also described numerous anatomical differences existing between the two taxa and these are summarised in Burger and Natsumo (1974).
Burger and Natsumo further divided the true seasnakes (family Hydrophiidae) into two subfamilies; the 'thick seasnakes', subfamily Ephalophiinae and the 'flat seasnakes', subfamily Hydrophiinae.

Table 1. Summarising Burger & Natsumo (1974) Hydrophiid subfamilies

	Ephalophiinae Thick seasnakes	**Hydrophiinae** Flat seasnakes
Body shape	Most primitive Rounded and colubrinae or slightly to moderately compressed	More advanced Strongly compressed and flattened
Ventral scales	Entire and almost as broad as those of colubrids or elapids	Small and often divided, smaller than, or only slightly larger than, neighbouring dorsals
Tail shape	Slightly to extremely compressed	Extremely compressed
New Guinea Genera	*Aipysurus* *Emydocephalus* *Hydrelaps* *Parahydrophis*	*Acalyptophis* *Hydrophis* (incl. *Astrotia, Disteira & Enhydrina*) *Lapemis* *Pelamis*

Twenty two species of true seasnakes, as distinct from the sea kraits, representing eleven genera (following Cogger [1975a, 1975b & 1992] in recognising *Astrotia, Disteira* and *Enhydrina*) have been recorded around the coastline of New Guinea (Table 2). Two of these species, *Emydocephalus annulatus* and *Parahydrophis mertoni*, have yet to be collected in PNG waters.

The distribution of seasnakes in PNG waters is poorly documented with only single records existing for many species. The area of greatest diversity appears to be along the Western Province coastline of the Arafura Sea from the mouth of the Pahoturi River to the Fly River and in the Gulf of Papua off Gulf and Central Provinces (17 species). Compared to this, only two species have been recorded further east along the Coral Sea coastline of Milne Bay Province. Seasnake species appear to be fairly poorly represented off the northern and island provinces. Three species, including a newly described *Enhydrina*, are recorded from the Sepiks, Morobe, Madang and the Manus archipelago while three others are reported, rather nebulously, from the seas around New Britain. This apparent imbalance in diversity may in part be a result of limited collecting in the eastern and northern regions while the Gulf of Papua and the Fly River mouth has long been home to an active prawn trawling industry with seasnakes regularly being taken in nets. In addition, the pioneer herpetologist and government officer Fred Parker spent much of his time investigating the snakes of the southern Trans-Fly of Western Provinces and his collections included many seasnakes captured by off-shore fishermen.

The prawn fishermen are also probably the people most at risk from seasnake bite. Since seasnakes are air-breathing reptiles, capable of holding their breath for a maximum of two hours, many of those trapped in the trawl nets have drowned by the time they are brought to the surface. However, bites do occur when semi-conscious specimens are roughly handled as the nets are retrieved and accidents of this kind, when the boats may be many hours, or even days, from port, may easily end tragically. Seasnake venoms are capable of causing death in three hours or less, so consequently many deaths at sea are probably not reported to the authorities with the result that the true scale of seasnake morbidity and mortality in PNG remains unknown.

Table 2: True seasnakes in PNG waters

Scientific name	Common name
Acalyptophis peronii	Spiny-headed seasnake
Aipysurus duboisii	Reef shallows seasnake
Aipysurus eydouxii	Eydoux's, stagger-banded or spine-tailed seasnake
Aipysurus laevis	Olive-brown seasnake
Astrotia stokesii	Stokes's or large-headed seasnake
Disteira kingi	Spectacled or King's seasnake
Disteira major	Olive-headed or greater seasnake
Enhydrina schistosa	Common or beaked seasnake
Enhydrina zweifeli	'Sepik or Zweifel's seasnake'
Hydrelaps darwiniensis	Black-ringed mangrove or Port Darwin smooth seasnake
Hydrophis atriceps	Black-headed seasnake
Hydrophis elegans	Elegant or bar-bellied seasnake
Hydrophis gracilis	Graceful small-headed or slender seasnake
Hydrophis macdowelli	Small-headed or McDowell's seasnake
Hydrophis melanosoma	Black-banded robust seasnake
Hydrophis ornatus ornatus	Ornate reef seasnake
Hydrophis pacificus	Pacific seasnake
Hydrophis vorisi	Estuarine seasnake
Lapemis hardwickii	Hardwicke's or spine-bellied seasnake
Pelamis platurus	Yellow-bellied or pelagic seasnake

Southern Irian Jayan and Aru Is. species not yet recorded from PNG waters

Emydocephalus annulatus	Turtle-headed or ringed seasnake
Hydrophis ornatus ocellatus	Reef seasnake
Parahydrophis mertoni	Arafura smooth or northern mangrove seasnake

Several species are primarily off-shore species, most notably the black and yellow pelagic seasnake *Pelamis platurus* which may be seen in huge aggregates drifting on the surface. Such species usually only appear inshore after storms when even an oceanic wanderer such as *Pelamis* may be forced close to land and cast up onto the beaches. Rough weather may also cause large numbers of inshore seasnakes to congregate in river mouths, estuaries and mangrove swamps.

Some seasnakes inhabit coral reefs but do not generally harass divers although it should be noted that large species such as *Astrotia stokesii* and *Aipyurus laevis* possess fangs which are long enough to penetrate a wet suit and while *A. laevis* is generally placid by nature, *Astrotia* can become aggressive if disturbed.

A number of species may also be encountered in river mouths and mangrove swamps. Species of the genera *Enhydrina* and *Hydrophis* may enter estuarine brackish or riverine freshwater and travel inland where they may pose a threat to river fishermen or bathers. Seasnakes of the genus *Enhydrina* have been implicated in snakebites in Papua New Guinea and are believed responsible for serious bites in the Ramu River system. Some seasnakes have become so adapted to freshwater than they have become confined to inland lakes eg. the seasnake *Hydrophis semperi* in Lake Taal, Luzon Island, Philippines and the sea krait *Laticauda crockeri* in Lake Te-Nggano, Rennel Island, Solomon Islands. As yet no seasnake has been found to dwell entirely in freshwater in PNG, although Fred Parker reported three rapidly fatal bites to young girls in the village of Wipim, Western Province in the mid-1970's. The pool at Wipim is connected by tributaries to the tidal Oriomo River where I collected a seasnake believed to be a specimen of *Enhydrina schistosa.* A land-locked seasnake may possibly have been the cause of the deaths which occurred over a period of 12 months. Neither Parker nor I were able find any aquatic snake in the immediate vicinity which could have caused these deaths (*see* Parker's snake - An enigma).

Although seasnake fatalities in PNG are rarely reported, with the only documented cases being the Ramu River bites caused by a specimen of *Enhydrina* (possibly the newly-described Sepik species *E. zweifeli*), several species have been implicated in fatalities elsewherewithin their geographical ranges, amongst them *Astrotia stokesii, Hydrophis elegans, H. ornatus* and *Lapemis hardwickii.*

A few seasnakes possess such tiny fangs that they are probably incapable of inflicting a bite to a human, and those species which have evolved to feed specifically on fish eggs, *Aipysurus eydouxi* and *Emydocephalus annulatus,* may actually be evolving away from being venomous. Due to the difficulty persons may experience in identifying seasnakes in the field and the possibility that unrecorded Australian or Indonesian, or even undescribed, species may also occur along the PNG coastlines, all seasnakes should be treated as potentially dangerous and handling, even of apparently dead specimens, should be strictly avoided.

Inshore seasnakes may be confused with other marine species eg. the sea kraits, *Laticauda* spp., the little file snake, *Acrochordus granulatus*, or the mangrove and mud snakes of the Homalopsinae. Most true hydrophiid seasnakes are entirely marine and unable to move on land but it should be noted that the two amphibious mangrove mudflat dwelling seasnakes *Parahydrophis mertoni* (southern Irian Jaya) and *Hydrelaps darwiniensis* (southern Western Province), are not entirely helpless on land and whilst rare, small and generally placid, they may be confused with the five species of mildly venomous mud snakes and mangrove snakes belonging to the Homalopsinae (Colubridae) which also occur in southern Western Province. The lack of a loreal scale should distinguish both of these seasnakes from all of the homalopsine mud snakes and mangrove snakes, with the exception of *Fordonia leucobalia.* In common with all mud snakes these two seasnakes also possess a divided anal plate but although the first two subcaudal scales are paired, the remainder are single, in contrast to the homalopsines which have paired subcaudals throughout. In addition the two seasnakes are crossbanded whilst the homalopsines, with the exception of *Cantoria annulata*, are unicolour or speckled. Seasnakes also possess paddle-shaped tails.

Key to the Seasnakes, family Hydrophiidae, in New Guinea waters

1a. Ventral scales as broad, or nearly as broad, as those of terrestrial snakes (Colubridae or Elapidae) and distinct from neighbouring dorsal scales (fig. 1); body either rounded or only slightly to moderately compressed. Subfamily Ephalophiinae. 2
1b. Ventral scales reduced in size, as wide as, or slightly wider than, neighbouring dorsal scales (fig. 2); body strongly compressed and flattened. Subfamily Hydrophiinae. 5

2a. 3 supralabials, 2nd largest, elongate and in contact with eye (fig. 3); four infralabials, 3rd largest. *Emydocephalus**
2b. 6 or more supralabials; more than 6 infralabials. 3

3a. Dorsal scales in 36-39 rows at midbody. *Parahydrophis**
3b. Dorsal scales in fewer than 30 rows at midbody. 4

4a. Preocular scale present, separate from prefrontal; anterior temporal small, not in contact with posterior supralabials (fig. 4); tail strongly compressed and 'paddle-like'. *Aipysurus*
4b. Preocular scale absent, fused with prefrontal; anterior temporal large, passing between 5th and 6th supralabials, resembling elapid temporolabial (fig. 5); tail slightly compressed. *Hydrelaps*

5a. Posterior head scales fragmented and irregular, those around the eye with raised spinous or tuberculate 'horn-like' projections (fig. 6). *Acalyptophis*
5b. All scales of head regular well-defined scutes without raised projections (fig. 7). 6

6a. Mental groove absent (fig. 8); head elongate; body and head generally black above and yellow below. *Pelamis*
6b. Mental groove present (fig. 9); head rounded or elongate; body not contrastingly patterned black and yellow. 7

7a. Mental scale narrow, dagger-shaped (fig. 10), much longer than broad, extending back level with the posterior edge of the 1st pair of infralabials though may be obscured by deep mental groove. *Enhydrina*
7b. Mental scale normal, short and triangular (fig. 9), not extending back between 1st pair of infralabials. 8

8a. Posterior ventral scales indistinct from neighbouring scales; dorsal scales square or hexagonal and non-overlapping (juxtaposed) (fig. 11), spinous or tuberculate in males. *Lapemis*
8b. Posterior ventral scales distinct from neighbouring dorsal scales though may be single or divided; dorsal scales normal, keeled and overlapping (imbricate) (fig. 12), at least dorsally. 9

9a. Ventral scales divided into a pair of small foliform (leaf-shaped) scales forming a raised mid-ventral keel (fig. 13). *Astrotia*
9b. Ventral scales mostly undivided (fig. 2), or if all divided never foliform or forming a raised mid-ventral keel. 10

10a. Fewer than four maxillary teeth behind fangs or if more than four maxillary teeth, then anterior chin shields reduced and separated from contact with mental groove by elongate 1st pair of infralabials. *Disteira*
10b. More than four maxillary teeth behind fangs; anterior chin shields not reduced, large and in contact with mental groove. *Hydrophis*

* *Emydocephalus annulatus* and *Parahydrophis mertoni* have been recorded from southern Irian Jaya but since representatives of

these genera have not been collected from Papua New Guinean waters they are not included in this guide and are included only in the key to enable them to be indentified should specimens be eventually collected from Western Province. The reader is referred to photographs and descriptions of these species in Cogger (1992) and Ehmann (1992).

The genera *Acalyptophis, Astrotia, Hydrelaps, Lapemis* and *Pelamis* are monotypic (contain only a single species). Keys to the genera *Aipysurus, Disteria, Enhydrina* and *Hydrophis* are included with their species accounts.

Keys to the hydrophiid genera are to be found on the following pages:

Genus	**Page**
Aipysurus	180
Disteria	185
Enhydrina	187
Hydrophis	189

Fig 1: Subfamily Ephalophiinae. Ventral scales (shaded) as broad, or nearly as broad, as ventrals of those of terrestrial colubrids and elapids.

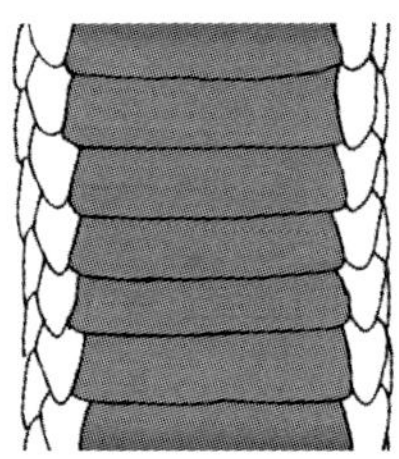

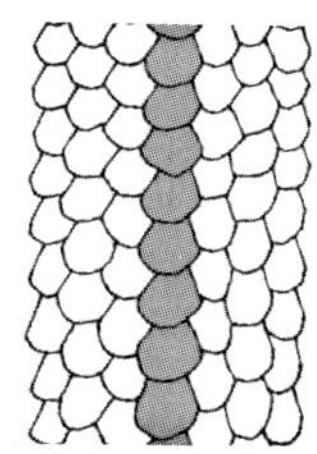

Fig 2: Subfamily Hydrophiinae. Ventral scales (shaded) reduced in width, only as broad as, or slightly broader than, neighboring scales.

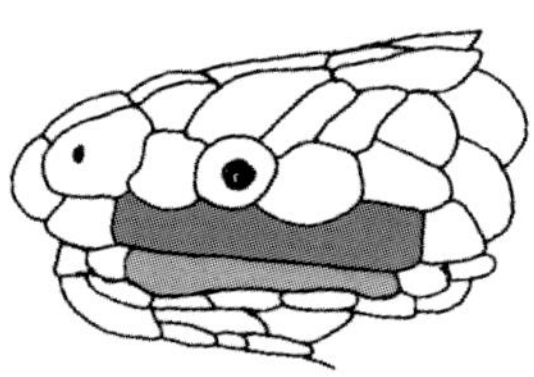

Fig 3: *Emydocephalus annulatus* (no locality data). Three supralabials with second (dark shaded) greatly enlarged; four infralabials with third (light shaded) greatly enlarged.

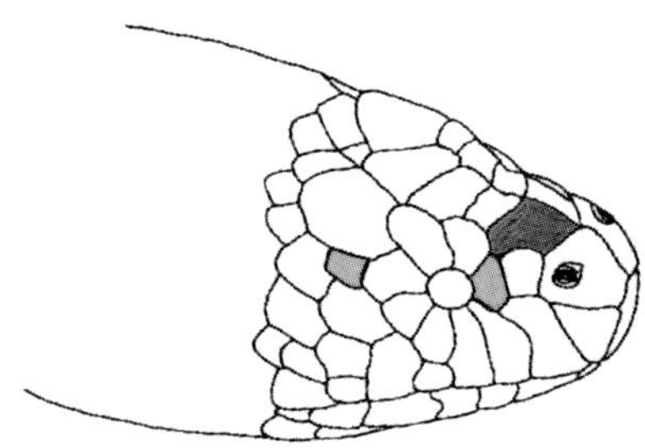

Fig 4: *Aipysurus laevis* (Ashmore Reef, Western Australia) (after Cogger 1975)
Preocular scale present (anterior light shaded), not fused with prefrontal (dark shaded); anterior temporal (posterior light shaded) small, excluded from contact with supralabials.

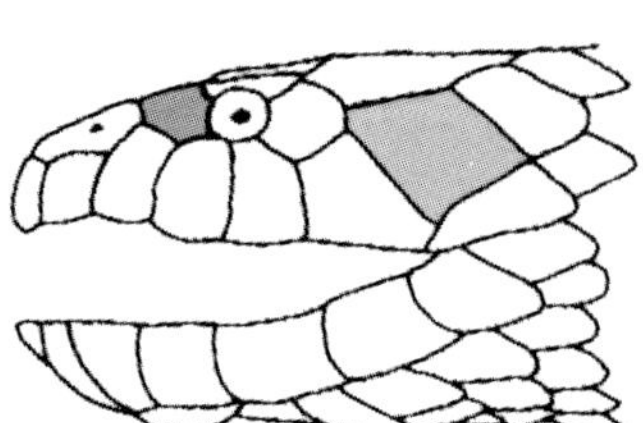

Fig 5: *Hydrelaps darwiniensis*
RMNH 10242 Maro River, Merauke, Irian Jaya.
(after Brongersma 1956).
Preocular scale absent, fused with prefrontal (dark shaded) which contacts eye; anterior temporal (light shaded) large, resembling elapid temperolabial, in contact with 5th and 6th supralabials.

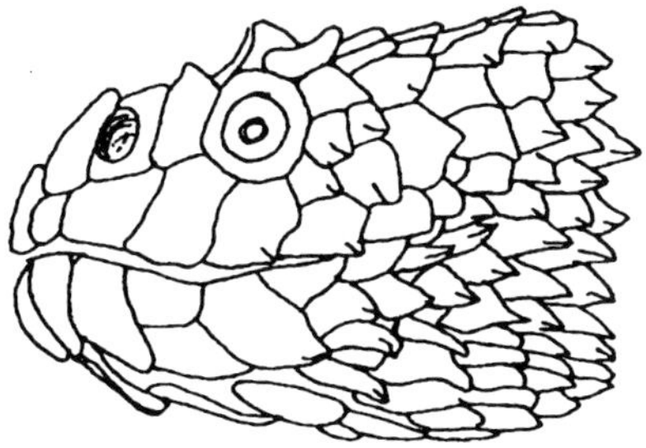

Fig 6: *Acalyptophis peronii*
RMNH 10243 Joes Island, Irian Jaya
(after Brongersma 1956)
Posterior head scales fragmented; anterior head scales with raised spinous or tubercular 'horn-like' projections

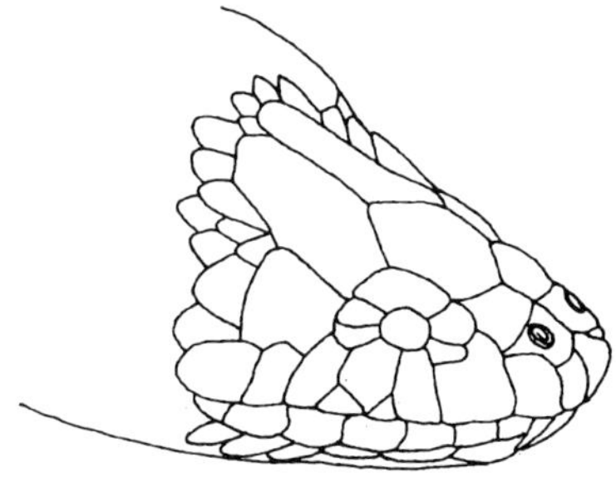

Fig 6: *Astrotia stokesii*
Darwin, Northern Territory, Australia
(after Cogger 1975)
Scales of head unfragmented and well defined, without raised projections.

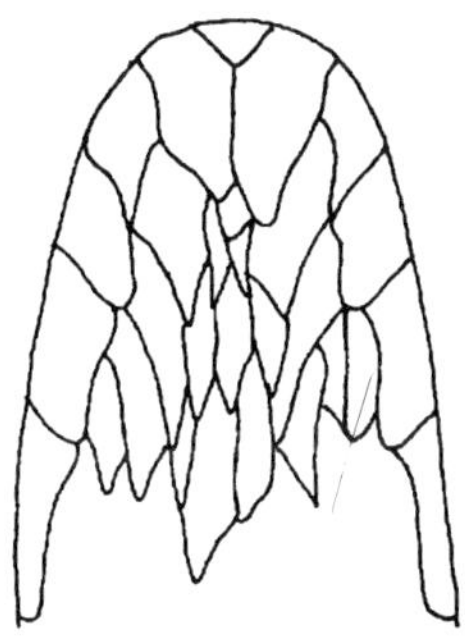

Fig 8: *Pelamis platurus*.
Mental groove absent, anterior chin shields in contact, posterior chin shields fragmented.

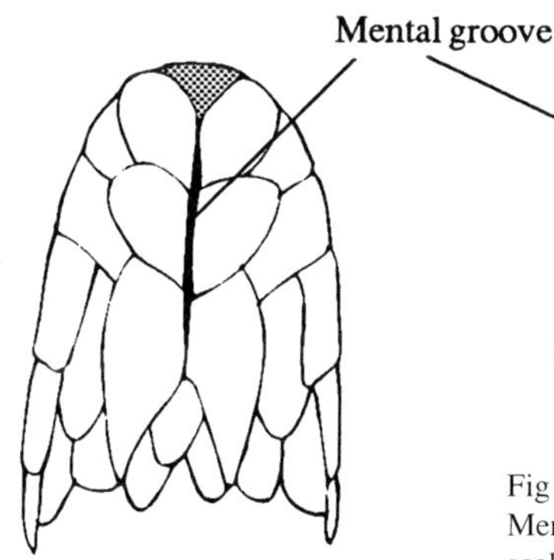

Fig 9:
Mental groove present; mental scale (shaded) normal, small and triangular, not extending between 1st infralabials.

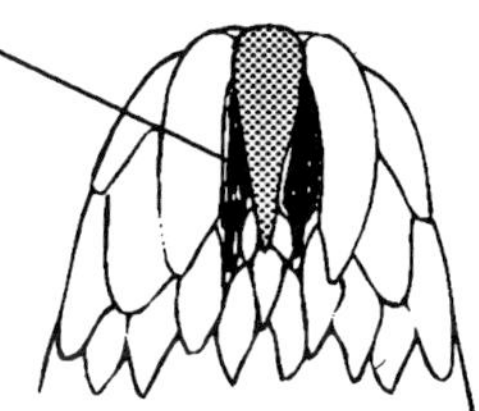

Fig 10: *Enhydrina schistosa*.
Mental groove present; mental scale (shaded) elongate and 'dagger-shaped', extending back between lst infralabials.

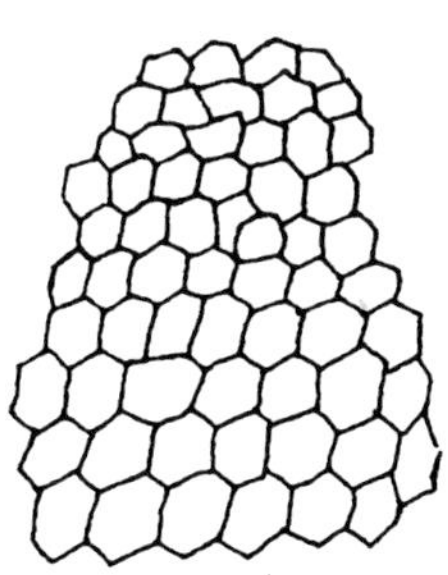

Fig 11: *Lapemis hardwickii*.
Dorsal scales squarish or hexagonal, larger laterally than dorsally, juxtaposed (non-overlapping).

Fig 12: Dorsal scales normal and imbricate (overlapping).

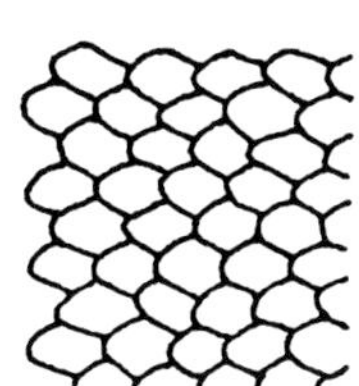

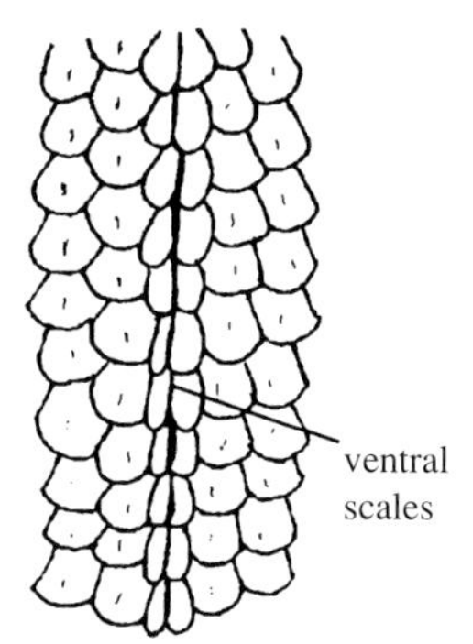

Fig 13: *Astrotia stokesii*.
Ventral scales small and foliform (leaf-shaped) forming a mid-ventral keel.

MIGHTY-TAILED or PIPE SEASNAKES
Aipysurus spp.
Map: 32a

Taken from the Greek *Aipys* meaning 'sheer, mighty, hanging straight down' and *-urus* for 'tail', the seasnakes of the genus *Aipysurus* are found throughout the Indo-Australian archipelago although they are centred on northern and western Australia. Three species occur in Papua New Guinea waters.

Key to the 'Mighty-tailed or Pipe seasnakes', genus *Aipysurus*, in Papua New Guinea waters

1a.	Fewer than 20 dorsal scale rows at midbody.	2
1b.	21 - 25 dorsal scale rows at midbody.	*Aipysurus laevis*
2a.	17 dorsal scale rows at midbody; ventrals 150 or fewer; 6 supralabials; head scales large, regular and unfragmented.	*Aipysurus eydouxii*
2b.	19 dorsal scale rows at midbody; ventrals 150 or more; 8 - 9 supralabials; head scales irregular and fragmented.	*Aipysurus duboisii*

REEF SHALLOWS SEASNAKE
Aipysurus duboisii

Average/Maximum length: 0.7m / 1.2m.

Description:

a) physique: Head broad, eye small; fore, mid and hind body stout; tail laterally compressed.

b) colouration: Cream to white dorsally with 32 - 38 purple-brown bands which may meet mid-dorsally and completely obscure lighter pigment from above; undersides cream; head dark brown, throat white with darker tips to scales.

c) scalation: DMB 19, smooth or weakly keeled, overlapping on anterior body but not overlapping on posterior body; V 150 - 165, with a small notch, as wide as adjacent dorsal scales; SC 23 - 35, all single; anal plate divided; head scales irregular and fragmented, except nasal and rostral scales; 8 - 9 supralabials with 4th, 5th and 6th in contact with the eye; 8 - 9 infralabials.

Habits: Inhabits shallow offshore waters over reefs 10cm - 3.5m deep, although may dive to 50m, often being forced to crawl on exposed coral debris at low tide. Active in the late afternoon and early evening. Forages for food around coral reefs. Viviparous.

Temperament: Will bite if provoked and an unprovoked attack on a diver is on record; highly venomous.

Prey: Reef fish, including eels, blennies, surgeon fish and parrot fish.

Seas: Timor, Arafura and Coral.

Distribution within PNG: CENT: Yule Is.

Distribution outside PNG: Australia (n.WA, NT, Qld, Gulf of Carpentaria) and New Caledonia.

EYDOUX'S, STAGGER-BANDED or SPINY-TAILED SEASNAKE
Aipysurus eydouxii

Average/Maximum length: 0.6m / 1.1m.

Description:

a) physique: Head moderately broad, eye small; neck slender; fore, mid and hind body stout; tail laterally compressed.

b) colouration: Cream to white dorsally with 28 - 35 alternating black-brown bands which may meet mid-dorsally; head yellow brown.

c) scalation: DMB 17, smooth; V 124 - 150, weakly keeled with a small notch, as wide as adjacent dorsal scales; SC 23 - 35, all single; anal plate divided; head scales large, regular and unfragmented; 6 supralabials with 4th in contact with the eye; 6 infralabials.

Habits: Poorly documented. Usually inhabits turbid offshore waters over mud or sand, between 2 - 50m deep, rather than clear water over reefs, and may congregate in large numbers inshore, in estuaries or river mouths during the summer. Forages for food in crevices and burrows. Viviparous with litter sizes between 1 - 12, usually fewer than six.

Temperament: This species does not appear to bite even when provoked, the venom glands are degenerate, the fangs small and the venom weak.

Prey: Fish eggs.

Seas: South China, Flores, Timor, Arafura and Coral.

Distribution within PNG: WEST: (Madiri, Fly R. delta, Mawatta, mouth of Binaturi R. and 60-80km ENE Daru).

Distribution outside PNG: Irian Jaya (80km N Kaap Valsch, off Prince Frederik Hendrik Is.); Australia (n.WA, NT, Qld incl. Thursday Is., Torres Strait Is. and Gulf of Carpentaria) and Indo-Malaysia.

OLIVE-BROWN SEASNAKE

Olive-brown seasnake – *Aipysurus laevis* (no locality data)
Photo: Hal Heatwole, North Carolina State University

Aipysurus laevis

Average/Maximum length: 1.2 - 1.7m / 2.2m.

Description:

a) physique: Head large and powerful, eye moderate; neck and body stout; tail laterally compressed.
b) colouration: Body and head uniform olive brown dorsally, may be paler lower on the flanks; tail brown to white with a brown dorsal ridge.
c) scalation: DMB 21 - 25, smooth and overlapping; V 135 - 155, weakly keeled with a small notch, as wide as adjacent dorsal scales; SC 25 - 35, all single; anal plate divided; head scales large, regular and unfragmented, except for some fragmentation of the parietal scales; 8 - 10 supralabials with 4th, 5th and 6th in contact with the eye.

Habits: A common and fairly sedentary reef species often seen repeatedly in the same location, resting on coral or on the sea floor. Inhabits coral reefs, lagoons or rocky shores at depths of 5 - 45m and occasionally enters estuaries and river mouths. Forages for food in coral crevices. Viviparous with clutch sizes between 1 - 5.

Temperament: Generally inoffensive and 'curious' but more aggressive in the breeding season (May to June) or if molested. Individual olive-brown seasnakes are commonly encountered in the same locality by divers, without incident.

Prey: Fish, crustaceans and fish eggs.

Seas: Timor, Arafura and Coral.

Distribution within PNG: WEST: Fly R. mouth 60-80 km ENE Daru, Daru, Bobo and Parama Is. Dagagota Reef, Gulf of Papua and 60-80km ENE Daru. M.BAY: Samarai Is.

Distribution outside PNG: Indonesia (Aru and Kei Is.); Australia (n.WA, NT, Qld, Gulf of Carpentaria) and New Caledonia. A former subspecies *A. l. pooleorum* from Shark Bay, WA, has been elevated to species status by some authors (Ehmann 1992).

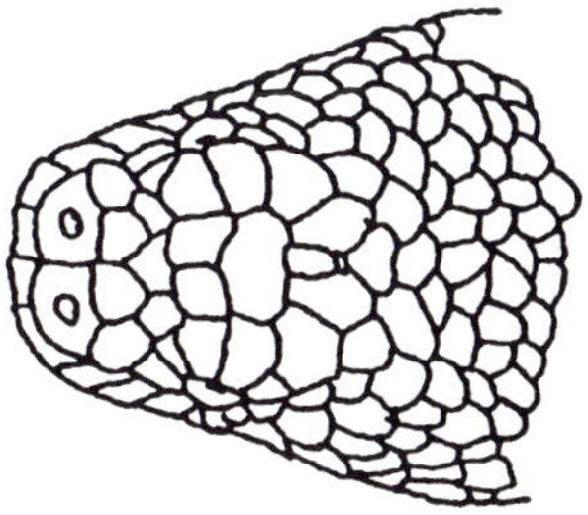

Aipysurus laevis
(after Smith, 1926)

BLACK-RINGED MANGROVE or PORT DARWIN SMOOTH SEASNAKE

Hydrelaps darwiniensis

Map: 32a

Average/Maximum length: 0.4m / 0.56m.

Description:

a) physique: Head elongate and narrow, indistinct from neck, eye small; neck slender; body slender, slightly stouter; tail laterally compressed.

b) colouration: Body and tail pale grey or yellow with 35 - 55 darker brown, black or grey rings which are broadest in the centre of the back; head coloured as darker rings with lighter patches.

c) scalation: DMB 25 - 29 smooth and overlapping; V 163 - 172, 2 - 3x wider than adjacent dorsal scales; anal plate divided; SC 27 - 39, mostly single; head scales large, regular and unfragmented; 6 supralabials with 3rd and 4th in contact with the eye; 7 infralabials preocular scale absent allowing the prefrontal to contact the eye; single postocular.

Habits: Inhabiting the inter-tidal zone, *Hydrelaps* is able to move onto land, due to the presence of overlapping moderately broad ventral scales, where it forages amongst mangrove roots and mudflat debris for prey. Typical habitats include tidal creeks and mangrove swamps. Viviparous with a clutch size of up to five.

Temperament: Generally inoffensive but may bite if molested. Nothing is known of its venom.

Prey: Small fish such as mudskippers, gobies such as the bearded goby *Scartelaos viridis*, and crabs.

Seas: Timor and Arafura.

Distribution within PNG: WEST: Daru Is. GULF: Kiamari.

Distribution outside PNG: Indonesia (Aru Is.); Irian Jaya (Merauke); Australia (WA, NT, w.Qld, Gulf of Carpentaria).

SPINY-HEADED SEASNAKE

Acalyptophis peronii

Map: 32a

Average/Maximum length: 1.0m / 1.2m.

Description:

a) physique: Head small, eye small; anterior body slender, mid and posterior body stout; tail laterally compressed.

b) colouration: Pale grey to cream dorsally with 25 - 30 darker bands around the body but broadest mid-dorsally; head pale.

c) scalation: DMB 21 - 31, all keeled, overlapping on anterior body but not overlapping on posterior body; V 140 - 210, unkeeled, as wide as adjacent dorsal scales; SC 44; anal plate entire; head scales irregular and fragmented, those around eye spinous and raised posteriorly to form 'horns'; 6 - 7 supralabials with 3rd and 4th in contact with the eye; 9 infralabials; a single preocular; 4 postoculars; prefrontals absent.

Habits: Poorly documented. Active in late aftenoon and at night in 11-19m deep offshore waters over reefs. Forages for food in coral sand and around coral reefs. Viviparous with clutch sizes of ten recorded.

Temperament: Generally inoffensive, biting only when severely provoked but highly venomous.

Prey: Fish, especially gobies.

Seas: Timor, Arafura and Coral.

Distribution within PNG: WEST: Fly R. mouth 60-80km ENE Daru.

Distribution outside PNG: Irian Jaya (Joes Is. near Salawatti Is.); Australia (n.WA, NT, Qld incl. Thursday Is., Torres Strait Is. and Gulf of Carpentaria) and Indo-China (possibly in error).

STOKES' SEASNAKE
Astrotia stokesii
Maps: 32a

Average/Maximum length: 0.8 - 1.2m / 1.6m.
Description:
a) physique: Head broad and distinct from the neck, eye small; neck and body especially stout; tail laterally compressed.
b) colouration: Dorsally yellow, grey or brown with 28 - 34 broad saddles which alternate with smaller dark bars or paired spots; head dark; tail paler than body.
c) scalation: DMB 45 - 63, keeled with a series of tubercles and strongly overlapping; V 226 - 286, small and divided to form a distinctive mid-ventral keel; anal scales divided and enlarged; head scales large, regular and unfragmented; 8 - 10 supralabials with 4th to 6th in contact with the eye.

Habits: Active by day or night. An inhabitant of turbid offshore waters or shelving or sloping coral reefs, between 3 - 22m deep, and coral sand sea beds. Forages for food amongst coral growths or in sea floor burrows. *A. stokesii* may be extremely abundant, large numbers congregating in huge drifting 'slicks', the largest recorded of which, measuring 3m by approximately 100km, may have contained millions of individual snakes. Viviparous with litter sizes between 1 - 5, occasionally up to 12. May move onto land at low tide.
Temperament: Inclined to bite with aggression if molested.
Prey: Slow-moving fish such as frogfish and banded batfish *Halophryne diemensis.*

Seas: South China, Flores, Timor, Arafura and Coral.
Distribution within PNG: WEST: Fly R. mouth 60-80km ENE Daru.
Distribution outside PNG: Indonesia (Aru Is.); Irian Jaya (80km N Kaap Valsch, off Prince Frederik Hendrik Is.); Australia (n.WA, NT, Qld incl. Torres Strait and Gulf of Carpentaria); Indo-Malaysia and Indo-China.

Stokes' seasnake – *Astrotia stokesii*
(Waters off Singapore)
Photo: Harold Voris, Chicago Field Museum

Stokes' seasnake – *Astrotia stokesii*
(Waters off Singapore)
Photo: Harold Voris, Chicago Field Museum

AUSTRALO-PAPUAN SEASNAKES

Disteria spp.

Map: 32a

Currently the genus *Disteria* is considered to contain four species from India to Australia. Two species occur in the Australo-Papuan region although they are often placed in the genus *Hydrophis* by taxonomists.

Key to the species of seasnakes, genus *Disteira,* in Papua New Guinea waters

1a. 324 - 342 ventrals; 50 or more dark saddles over the body and tail; anterior chin shields large and bordering the mental groove. *Disteira kingi*

1b. 230 - 266 ventrals; fewer than 40 dark saddles over the body and tail; reduced anterior chin shields separated from mental groove by elongate 1st infralabials. *Disteira major*

SPECTACLED or KING'S SEASNAKE

Disteira kingi

Average/Maximum length: 0.6m / 1.5m.

Description:

a) physique: Head small, elongate and pointed, eye small; neck and body fairly slender; tail laterally compressed.

b) colouration: Body and tail light grey above, cream below with 50 - 65 darker rounded saddles which are blackest on the neck; head black except for a white ring around the eye.

c) scalation: DMB 37 - 39, keeled and overlapping dorsally, smooth and barely overlapping laterally; V 324 - 342, twice as wide as adjacent dorsal scales; anal scales not enlarged; head scales large, regular and unfragmented; 7 - 8 supralabials with 3rd and 4th or 5th in contact with the eye; chin shields large, bordering mental groove.

Habits: Diurnal but little documented. Usually inhabits turbid offshore waters over mud or muddy sand, between 7 - 22m deep. Viviparous.

Temperament: Not known.

Prey: Eels including morays i.e. *Muraena reticularis.*

Taxonomic note: This species is included in the genus *Hydrophis* by some authors (Smith 1926; Ehmann 1993).

Seas: Timor, Arafura and Coral.

Distribution within PNG: WEST: Fly R. mouth 60-80km ENE Daru.

Distribution outside PNG: Australia (n.WA, NT, Qld incl. Torres Strait and Gulf of Carpentaria).

OLIVE-HEADED or GREATER SEASNAKE

Disteira major

Average/Maximum length: 0.8m / 1.5m.

Description:

a) physique: Head narrow, indistinct from neck, eye small; neck and body slender but robust; tail laterally compressed.

b) colouration: Body and tail pale grey to yellow with 25 - 38 olive to black oval saddles, alternating with fine cross-bars and lateral spots; head olive brown with darker flecking.

c) scalation: DMB 37 - 43, faintly keeled and barely overlapping; V 230 - 266, twin keeled and slightly wider than adjacent dorsal scales; anal scales enlarged; head scales large, regular and unfragmented; 7 - 8 supralabials with 3rd and 4th in contact with the eye; chin shields separated from mental groove by first infralabials.

Habits: Diurnal but little documented. Usually inhabits turbid inshore waters, such as estuaries, over mud or muddy sand, between 3 - 30m deep. Viviparous with a litter size of up to five.
Temperament: Not known.
Prey: Eels and long-bodied fish such as the eel-tailed catfish *Cnidoglanus* sp.
Taxonomic note: This species is included in the genus *Hydrophis* by some authors (Smith 1926; Ehmann 1993).

Seas: Timor, Arafura and Coral.
Distribution within PNG: WEST: Fly R. mouth 60-80km ENE Daru.
Distribution outside PNG: Irian Jaya (80km N Kaap Valsch, off Prince Frederik Hendrik Is.); Australia (WA, NT, Qld, NSW and Gulf of Carpentaria) and New Caledonia.

BEAKED SEASNAKES
Enhydrina spp.

Map: 32 a+c

Formerly thought to contain a single species, a second species of *Enhydrina* was described by the Russian taxonomist V. E. Kharin in 1985.

Key to the Beaked seasnakes, genus *Enhydrina,* in Papua New Guinea waters

1a. Rostral scale undivided; dorsal scales weakly keeled; preocular scale present.
Enhydrina schistosa

1b. Rostral scale divided; dorsal scales strongly keeled or spinose; preocular scale absent.
Enhydrina zweifeli

COMMON or BEAKED SEASNAKE

Enhydrina schistosa

Average/Maximum length: 0.9 - 1.2m / 1.4m.

Description:

a) physique: Head elongate, slightly distinct from neck with a distinctive curved snout, eye small; neck and forebody slender, remainder of body stouter and deeper; tail laterally compressed.
b) colouration: Body and tail pale grey or blue-grey with 45 - 55 darker saddles which fade out on the lower flanks and tail; undersides are white; head mottled grey.

c) scalation: DMB 49 - 66, faintly keeled and overlapping; V 239 - 322, slightly wider than adjacent dorsal scales; anal scales enlarged; head scales large, regular and unfragmented; 7 - 8 supralabials with 3rd and 4th in contact with the eye; preocular present; undivided rostral scale extended into a 'beaklike' projection; chin shields separated from mental groove by first infralabials; single elongate 'dagger-like' mental scale present, extending back between the chin shields and surrounded by loose, extensible scale-less skin.

Habits: Active diurnally or nocturnally. Usually inhabits turbid inshore waters less than 20m deep and often shallower than 5m, such as estuaries or shallow bays, where it forages on the sea bed, between 3 - 30m deep. Viviparous with a litter size of up to 12.

Common beaked seasnake
– *Enhydrina schistosa*
(mouth of Muar River, Malaysia)
Photo: Harold Voris, Chicago Field Museum

Temperament: The beaked seasnake bites quickly and easily if handled. Snakebites in the Ramu River, Madang Province, are believed to have been caused by snakes of this genus which are believed to be responsible for 50% of all seasnake bites and 90% of all fatalities. An antivenom is produced for this species at the Commonwealth Serum Laboratories in Victoria, Australia.
Prey: Large fish such as pufferfish and catfish.
Taxonomic note: This species is included in the genus *Disteria* by some authors (McDowell 1972).

Seas: Arafura and Coral.
Distribution within PNG: WEST: Gulf of Papua and Fly R., s. of Balimo. A specimen of seasnake I collected in 1986, from the Oriomo River 60 km inland at Old Zim, is believed to belong to this species. Unfortunately the specimen was lost. (*See* Parker's snake - an enigma).
Distribution outside PNG: Irian Jaya (80km N Kaap Valsch, off Prince Frederik Hendrik Is.); Australia (NT, Qld and Gulf of Carpentaria).

'SEPIK or ZWEIFEL'S SEASNAKE'
Enhydrina zweifeli
Known length: 0.8m.
Description:
a) physique: As *E. schistosa*.
b) colouration: Dorsally grey with dark transverse saddles which are fairly well defined posteriorly; underside of body and throat white, lower labials grey.
c) scalation: DMB 48, strongly keeled or spinous and juxtaposed; V 272, no wider than adjacent dorsal scales with those of the anterior body divided by a groove; eight anal scales enlarged; SC 52+; head scales large, regular and unfragmented; 6 - 7 supralabials with 3rd or 4th to 5th in contact with the eye; preocular absent; rostral triangular and divided into two by a vertical suture; 9 - 11 infralabials; single elongate 'v'-shaped mental scale present but hidden in the skin of the gular fold.

Habits: Only known from the holotype which was collected from the mouth of the Sepik River in 1966. With such a distribution it is possible that this snake enters rivers.
Temperament: Nothing known but must be considered a dangerous species.
Prey: Probably fish.
Taxonomic note: Described by Kharin (1985) from a single male specimen in the American Museum of Natural History (AMNH 104340 formerly labelled *E. schistosa*) on the basis of the divided rostral, the absence of preoculars, heavily keeled or spinous dorsal scales and the condition of the hemipenes.

Seas: Bismarck.
Distribution within PNG: E.SPK: mouth of Sepik R.
Distribution outside PNG: None. Possibly a PNG endemic.

BANDED SEASNAKES

Hydrophis spp.

Maps: 32 b+c

This, the largest genus of seasnakes, is represented in the Australasian region by upwards of a dozen species of which eight have been reported from Papua New Guinean waters. Most are in some way banded, either partially or completely around the body, although in some species the banding is less evident in adults than it is in juveniles.

Key to the banded seasnakes, genus *Hydrophis,* in Papua New Guinean waters

Some of the characters required to determine species may be impossible to examine under field conditions

1a. Posterior ventrals all undivided. *Hydrophis gracilis*
1b. Posterior ventrals distinct and undivided. 2

2a. 10 or more solid maxillary teeth behind fang. *Hydrophis ornatus*
2b. Fewer than 10 solid maxillary teeth behind fang. 3

3a. Usually 3rd to 5th supralabials in contact with the eye; 324 - 421 ventrals; 45 - 49 dorsal scale rows at midbody. *Hydrophis pacificus*
3b. Only 3rd and 4th supralabials in contact with the eye. 4

4a. Head completely black in all ages, followed by a light ring; ventrals 323 -514. *Hydrophis atriceps*
4b. Head black in juveniles becoming lighter in adults; ventrals fewer than 421. 5

5a. Body pattern of dark edged, light centred round markings which are confined to the upper flanks and vertebral region and do not circle the body; ventral fewer than 275. *Hydrophis macdowelli*
5b. Body pattern comprised of dark bands which circle the body or are broken mid-ventrally or vertebrally. 6

6a. 37 or fewer dorsal scale rows at midbody; 70 - 80 dark crossbands or rings around the body. *Hydrophis vorisi*
6b. 37 or more dorsal scale rows at midbody; 35 - 70 dark crossbands or rings around the body. 7

7a. 37 - 49 dorsal scale rows at midbody; 345 - 432 ventrals; 55 or fewer dark crossbands. *Hydrophis elegans*
7b. 37 - 43 dorsal scale rows at midbody; 266 - 370 ventrals; 50 or more dark crossbands. *Hydrophis melanosoma*

BLACK-HEADED SEASNAKE

Hydrophis atriceps

Average/Maximum length: 1.0m / 1.3m.

Description:

a) physique: Head small and indistinct from neck, eye small; neck and forebody slender; mid and hindbody stout and compressed; tail laterally compressed.
b) colouration: Body and tail yellow or light brown with 50 - 70 dark edged, lighter centred, crossbands which are broadest along the centre of the back, dark spots may also be present in the lighter gaps; head

dark brown or black.

c) scalation: DMB 35 - 49 slightly keeled and juxtaposed; V 323 - 514, barely wider than adjacent dorsal scales and mostly undivided; head scales large, regular and unfragmented; 6 - 7 supralabials with 3rd and 4th in contact with the eye.

Habits: Little documented. Found offshore where it forages for prey on the sea bed. Viviparous.
Temperament: Not known but potentially highly dangerous.
Prey: Probably eels.
Taxonomic note: Formerly recognised as a subspecies of *H. fasciatus* (McDowell 1972 and Cogger 1975).

Seas: Timor and Arafura.
Distribution within PNG: GULF: Gulf of Papua.
Distribution outside PNG: Indonesia (Java and Aru Is.); Australia (NT); Indo-China and the Philippines.

ELEGANT or BAR-BELLIED SEASNAKE
Hydrophis elegans
Average/Maximum length: 1.3m / 2.1m.
Description:

a) physique: Head small and indistinct from neck, eye small; neck and forebody slender; mid and hindbody much stouter and more compressed; tail laterally compressed.
b) colouration: Body and tail brown or light grey above, whiter below, with 35 - 55 black crossbands and dark spots in the lighter spaces between; head dark brown or black in juveniles, lighter in adults.
c) scalation: DMB 35 - 49 slightly keeled and weakly imbricate; V 323 - 514, smooth and fractionally wider than the adjacent dorsal scales and mostly undivided; head scales large, regular and unfragmented; 6 - 7 supralabials with 3rd and 4th in contact with the eye.

Elegant seasnake – *Hydrophis elegans*
(Gulf of Carpentaria)
Photo: Hal Heatwole, North Carolina State University

Hydrophis elegans
Gulf of Carpentaria

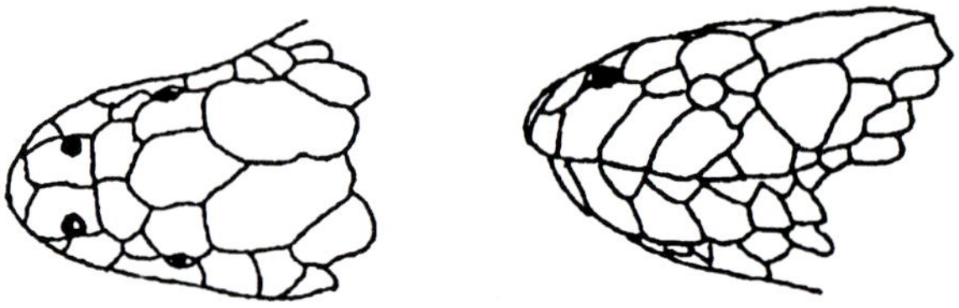

Habits: Active diurnally or nocturnally. Generally offshore in water up to 20m deep, occasionally as deep as 80m, but moving inshore into estuaries and river mouths in the summer months where it forages for prey on the sea bed, and occasionally seen basking on land. Preferred habitats are turbid waters over muddy sea beds. Viviparous with a litter size numbering between 8 and 23.
Temperament: Will bite easily if molested.
Prey: Long bodied fish and eels including the moray eel *Muraena reticularis*.

Seas: Timor, Arafura and Coral.
Distribution within PNG: WEST: Fly R. mouth 60-80km ENE Daru. GULF: Deception Bay. E.SPK: Sepik R. mouth.

Distribution outside PNG: Irian Jaya (80km N Kaap Valsch, off Prince Frederik Hendrik Is. and McCluer Gulf, off Fak Fak); Indonesia (Aru Is.); Australia (WA, NT, Qld, Gulf of Carpentaria, Torres Strait, NSW).

SLENDER or GRACEFUL SMALL SEASNAKE
Hydrophis gracilis
Average/Maximum length: 0.7m / 1.0m.
Description:
a) physique: Head small and indistinct from neck, eye small; neck and forebody extremely slender in contrast to much thicker and deeper mid and hindbody; tail laterally compressed.
b) colouration: Body and tail grey above, white below, with 45 - 70 black or grey crossbands, incomplete anteriorly but ringing the body posteriorly; head black above, light brown laterally.
c) scalation: DMB 30 - 36, with small tubercules dorsally and a larger central tubercule or prominent keel ventro-laterally, juxtaposed; V 220 - 287, all divided (characteristically of this species); head scales large, regular and unfragmented; 6 supralabials with 3rd and 4th in contact with the eye.
Habits: Generally offshore in deep turbid water of bays over muddy bottoms. Viviparous.
Temperament: Will bite easily if molested and has caused fatalities.
Prey: Long bodied fish and eels.
Taxonomic note: Formerly placed in the genus *Microcephalophis* (Smith 1926).

Seas: South China, Flores and Arafura.
Distribution within PNG: WEST: Gulf of Papua.
Distribution outside PNG: Persian Gulf; Indo-China, Indo-Malaysia and Australia (n.e.Qld, Torres Strait).

SMALL-HEADED or McDOWELL'S SEASNAKE
Hydrophis macdowelli
Average/Maximum length: 0.8m / 1.0m.
Description:
a) physique: Head extremely small and indistinct from neck, eye small; neck and forebody extremely slender, becoming much thicker and deeper by mid and hindbody; tail laterally compressed.
b) colouration: Body whitish above, tail and undersides yellowish, with 30 large rounded, black edged, lighter grey centred spots along the back and three series of smaller black spots along the flanks, all becoming fainter posteriorly; tail with several faint grey bands; head black in juveniles, brown in adults.
c) scalation: DMB 35 - 39; V 252 - 274, mostly undivided; head scales large, regular and unfragmented; 6 supralabials with 3rd and 4th in contact with the eye.
Habits: Nocturnal and diurnal, this seasnake usually occurs around reefs over sandy bottoms, from depths of 7 - 30m, but may also be found in estuaries and muddy bottomed river mouths. Viviparous.
Temperament: Nothing known.
Prey: Long bodied fish and eels.
Taxonomic note: Reported by previous authors as *Hydrophis* sp. (Cogger 1975 and Parker 1982) and believed to have been reported previously under the name *Disteira mjobergi*, this hydrophiid was described by Kharin (1983) and named in honour of the eminent American herpetologist Samuel B. McDowell.

Seas: Timor, Arafura and Coral.
Distribution within PNG: WEST: Fly R. mouth 60-80km ENE Daru.
Distribution outside PNG: Australia (WA, NT, Qld, Gulf of Carpentaria and Torres Strait).

BLACK-BANDED or ROBUST SEASNAKE

Hydrophis melanosoma

Average/Maximum length: 1.0m / 1.55m.

Description:

a) physique: Head indistinct from neck, eye small; body fairly stout; tail laterally compressed.

b) colouration: Body and tail olive above, yellow below, with 50 - 70 broad black bands which may be incomplete or form complete rings around the belly; head black above with yellow patches on the snout and labials.

c) scalation: DMB 37 - 43, imbricate and smooth in females, keeled in males; V 266 - 370, mostly undivided and wider than adjacent dorsal scales, each with two small keels; head scales large, regular and unfragmented; 6 - 7 supralabials with 3rd and 4th in contact with the eye.

Habits: Little known but believed to inhabit deep offshore waters. Viviparous.

Temperament: Nothing known.

Prey: Bottom-dwelling eels.

Seas: Flores, Timor, Arafura and Coral.

Distribution within PNG: M.BAY: Orangerie Bay.

Distribution outside PNG: Australia (WA, e.Qld) and Indo-Malaysia.

ORNATE or REEF SEASNAKE

Hydrophis ornatus

Average/Maximum length: 1.0m / 1.2m.

Description:

a) physique: Head slightly distinct from neck, eye small; body fairly heavy; tail laterally compressed.

b) colouration: Body and tail blue-grey above, off-white below, with 30 - 60 broad dark bars and small rounded markings, faint in adults, which do not continue onto the ventrolateral surfaces which are largely immaculate; head darker than body.

c) scalation: DMB 39 - 59; V 240 - 340, undivided and twice as wide as adjacent dorsal scales, broadest anteriorly; head scales large, regular and unfragmented; 7 - 8 supralabials with 3rd and 4th in contact with the eye.

Ornate or reef seasnake – *Hydrophis ornatus* (waters off Mersing, Malaysia)
Photo: Harold Voris, Chicago Field Museum

Habits: Found in clear offshore waters over coral reefs and turbid riverine and estuarine waters. Viviparous.

Temperament: Nothing known.

Prey: Eels and fish including pony fish *Leiognathus* sp. and anchovies *Setipinna papuensis*.

Seas: South China, Flores, Timor, Arafura and Coral.

Distribution within PNG: WEST: Fly R. mouth 60-80km ENE Daru, Mabaduwane, Pahoturi R. mouth and Mawatta, Binaturi R. mouth; WNB/ENB: no precise locality data.

Distribution outside PNG: Irian Jaya (southwest and west without precise locality data); Australia (WA, NT, Qld, Gulf of Carpentaria, NSW); Solomon Is.; Persian Gulf; Indo-China and Indo-Malaysia. Smith (1926) recognised two subspecies; *H. o. ornatus* for Indo-Chinese, Indo-Malaysian and New Guinea populations and *H. o. ocellatus* for populations from west New Guinea, the Aru Is. and Australia. A further subspecies, *H. o. maresinensis*, may be recognised for Far East Asian populations.

PACIFIC SEASNAKE

Hydrophis pacificus

Average length: 1.4m.

Description:

a) physique: Head large, but short, and slightly distinct from neck, eye small; body anteriorly slender, posteriorly heavily built and laterally compressed; tail laterally compressed.

b) colouration: Body and tail dark grey above, off-white below, with 49 - 72 dark, (black in juveniles) bands which may continue under the venter but may be mis-aligned on the centre of the back; head black with yellow spots on the snout and behind the eyes in juveniles but much paler in adults.

c) scalation: DMB 45 - 49, imbricate; V 324 - 421, mostly undivided and slightly wider than the adjacent dorsal scales; head scales large, regular and unfragmented; 7 - 8 supralabials with 3rd to 5th in contact with the eye.

Habits: Found offshore in deep water. Viviparous with a litter size of up to 17.

Temperament: Nothing known.

Prey: Probably eels.

Taxonomic note: Placed in synonymy with *H. belcheri* by Smith (1926), in *H. elegans* by McDowell (1972) but resurrected as a valid species by Cogger (1975).

Seas: Arafura, Coral and Bismarck.

Distribution within PNG: WEST: Fly R. mouth 60-80km ENE Daru. GULF: Purari R. mouth, Orokolo Bay. ENB/WNB without precise locality data.

Distribution outside PNG: Irian Jaya (80km N Kaap Valsch, off Prince Frederik Hendrik Is.); Australia (NT, w.Qld, Gulf of Carpentaria).

ESTUARINE SEASNAKE

Hydrophis vorisi

Average length: 0.5 - 0.6m.

Description:

a) physique: Head very small, indistinct from neck, eye small; body elongate, very slender anteriorly, posteriorly heavier, deeper and more compressed; tail laterally compressed.

b) colouration: Body and tail off-white with 70 - 80 dark bands around the body which may be mis-aligned on the centre of the back.

c) scalation: DMB 29 - 37, imbricate; V 330 - 350, mostly undivided and slightly wider than the adjacent dorsal scales, broadest anteriorly; anal scales slightly enlarged; head scales large, regular and unfragmented; 7 supralabials with 3rd and 4th in contact with the eye.

Habits: Estuarine and shallow inshore habitats. Viviparous.

Temperament: Nothing known.

Prey: Probably eels.

Taxonomic note: Formerly the two Western Province specimens were tentatively assigned to *H. obscurus* by McDowell (1972), Cogger (1975) and Parker (1982), a species otherwise confined to Indo-China. Kharin (1984) created a new species for the PNG specimens, *H. vorisi,* in honour of the American herpetologist Harold Voris.

Seas: Timor and Arafura.

Distribution within PNG: WEST: Sturt Is., Fly R. and Balimo, Aramia R.

Distribution outside PNG: None. A PNG endemic.

HARDWICKE'S or SPINE-BELLIED SEASNAKE

Lapemis hardwickii

Maps: 32 a+c

Average/Maximum length: 1.0m / 1.5m.

Description:

a) physique: Head large and slightly distinct from neck, eye small; neck stout; body heavier, deeper and more compressed; tail laterally compressed.

b) colouration: Body and tail light tan to cream with 30 - 55 large, darker brown circular blotches which are fused along the back, overall impression may be of a lateral zig-zag between pale undersides and darker upper surfaces; head as back.

c) scalation: DMB 23 - 45, hexagonal and juxtaposed, smooth dorsally, keeled ventrally and heavily tuberculate or spinous in adult males; V 110 - 240, mostly undivided and almost indistinguishable from dorsal scales posteriorly; anal scales small and barely enlarged; head scales large, regular and unfragmented; 7 - 8 supralabials with 3rd and 4th in contact with the eye; chin shields much reduced and separated by 1st infralabials.

Habits: Diurnal and nocturnal. An inhabitant of both clear offshore coral reef waters and turbid inshore estuarine habitats and the tidal zone over mud or sandy mud bottoms. Found between 6 - 15m but occasionally deeper to 55m. Viviparous with a litter size of 1 - 7.

Temperament: Inclined to bite if handled and recorded as causing human fatalities.

Prey: Bottom-dwelling, midwater swimming and reef crevice-dwelling fish, especially gobies and rabbitfish. Species recorded from stomach contents include clupeid fish, gobies, a moray eel and a crab cheliped.

Taxonomic note: Smith (1926) recognised a second species *L. curtus*, but this species was synonymised with *L. hardwickii* by Voris (1969). Some authors prefer to use the name *L. curtus* for Australian populations (Parker 1982).

Seas: South China, Flores, Timor, Arafura, Coral and Bismarck.

Distribution within PNG: WEST: Gulf of Papua. NCD: Port Moresby. E.SPK: Sepik R. mouth. MOR: no precise locality data. MAN: SW of Manus Is.

Distribution outside PNG: Irian Jaya: 80km N Kaap Valsch, off Kolepom (Prince Frederik Hendrik Is.), McCluer Gulf, off Fak Fak. Australia: NT, w.Qld, Gulf of Carpentaria. Indo-Malaysia, Indo-China and Philippines.

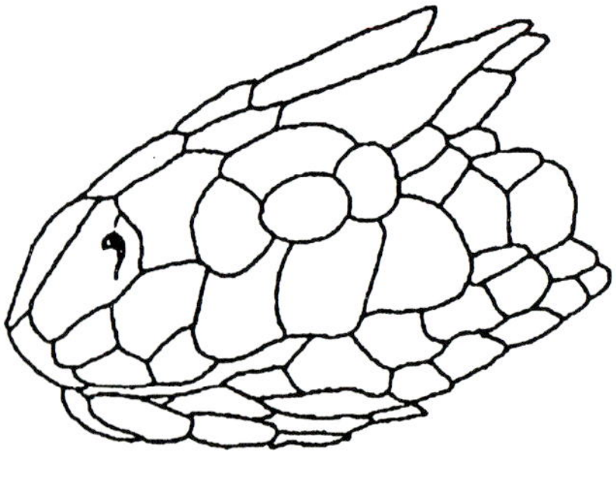

Lapemis hardwickii
Gulf of Papua

Hardwicke's seasnake – *Lapemis hardwickii*
(waters off Parit Botak, Malaysian Pen.) Photo: Harold Voris, Chicago Field Museum.

YELLOW-BELLIED or PELAGIC SEASNAKE
Pelamis platurus
Map: 32c

Average/Maximum length: 0.7m / 1.0m.
Description:
a) physique: Head elongate and distinct from neck, eye small; neck and body moderately stout, and laterally compressed; tail ovoid and laterally compressed.
b) colouration: Dorsum of head and body black or dark brown, venter bright yellow with transition between the two on the flanks either a clean straight line or a much fragmented zig-zag line; tail pale with darker blotches.
c) scalation: DMB 47 - 69, smooth, hexagonal and juxtaposed, lowest rows with small tubercules most evident in adult males; V 264 - 406, mostly divided by a slight median furrow or broken and indistinguishable from the dorsal scales; anal scales moderately enlarged; head scales large, regular and unfragmented; 7 - 8 supralabials with 4th and 5th separated from the eye by subocular scales.

Habits: Diurnal and nocturnal. Usually an inhabitant of open ocean 100m deep many kilometres from land, specimens may be beached by storms. On land they are totally helpless and unable to move. Although specimens may enter esturaries they are not found in brackish or freshwater. Viviparous with a litter size of 2 - 6. *Pelamis* rids itself of marine parasites by continually twisting and entwining its body in the water (a behaviour frequently observed and mis-understood) and regularily sloughing its skin. Occasionally huge 'slicks' consisting of thousands of yellow-bellied seasnakes may be encountered drifting with the current.
Temperament: Inclined to bite if handled and possessing a highly toxic venom.
Prey: Small surface-dwelling fish.

Seas: Indian and Pacific Oceans: South China, Flores, Timor, Arafura, Coral and Bismarck.
Distribution within PNG: ENB: no precise locality data
Distribution outside PNG: Irian Jaya (Kolepom - Prince Frederik Hendrik Is. and Fak Fak); Australia (WA, NT, Qld incl. Thursday Is., Torres Strait, NSW, Tas., Gulf of Carpentaria); East African coast, Madagascar, Persian Gulf, Indo-China, Indo-Malaysia, Philippines, Solomon Is., New Caledonia, New Zealand, Taiwan, Japan, se. Russia and west coast of Mexico, Panama and Ecuador.

Yellow-bellied seasnake – *Pelamis platurus*
(Port Elizabeth, S. Africa) Photo: Bill Branch, Port Elizabeth Museum

Yellow-bellied seasnake – *Pelamis platurus*
(Port Elizabeth, S. Africa) Photo: Bill Branch

Comparison between the Papuan taipan and the Papuan blacksnake

	PAPUAN TAIPAN	PAPUAN BLACKSNAKE
1. Dorsal colouration	Olive brown (Western), brown or black (Central) with broad orange vertebral stripe	Jet black, rarely brown
2. Ventral colouration	Cream with or without orange speckling	Black or gunmetal without speckling, throat and chin white
3. Head shape	Elongate, angular and distinct from neck	Short, rounded and barely distinct from neck
4. Body shape	Slender and triangular	Stout and rounded
5. Maximum length	3.4 m	2.4 m
6. Dorsal scales	Matt, keeled anteriorly, 21-23 rows at midbody	Glossy, smooth 19, rarely 21, rows at midbody
7. Ventral scales	220-250	221-230
8. Subcaudal scales	45-80, all paired	49-63, first 25-45 single, remainder paired
9. Anal plate	Single	Divided
10. Supraocular scale	Shelved over eye	Not shelved
11. Eye	Moderately large, round pupil	Small, round pupil
12. Habitat	Savanna and savanna woodland	Savanna woodland, swampland and forest
13. Activity	Diurnal, especially at the end of the wet season	Diurnal, especially at the end of the dry season
14. Abundance	Common	Rare
15. Distribution	Western, Central and Milne Bay, (Gulf?)	Western, Milne Bay and parts of Central, (Gulf?)
16. Prey preferences	Small mammals	Frogs and small mammals
17. Venom	Strongly neurotoxic and coagulant, weakly haemolytic	Strongly neurotoxic and haemolytic possibly haemorrhagic, probably not coagulant

Papuan blacksnake – *Pseudechis papuanus*
Rigo, Central Province
(BMNH 97.12.10.127)

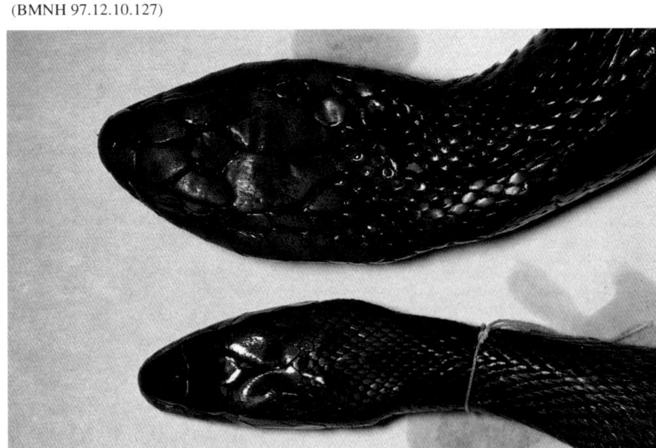

Papuan taipan – *Oxyuranus scutellatus canni*
Senggo, Irian Jaya
(BMNH 1992.542)

Comparison between the Papuan whip snakes and the Papuan blacksnake

		WHIP SNAKES	BLACKSNAKE
1.	Dorsal colouration	Olive brown or dark brown to black lightening towards tail which may be red-brown	Jet black, rarely brown
2.	Ventral colouration	Blue-grey with white to yellow throat and chin	Black or gunmetal without speckling throat and chin white
3.	Head shape	Short, rounded and barely distinct from neck	Short, rounded and barely distinct from neck
4.	Body shape	Elongate and rounded	Stout and rounded
5.	Maximum length	1.5 m	2.4 m
6.	Dorsal scales	Matt, to moderately glossy, smooth 15 rows at midbody	Glossy, smooth, 19 rarely 21, rows at midbody
7.	Ventral scales	160-225	221-230
8.	Subcaudal scales	69-105, all paired	49-63, first 25-45 single, remainder paired
9.	Anal plate	Divided	Divided
10.	Supraocular scale	Not shelved	Not shelved
11.	Eye	Very large, round pupil	Small, round pupil
12.	Habitat	Savanna and savanna woodland	Savanna woodland, swampland and forest
13.	Activity	Diurnal, even during hottest part of day	Diurnal, especially at the end of the dry season
14.	Abundance	Very common	Rare
15.	Distribution	Western and Central, Milne Bay (and Gulf?)	Western, Milne Bay and parts of Central (and Gulf?)
16.	Prey preferences	Lizards	Frogs and small mammals
17.	Venom	Neurotoxic, possibly haemorrhagic and coagulant	Strongly neurotoxic and haemolytic, possibly haemorrhagic, probably not coagulant

COMPARISON BETWEEN THE GROUND BOA AND THE DEATH ADDER

	GROUND BOA	DEATH ADDER
1. Dorsal colouration	Dark brown or red brown with darker brown square blotches which are united to form a zig-zag pattern	Highly variable, either uniform or with pale and dark bands across the body
2. Ventral colouration	Yellow or light brown with black or red mottling	White with black speckling, black and white markings especially evident on lower lips and throat
3. Head shape	Broad and angular, flattened and concave above, covered in small granular scales, no 'horns' above the eyes	Broad and angular, 'viper-like', at least some of the scales on top of the head enlarged, not granular, a raised 'horn-like' projection above each eye
4. Body shape	Extremely stout	Moderately stout
5. Tail	Stout and extremely short, <2x head length	Long and slender with a yellow tip, especially in juveniles
6. Maximum length	1.0 m	Usually less than 0.5 m, except Markham and Sepik populations - 1.0 m
7. Dorsal scales	Strongly keeled, 30-45 rows at midbody	Weakly keeled or smooth, 21-23 rows at midbody
8. Ventral scales	127-153	110-134
9. Subcaudal scales	11-22, all single	36-57, anteriorly single, posteriorly paired
10. Anal plate	Single	Single
11. Pelvic spurs	Spurs present in males either side of the cloaca (anus), absent in females	Absent
12. Eye	Very small, vertically elliptical pupil	Small, vertically elliptical pupil
13. Habitat	Damp woodlands and forests, coconut plantations, gardens	Monsoon woodlands, forests, kunai grasslands, gardens, coffee plantations
14. Activity	Nocturnal	Nocturnal
15. Abundance	Locally common	Very common
16. Distribution	Mainland PNG, especially northern provinces, altitude record 1200 m, also Bismarck and Manus provinces	Mainland PNG; only altitude record 1800 m
17. Prey preferences	Frogs, lizards and small mammals	Frogs, lizards and small mammals
18. Defensive posture	May roll into a ball with head protected in centre	Coiling back on itself in concentric curves with the head in the centre and striking from this position, but also able to strike from other positions
19. Teeth and fangs	Teeth present, no enlarged fangs	Teeth present, a pair of enlarged fangs present in the sheaths of skin in the front of the upper jaw
20. Venom	Non-venomous	Strongly neurotoxic and weakly haemolytic, possible anticoagulant and cytotoxic.

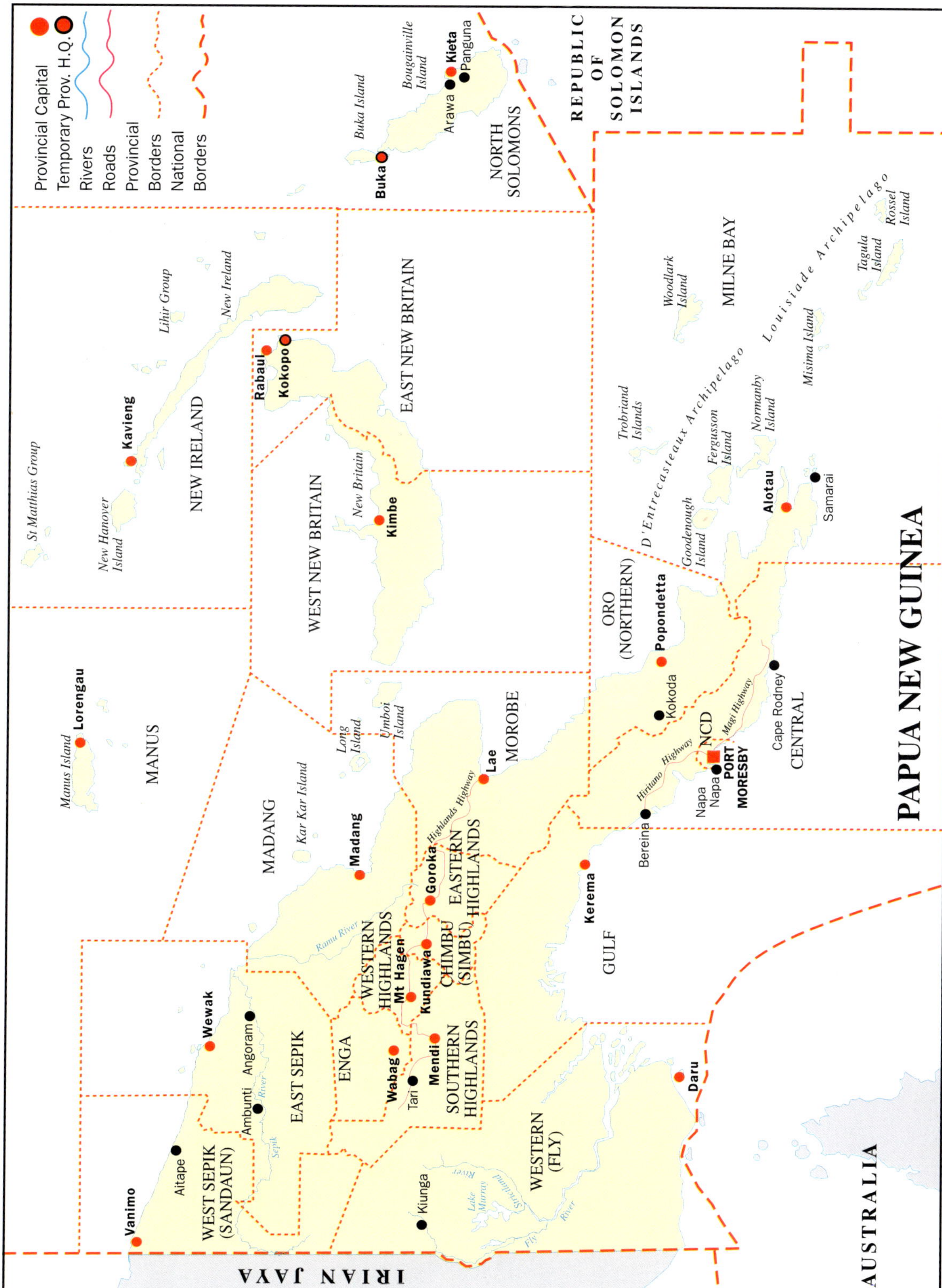

Provincial Capital
Temporary Prov. H.Q.
Rivers
Roads
Provincial Borders
National Borders
PAPUA NEW GUINEA
REPUBLIC OF SOLOMON ISLANDS
IRIAN JAYA
AUSTRALIA
NORTH SOLOMONS
Buka Island
Bougainville Island
Buka
Kieta
Panguna
Arawa
NEW IRELAND
St Matthias Group
New Hanover Island
Kavieng
Lihir Group
New Ireland
EAST NEW BRITAIN
Rabaul
Kokopo
WEST NEW BRITAIN
New Britain
Kimbe
MANUS
Manus Island
Lorengau
MILNE BAY
Woodlark Island
Trobriand Islands
D'Entrecasteaux Archipelago
Louisiade Archipelago
Goodenough Island
Fergusson Island
Normanby Island
Misima Island
Tagula Island
Rossel Island
Alotau
Samarai
ORO (NORTHERN)
Popondetta
Kokoda
NCD
PORT MORESBY
Napa Napa
Magi Highway
Hiritano Highway
Cape Rodney
CENTRAL
Bereina
MOROBE
Lae
Highlands Highway
Umboi Island
Long Island
Kar Kar Island
MADANG
Madang
Ramu River
EASTERN HIGHLANDS
Goroka
CHIMBU (SIMBU)
Kundiawa
WESTERN HIGHLANDS
Mt Hagen
ENGA
Wabag
SOUTHERN HIGHLANDS
Mendi
Tari
GULF
Kerema
EAST SEPIK
Wewak
Angoram
Ambunti
Sepik River
WEST SEPIK (SANDAUN)
Aitape
Vanimo
WESTERN (FLY)
Kiunga
Lake Murray
Strickland River
Fly River
Daru

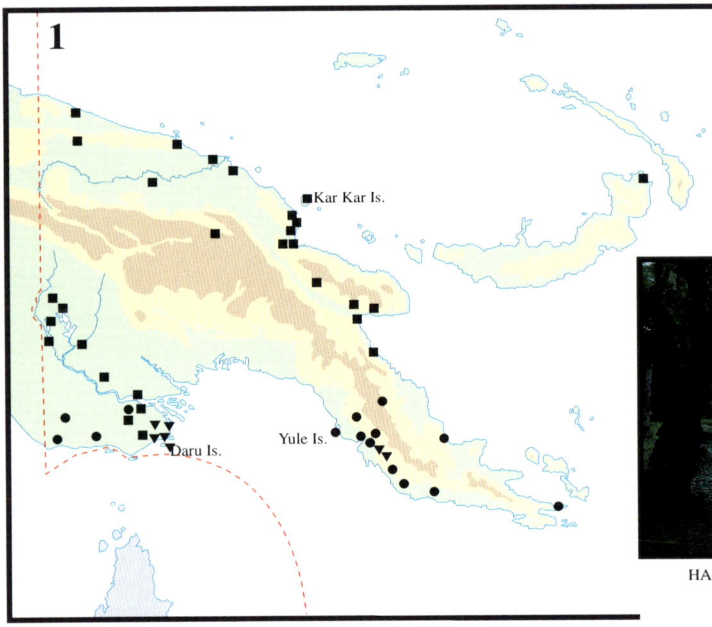

HABITAT: Coastal tidal river (Kunini, Western Province)
Morelia amethistina; Apodora papuana; Acrochordus arafurae; Tropidonophis

MAP 1

Distribution of the pygopodid snake-lizards, genus ***Lialis.***

Collection localities for *Lialis* spp. ● *L. burtonis* ■ *L. jicari* ▼ both spp.

MAP 2a

Distribution of the blind snakes, genus ***Ramphotyphlops***.

Collection localities for *Ramphotyphops* spp. ● *R. braminus*; ■ *R. erycinus*; ▲ *R. flaviventer*; ▼ *R leucoproctus*; ✖ *R. polygrammicus*; ★ *R. subocularis*

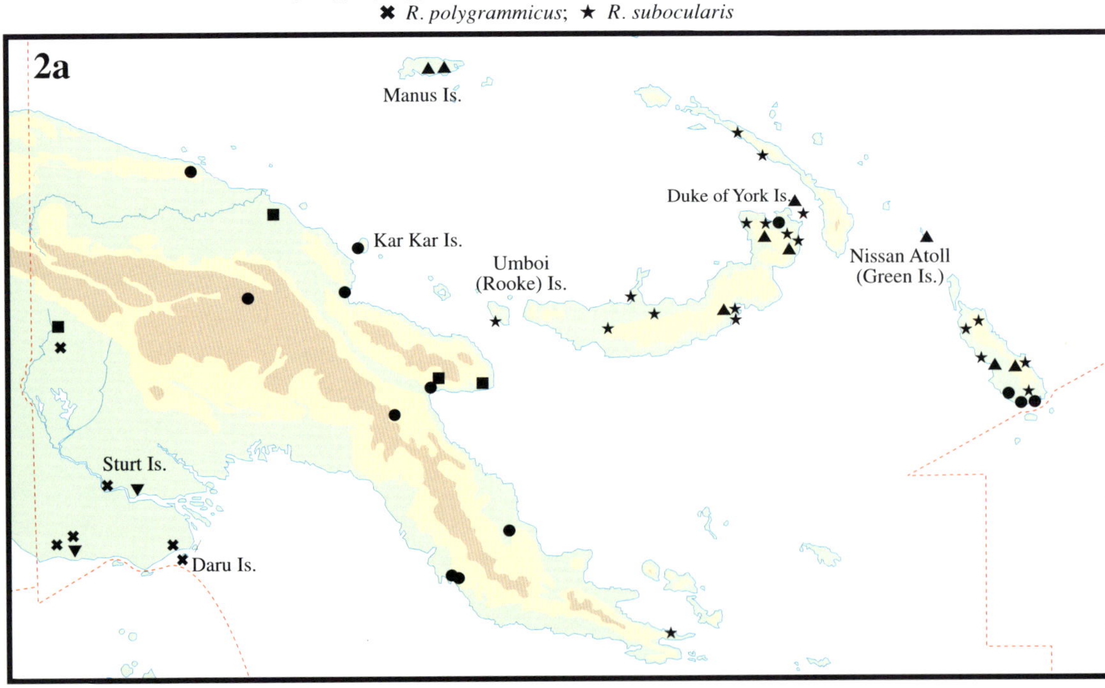

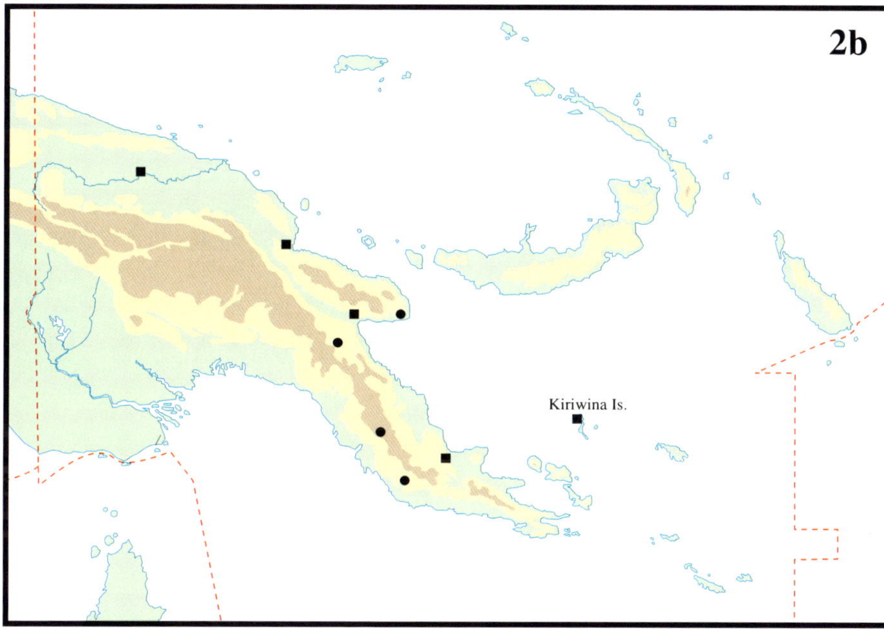

MAP 2b
Distribution of the blind snakes, genus ***Typhlops***.
Collection localities for *Typhlops* spp. ● *T. inornatus*;
■ *T. depresiceps*

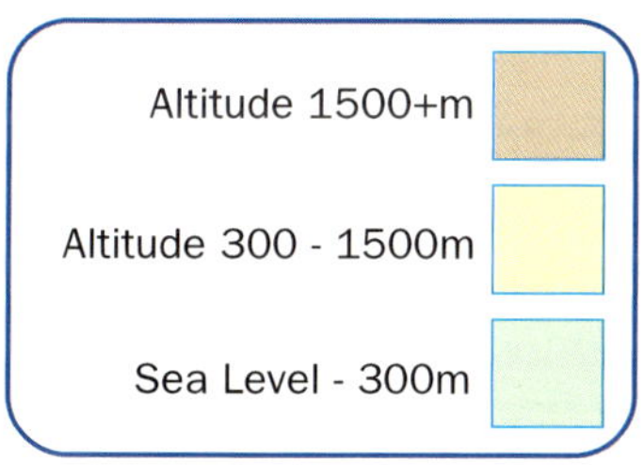

MAP 2c
Distribution of the blind snakes, genera ***Ramphotyphlops*** and ***Typhlops***.
Collection localities for West Irian spp. ▲ *R. multilineatus*; ▼ *T. ater*;
✛ *T. diardi* (questionable); ✖ *T. kraali*

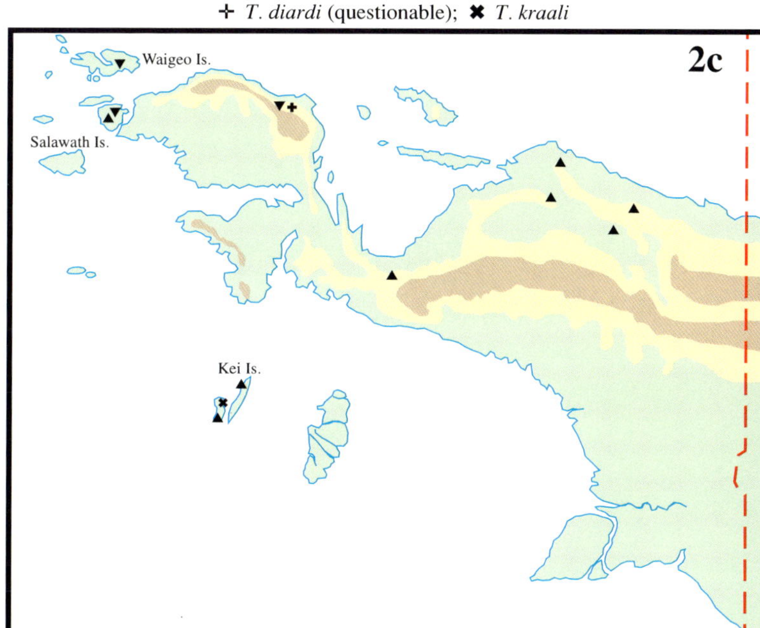

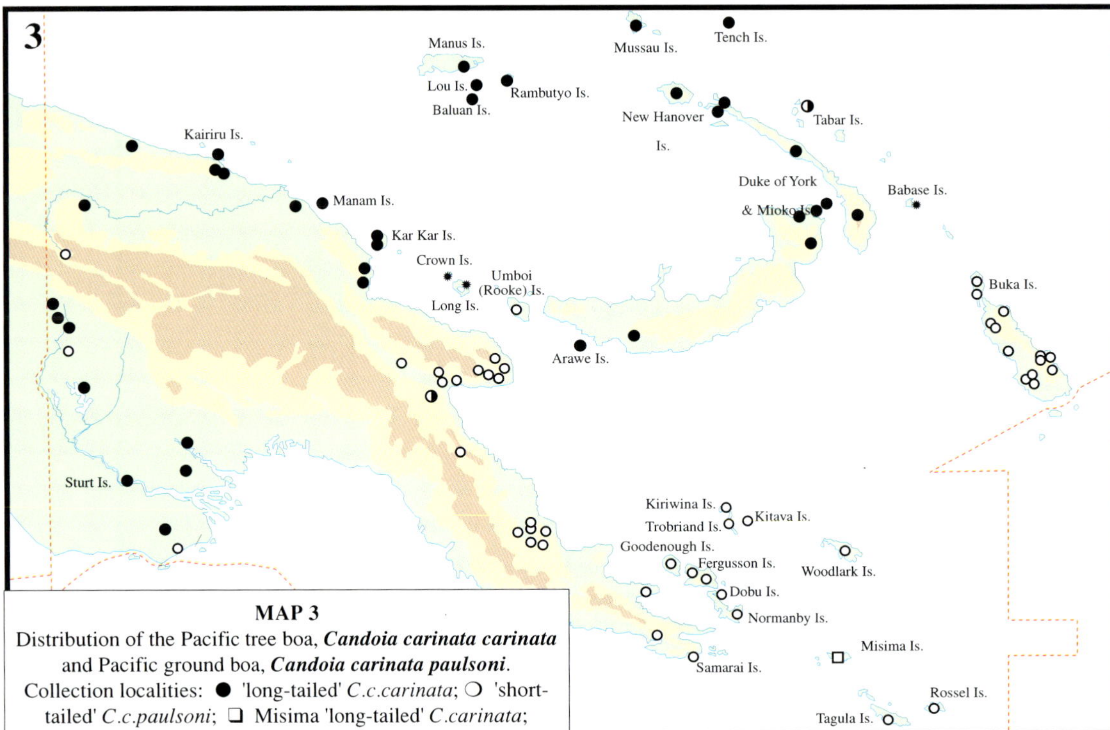

MAP 3
Distribution of the Pacific tree boa, ***Candoia carinata carinata*** and Pacific ground boa, ***Candoia carinata paulsoni***. Collection localities: ● 'long-tailed' *C.c.carinata*; ❍ 'short-tailed' *C.c.paulsoni*; ❑ Misima 'long-tailed' *C.carinata*; ◑ intermediate populations; ✷ *C.carinata* spp. (undetermined populations).

MAP 4
Distribution of the New Guinea ground boa, *Candoia aspera*. Collection localities ▲

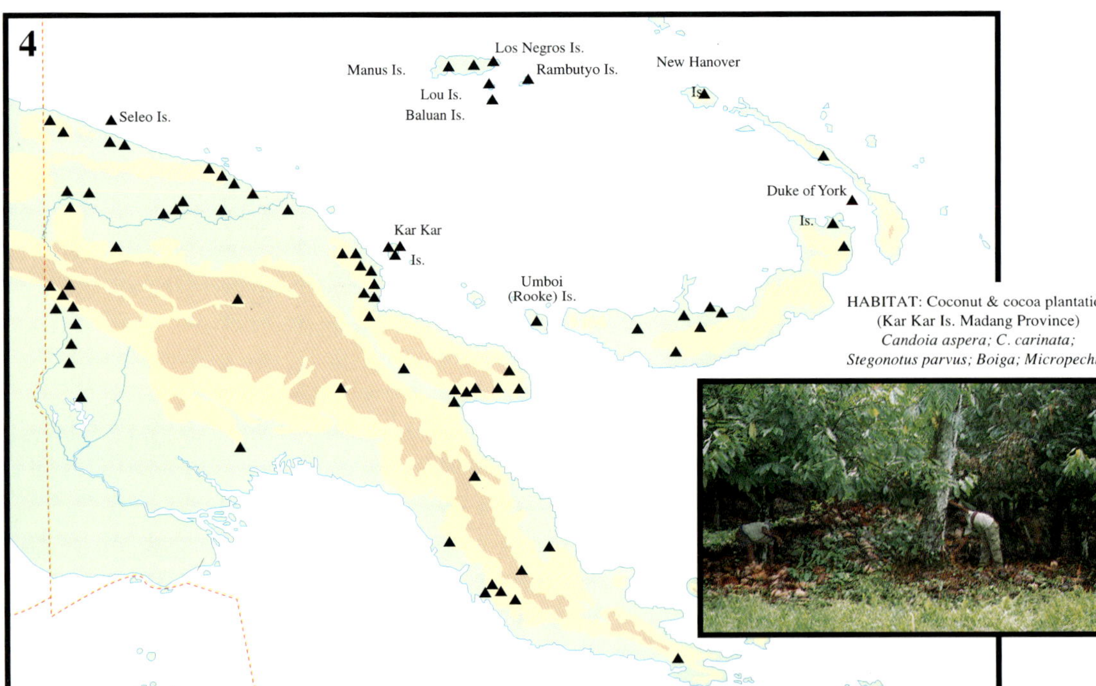

HABITAT: Coconut & cocoa plantation (Kar Kar Is. Madang Province) *Candoia aspera; C. carinata; Stegonotus parvus; Boiga; Micropechis.*

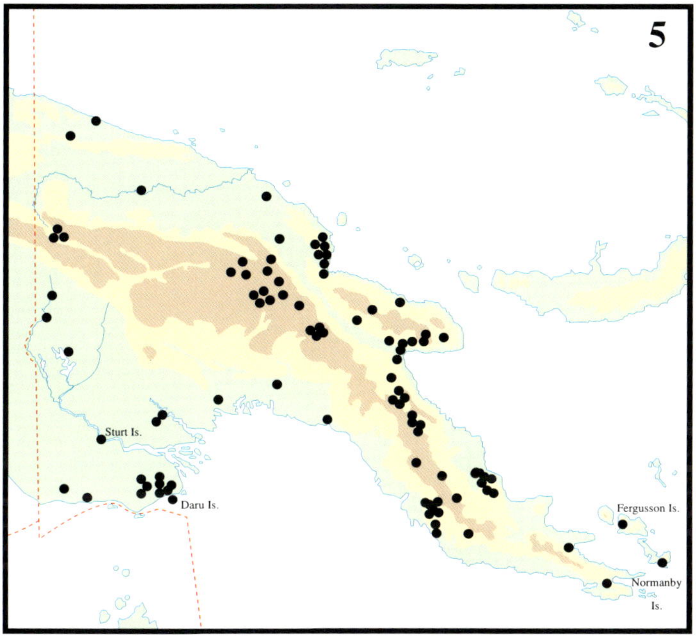

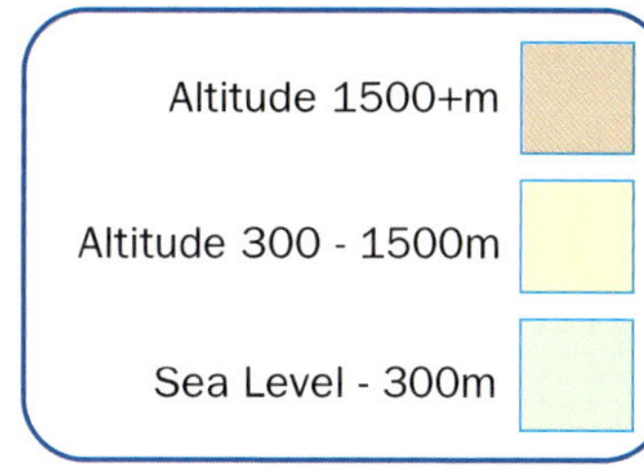

MAP 5
Distribution of the green tree python, ***Morelia viridis***. ● Collection localities

MAP 6
Distribution of the carpet python, ***Morelia spilota variegata***.
■ Collection localities

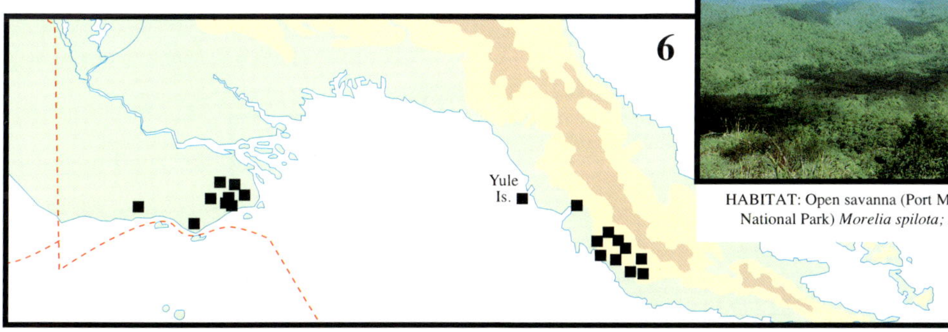

HABITAT: Open savanna (Port Moresby from Varirata National Park) *Morelia spilota; Boiga; Oxyuranus*.

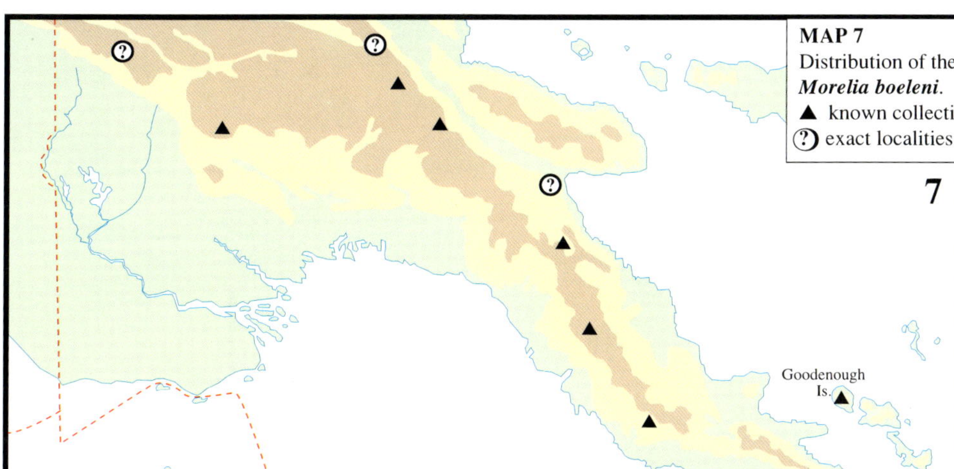

MAP 7
Distribution of the Boelen's python, ***Morelia boeleni***.
▲ known collection localities;
ⓘ exact localities undetermined

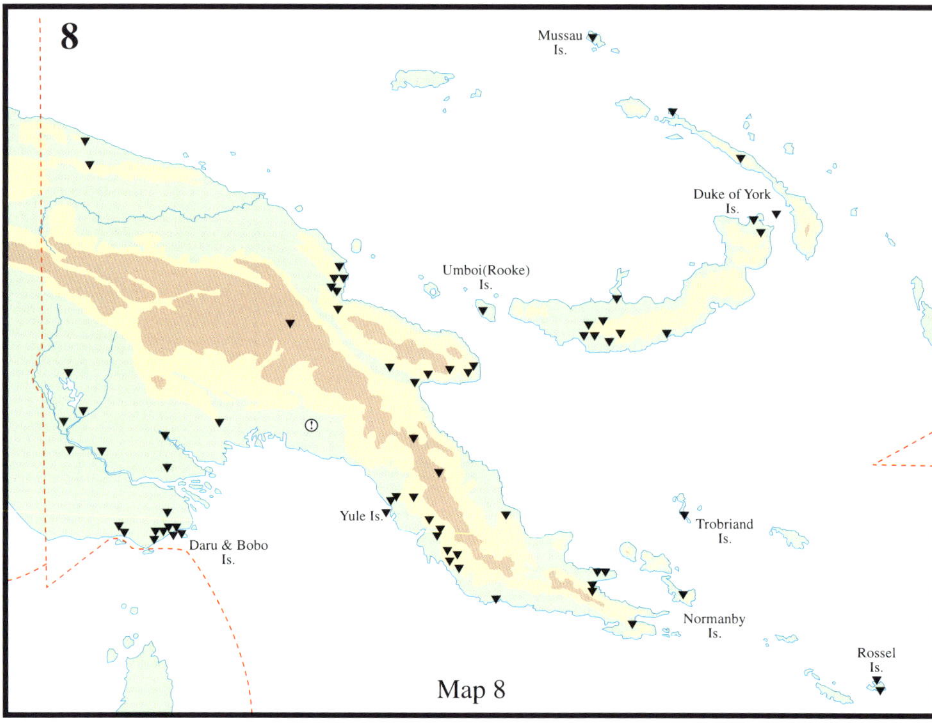

MAP 8

Distribution of the amethystine python, ***Morelia amethistina***. ▼ Collection localities; ⓘ provincial reports with imprecise locality data.

MAP 9

Distribution of the D'Albertis python, ***Leiopython albertisi***. ▼ Collection localities

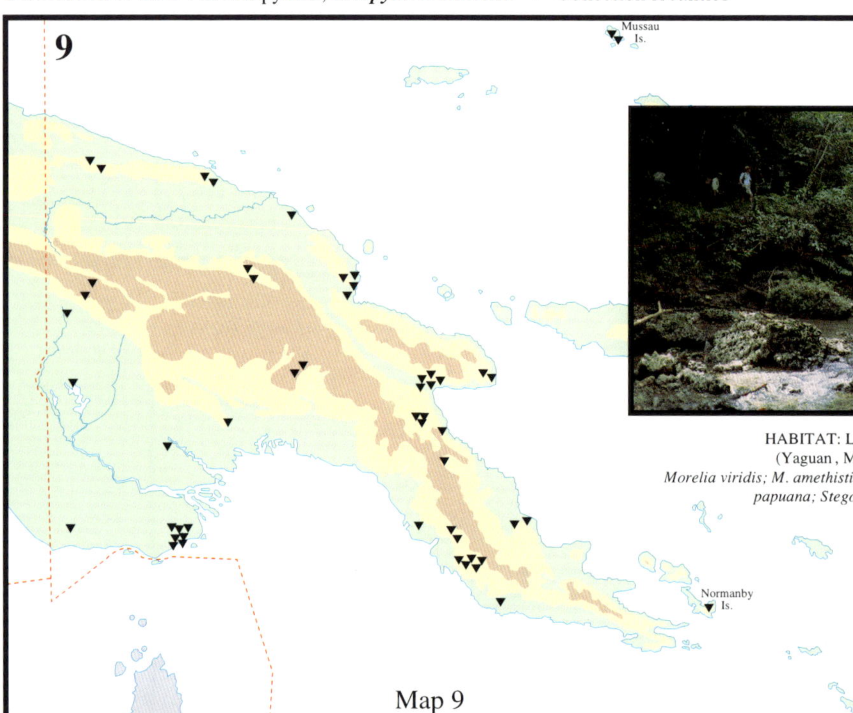

HABITAT: Low montane forest
(Yaguan, Madang Province)
Morelia viridis; M. amethistina; Leiopython albertisi; Apodora papuana; Stegonotus; Acanthophis.

MAP 10
Distribution of the Bismarck ringed python, ***Bothrochilus boa***.
▲ Collection localities

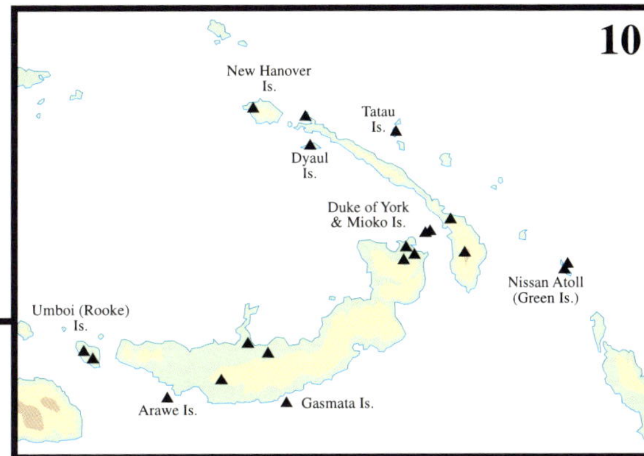

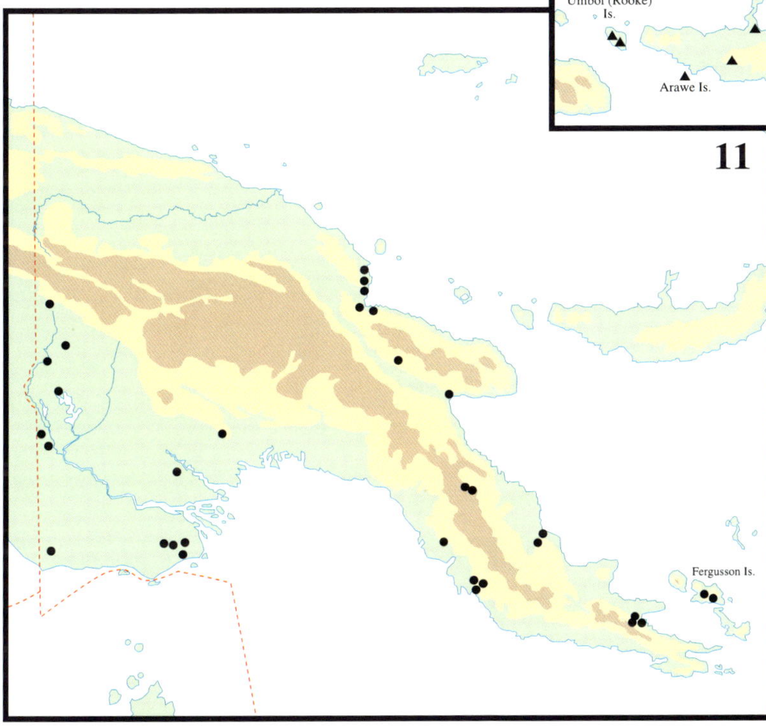

MAP 11
Distribution of the papuan olive python, ***Apodora papuana***.
● Collection localities

Altitude 1500+m
Altitude 300 - 1500m
Sea Level - 300m

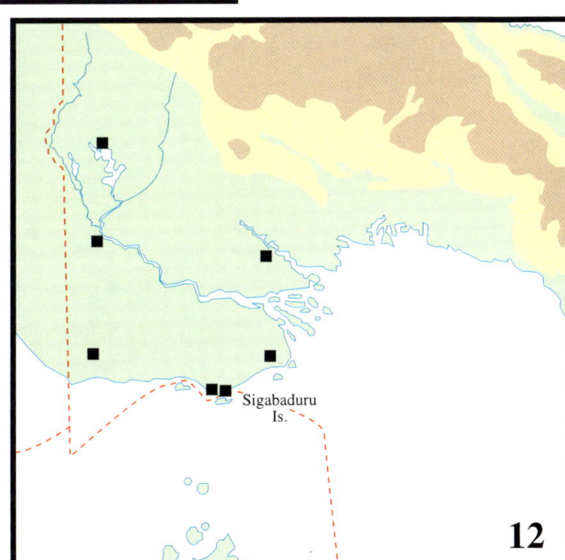

MAP 12
Distribution of the brown water python, ***Liasis fuscus***.
■ Collection localities

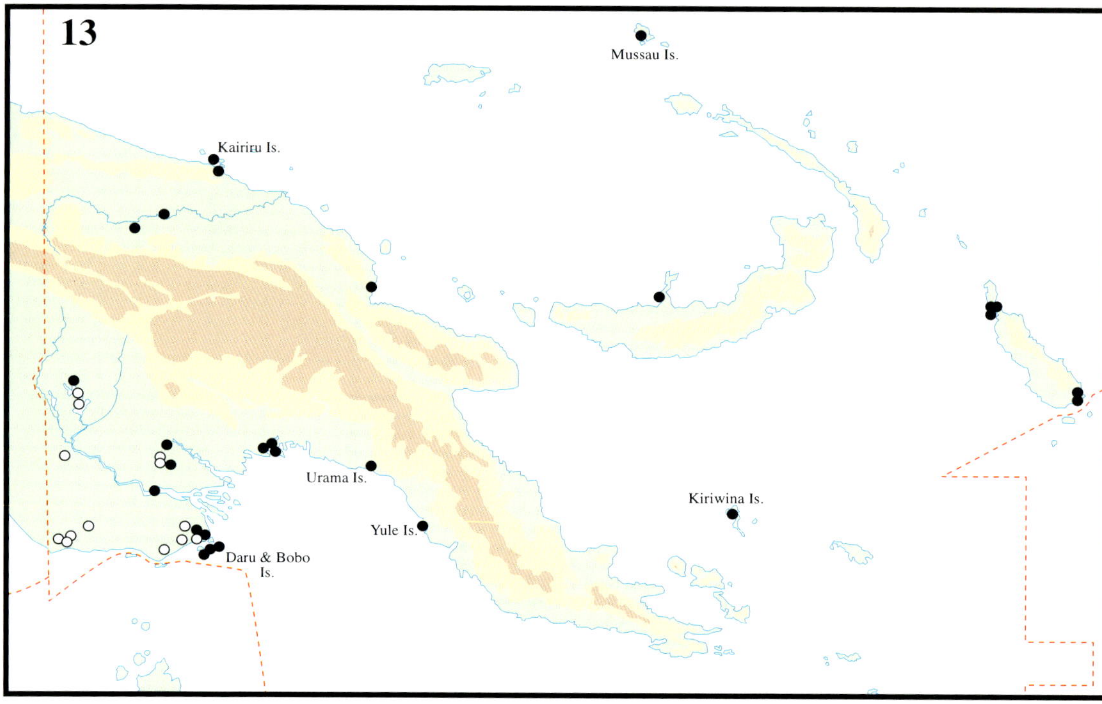

MAP 13
Distribution of the wart or file snakes, genus ***Acrochordus***.
Collection localities : ● *A. granulatus*; ❍ *A. arafurae*

MAP 14a
Distribution of the keelbacks, genus ***Tropidonophis***.
Collection localities: ■ *T. multiscutellatus*; ▲ *T. mairii mairii*; ▼ *T. mairii plumbea*; ✛ T. *picturatus*; ❑ *T. novaeguineae*; ✶ *T. mcdowelli*

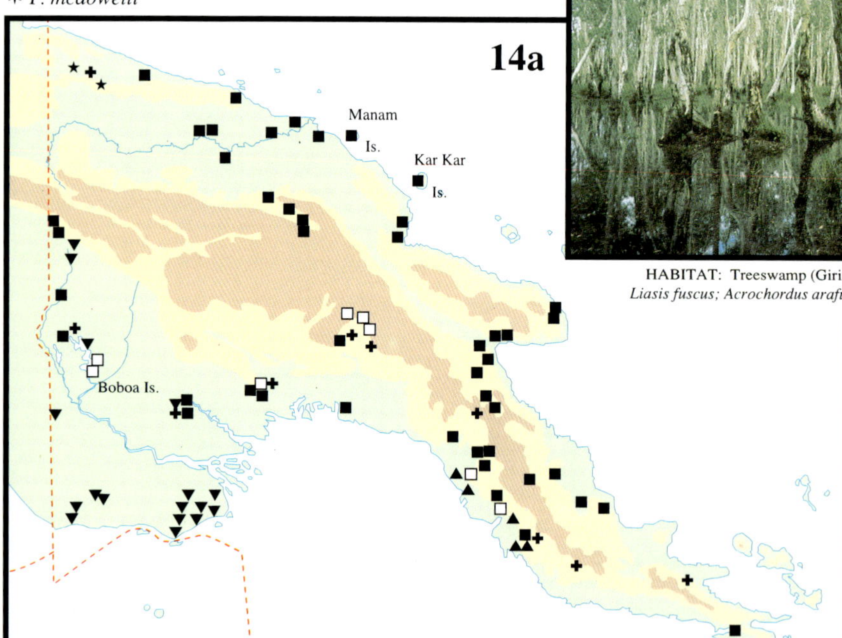

HABITAT: Treeswamp (Giringarande, Western Province)
Liasis fuscus; Acrochordus arafurae; Tropidonophis; Enhydris.

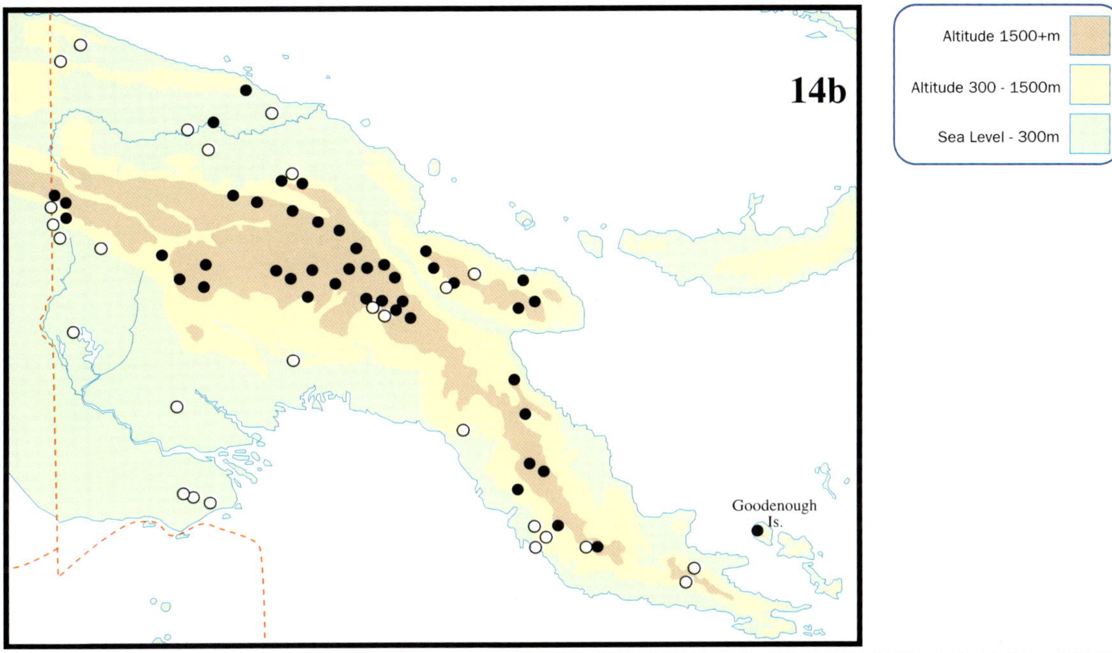

MAP 14b
Distribution of the keelbacks, genus ***Tropidonophis***.
Collection localities: ● *T. statisticus*; ❍ *T. doriae*

MAP 14c
Distribution of Irian Jayan keelbacks, genus ***Tropidonophis***.
Collection localities: ◆ *T. montanus*; ▲ *T. elongatus*

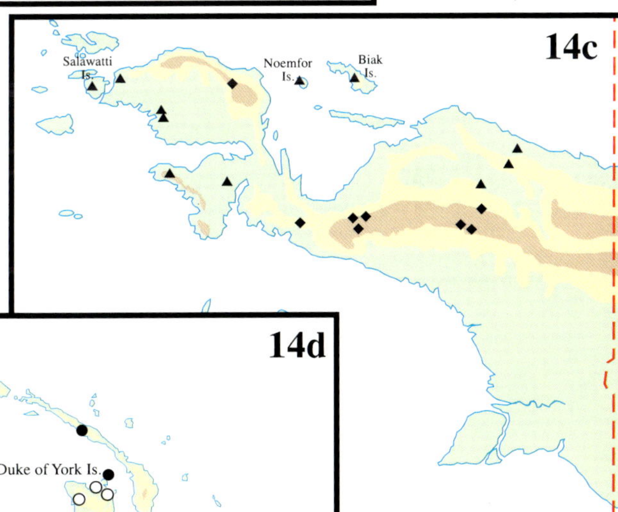

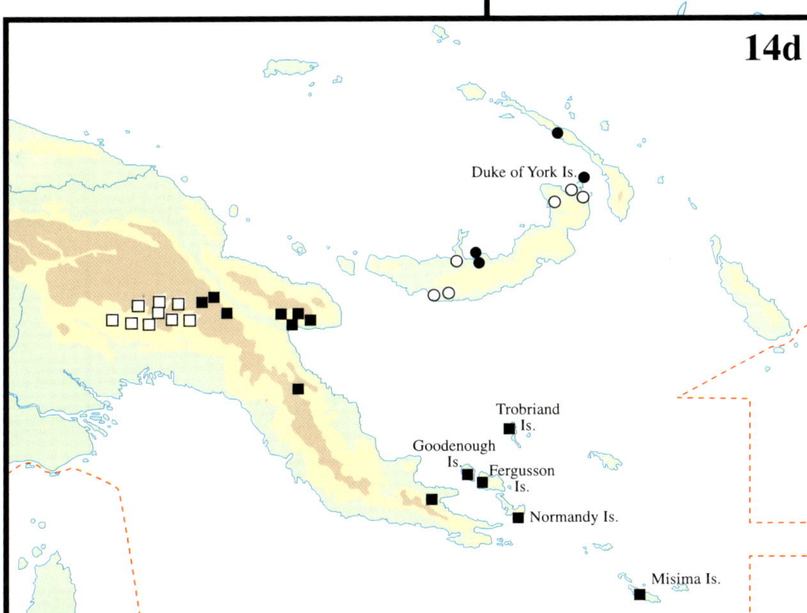

MAP 14d
Distribution of eastern PNG, Milne Bay & Bismarck keelbacks, genus ***Tropidonophis***
Collection localities:
❑ *T. parkeri*;
■ *T. aenigmaticus*;
❍ *T. dahlii*;
● *T. hypomelas*

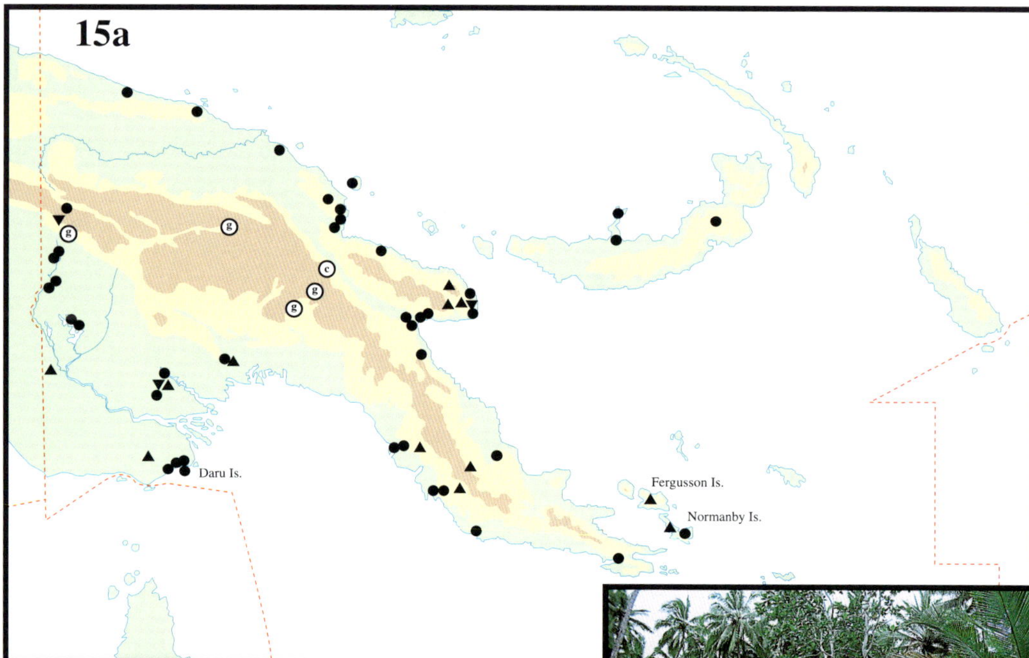

MAP 15a
Distribution of the treesnakes, genus ***Dendrelaphis***.
Collection localities: ● *D. calligastra*; ⓒ *D. calligastra* records with imprecise locality data; ▲ *D. gastrostictus*; ⓖ *D. gastrostictus* records with imprecise locality data; ▼ *D. lorentzi*

MAP 15b
Distribution of the treesnakes, genus ***Dendrelaphis***.
Collection localities: ■ *D. punctulatus*; ⓟ *D. punctulatus* records with imprecise locality data; ❑ *D. papuensis*; ✷ *D. salomonis*

HABITAT: Peripheral village (Kunini, Western Province)
Dendrelaphis; Tropidonophis; Fordonia; Enhydris; Boiga; Pseudechis papuanus

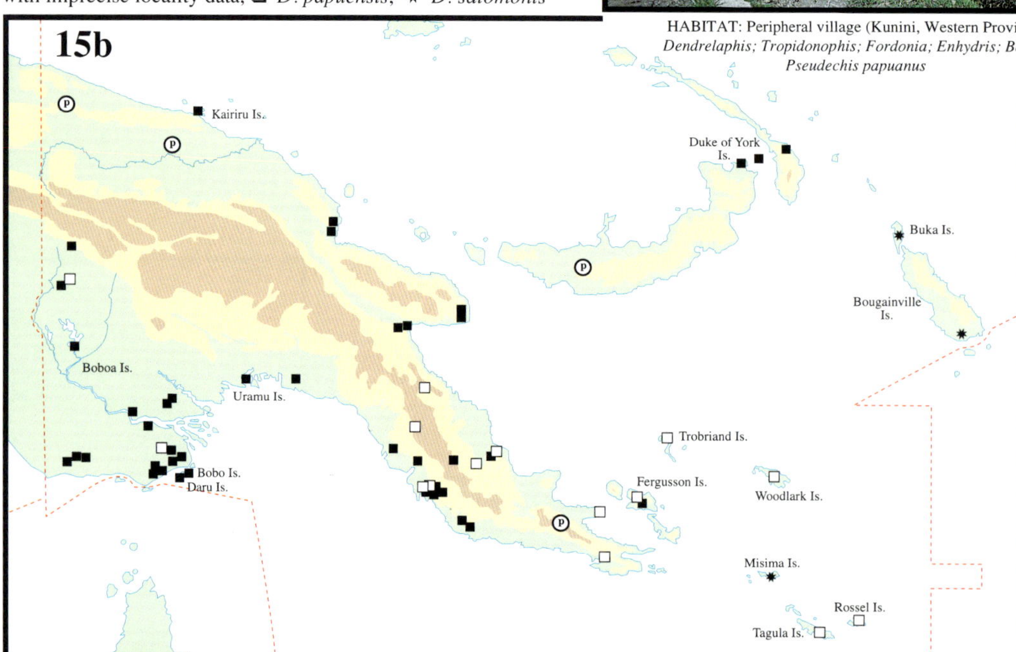

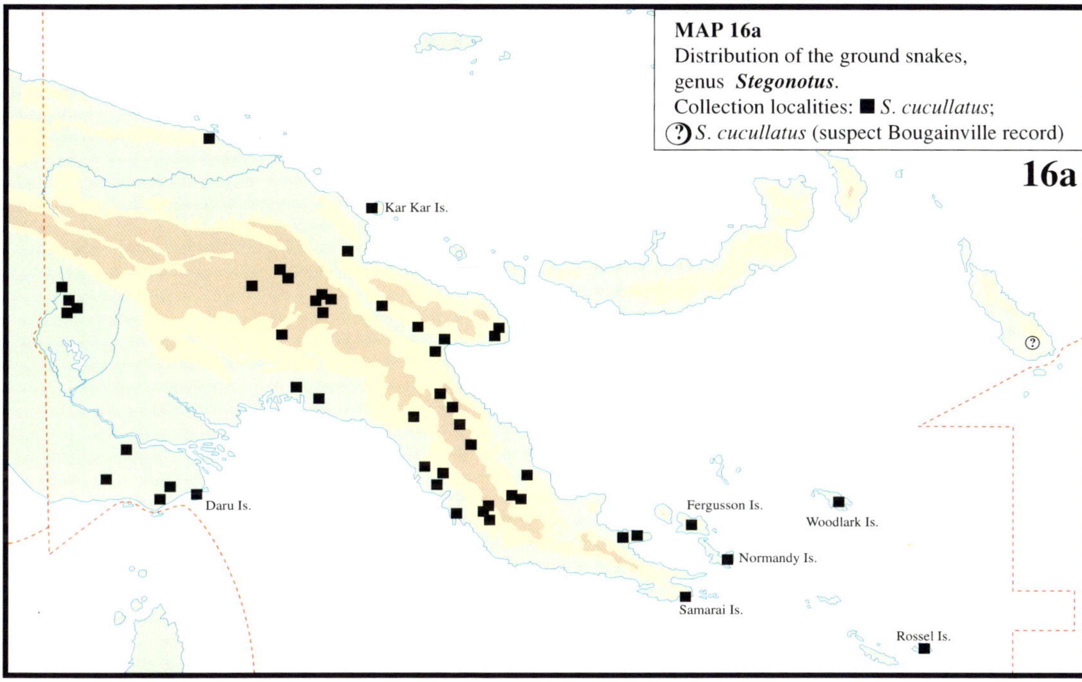
MAP 16a
Distribution of the ground snakes, genus **Stegonotus**.
Collection localities: ■ *S. cucullatus*; ? *S. cucullatus* (suspect Bougainville record)
16a
Kar Kar Is.
Daru Is.
Fergusson Is.
Woodlark Is.
Normandy Is.
Samarai Is.
Rossel Is.

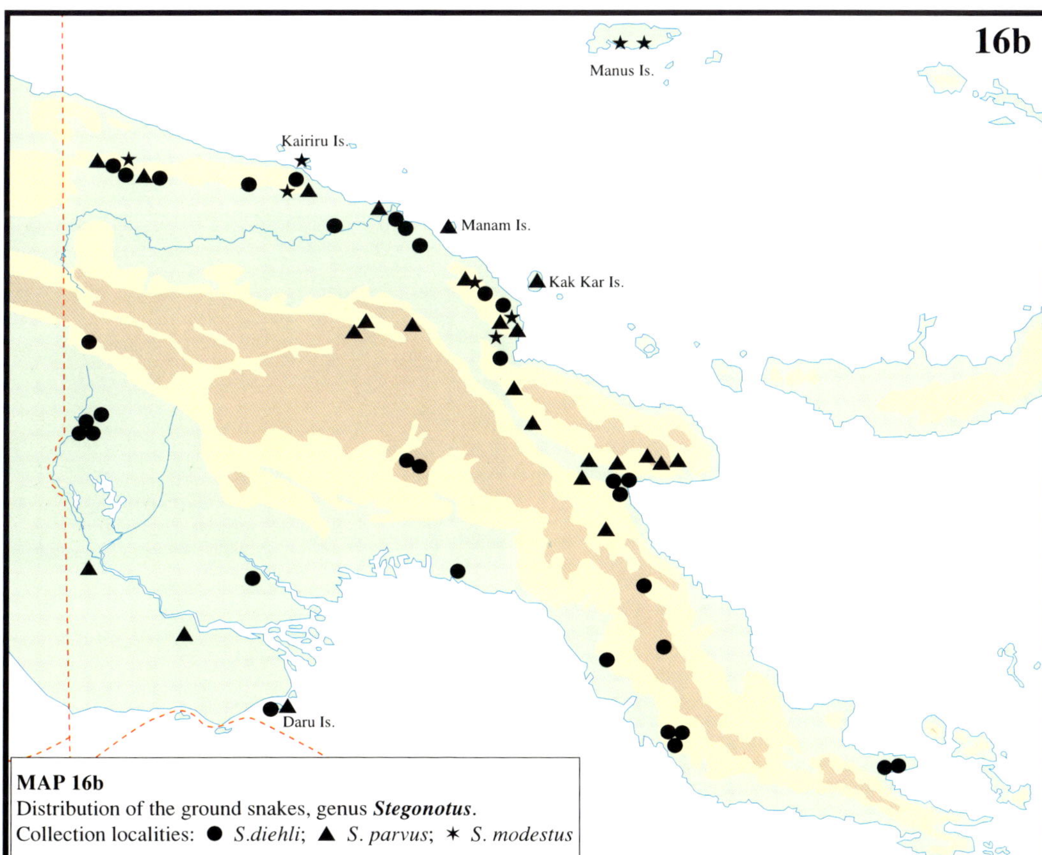
16b
Manus Is.
Kairiru Is.
Manam Is.
Kak Kar Is.
Daru Is.
MAP 16b
Distribution of the ground snakes, genus **Stegonotus**.
Collection localities: ● *S.diehli*; ▲ *S. parvus*; ✶ *S. modestus*

MAP 16c
Distribution of the Bismarck & Milne Bay ground snakes, genus ***Stegonotus***.
Collection localities:
▼ *S.*cf. *parvus* (McDowell 1985);
✛ *S. heterurus*;
✖ *S. guentheri*

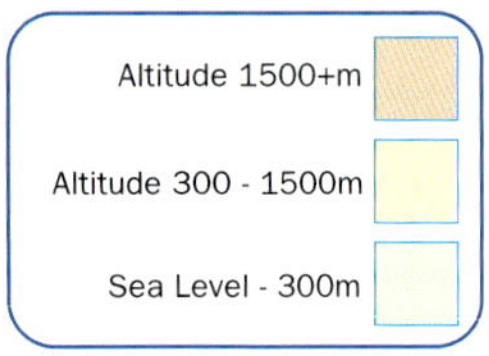

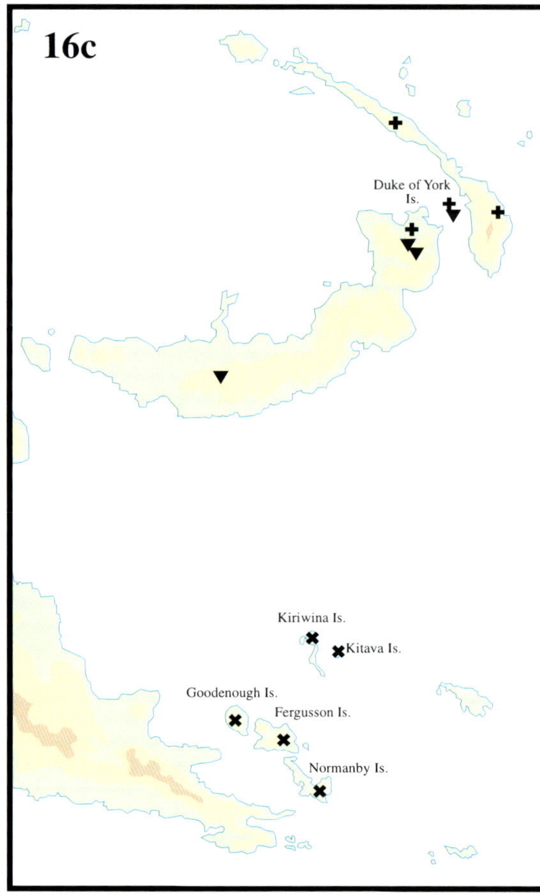

MAP 17a
Distribution of the mud and mangrove snakes, subfamily **Homalopsinae**.
Collection localities:
● *Fordonia leucobalia*;
▲ *Myron richardsoni*;
■ *Enhydris polylepis*

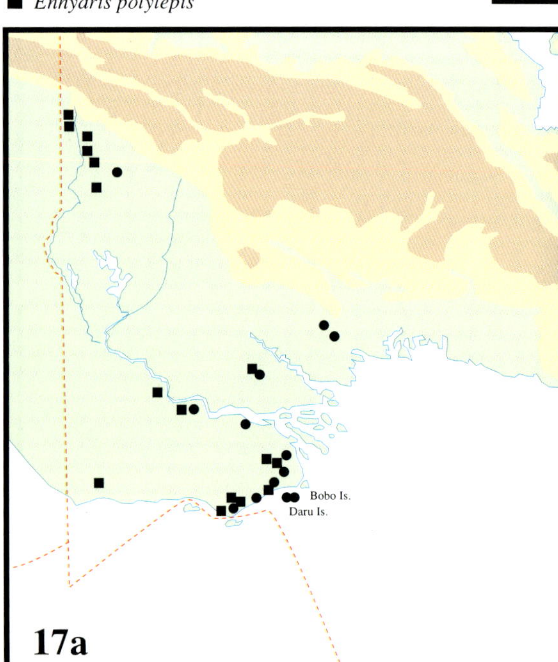

HABITAT: Lowland swamp forest (Oriomo, Western Province)
Morelia viridis; Leiopython albertisi; Tropidonophis; Fordonia; Enhydris; Boiga.

MAP 17b
Distribution of the mud and mangrove snakes, subfamily **Homalopsinae**, of Western Province.
Collection localities:
▲ *Myron richardsoni*;
✶ *Cantoria annulata*;
❍ Wipim, location of 1970's aquatic snake fatalities (*see* 'Parker's snake' - an enigma).

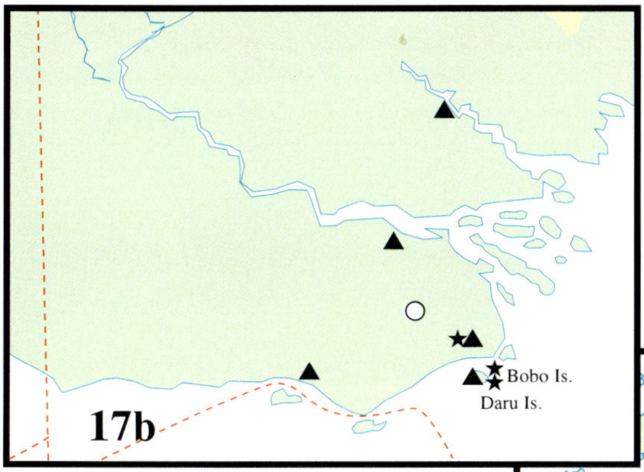

MAP 17c
Distribution of the mud and mangrove snakes, subfamily **Homalopsinae**, of Irian Jaya
(PNG species not shown)
Collection localities:
✛ *Cerberus rynchops novaeguineae*;
✖ *Heurnia ventromaculata*

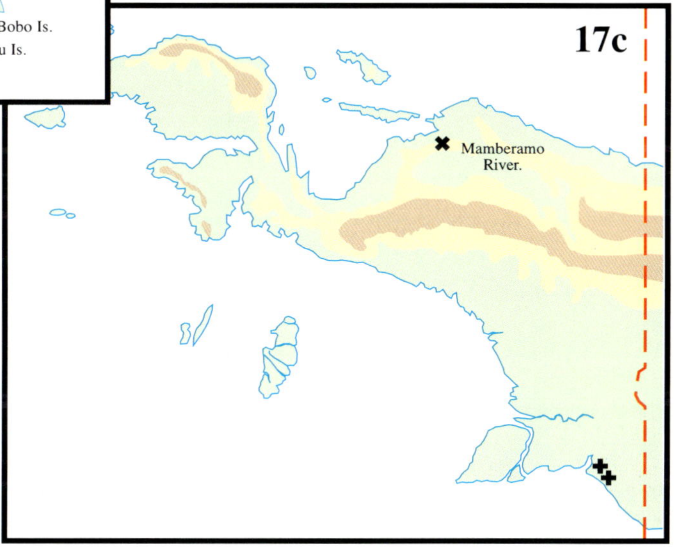

MAP 18
Distribution of the brown catsnake, ***Boiga irregularis***.
● Collection localities;
(!) provincial records with imprecise locality data
NB: The extremely wide distribution of this species is not fully illustrated by this map which is based on a limited number of collection localities and literature sources.

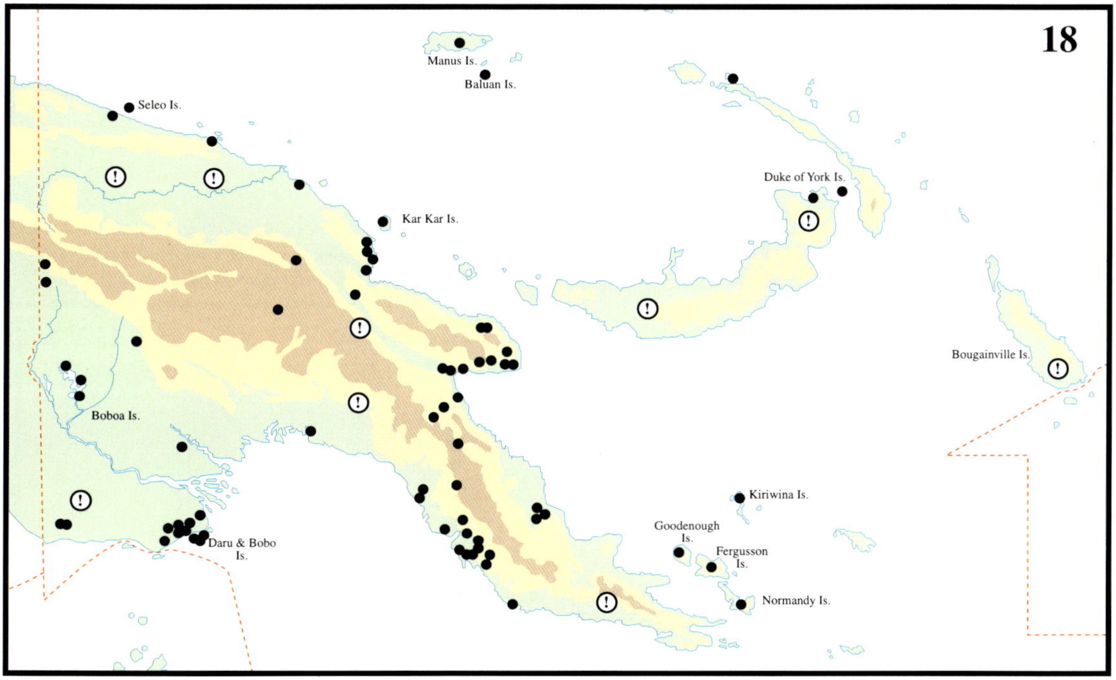

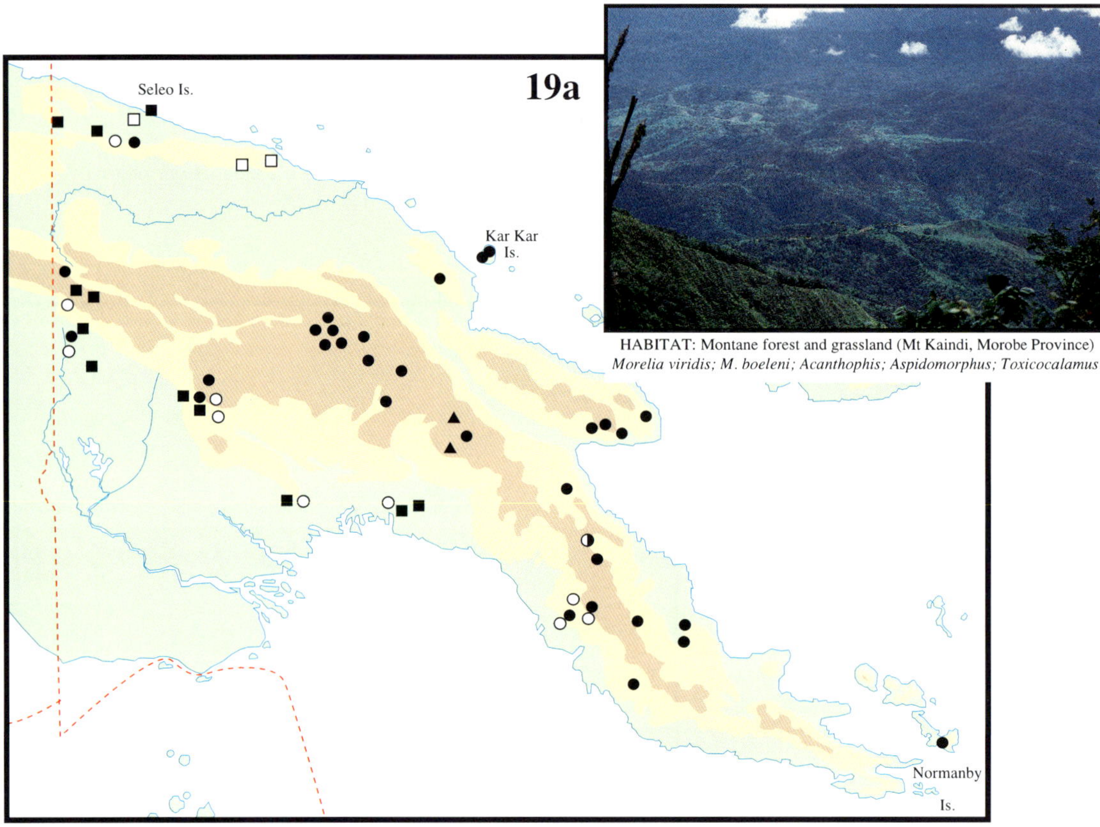

HABITAT: Montane forest and grassland (Mt Kaindi, Morobe Province)
Morelia viridis; M. boeleni; Acanthophis; Aspidomorphus; Toxicocalamus.

MAP 19a
Distribution of the mainland PNG forest snakes, genus ***Toxicocalamus***
Collection localities:
● *T. loriae*; ❍ *T. stanleyanus*;
◑ *T. loriae* x *T. stanleyanus* (Garaina pop.);
▲ *T. spilolepidotus*; ■ *T. preussi*;
❐ *T. buergersi*

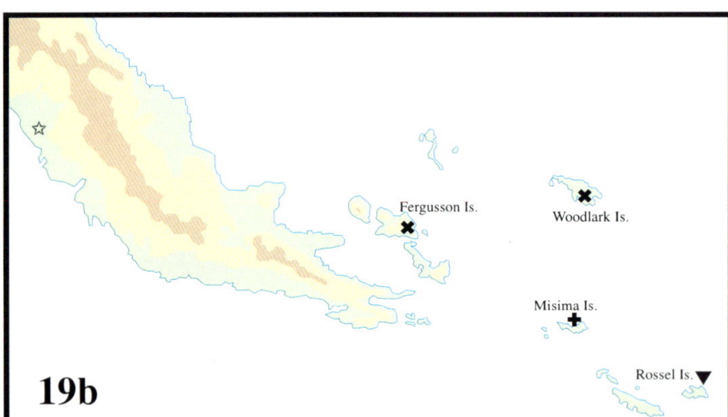

MAP 19b
Distribution of the insular Milne Bay forest snakes, genus ***Toxicocalamus***
Collection localities: ✖ *T. longissimus*; ✚ *T. misimae*;
☆ *T.* cf. *misimae* (Mekeo); ▼ *T. holopelturus*.

MAP 19c
Collection locality of the endemic Irian Jayan forest snake, genus ***Toxicocalamus***
✶ *T. grandis* (PNG Species not shown)

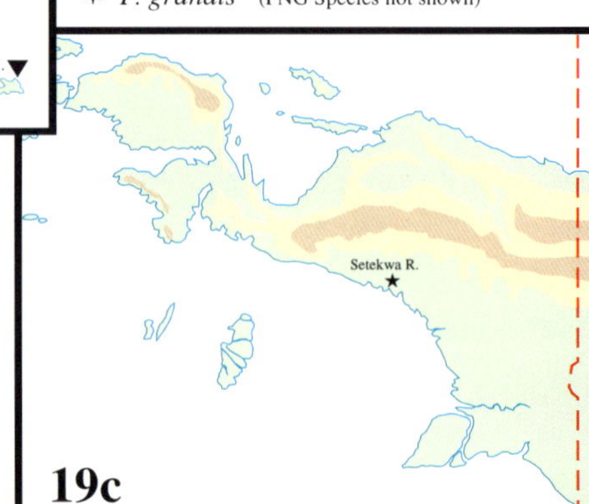

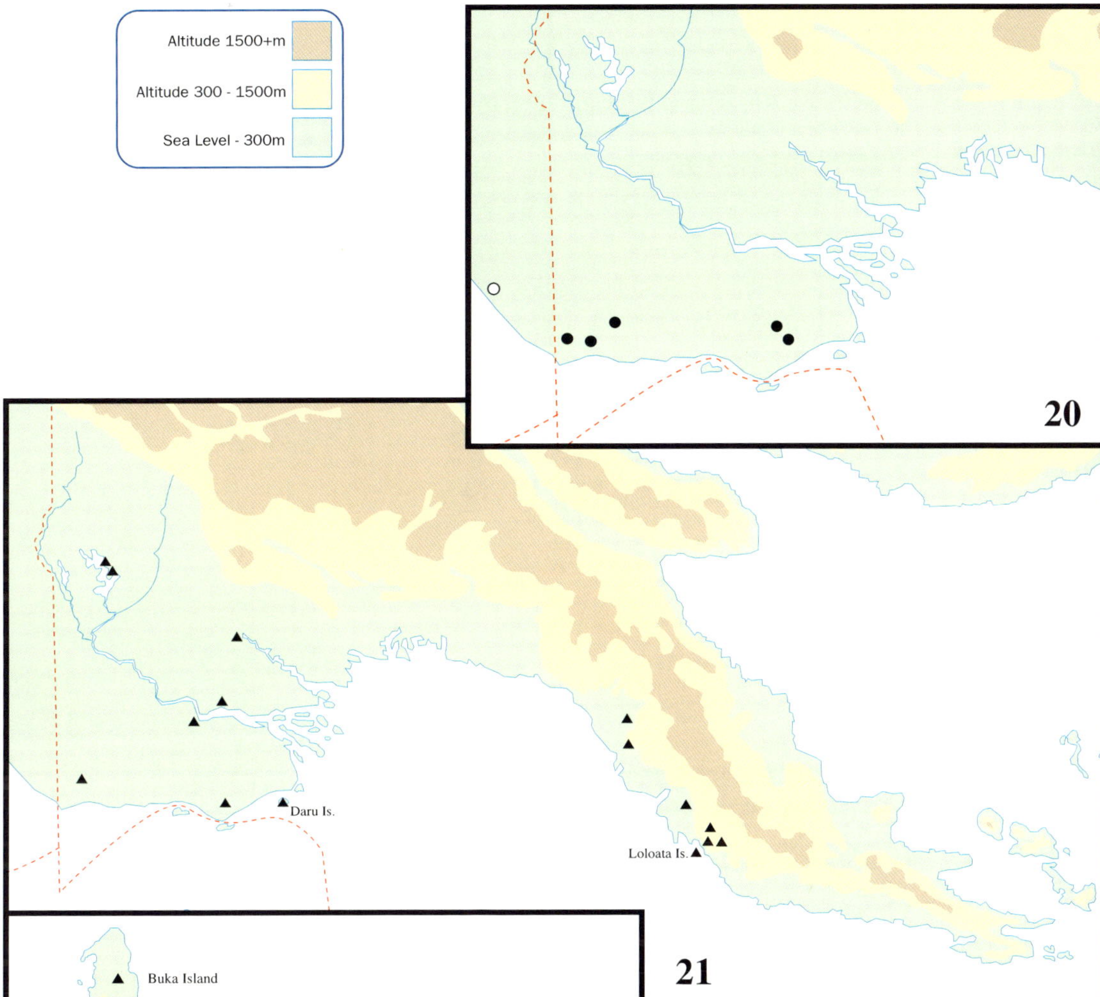

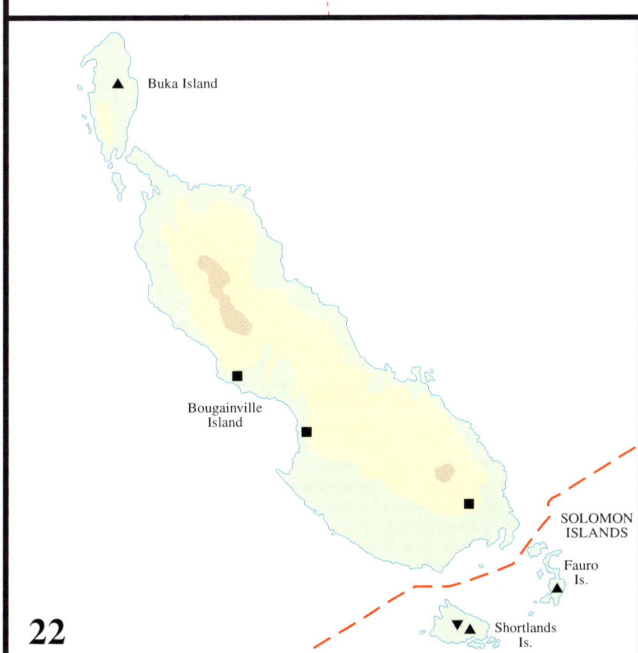

MAP 20
Distribution of the Australian small-eyed snakes, genus ***Rhinoplocephalus*** in Western Province and Irian Jaya.
Collection localities:
● *R. nigrostriatus*; ❍ *R. boschmai*

MAP 21
Distribution of the brown-headed snake, ***Furina tristis*** in Western and Central Provinces.
▲ Collection localities

MAP 22
Distribution of the North Solomons Province elapids. Collection localities:
▲ *Salomonelaps par*;
■ *Parapistocalamus hedigeri*;
▼ *Loveridgelaps elapoides*

MAP 23
Distribution of the Whipsnakes, genus ***Demansia***, in Western and Central Provinces. ● Collection localities.

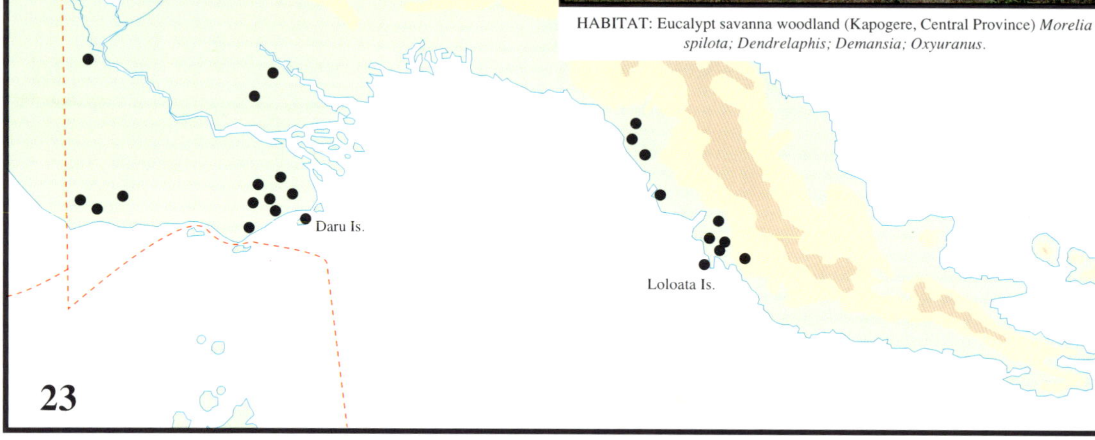

HABITAT: Eucalypt savanna woodland (Kapogere, Central Province) *Morelia spilota; Dendrelaphis; Demansia; Oxyuranus.*

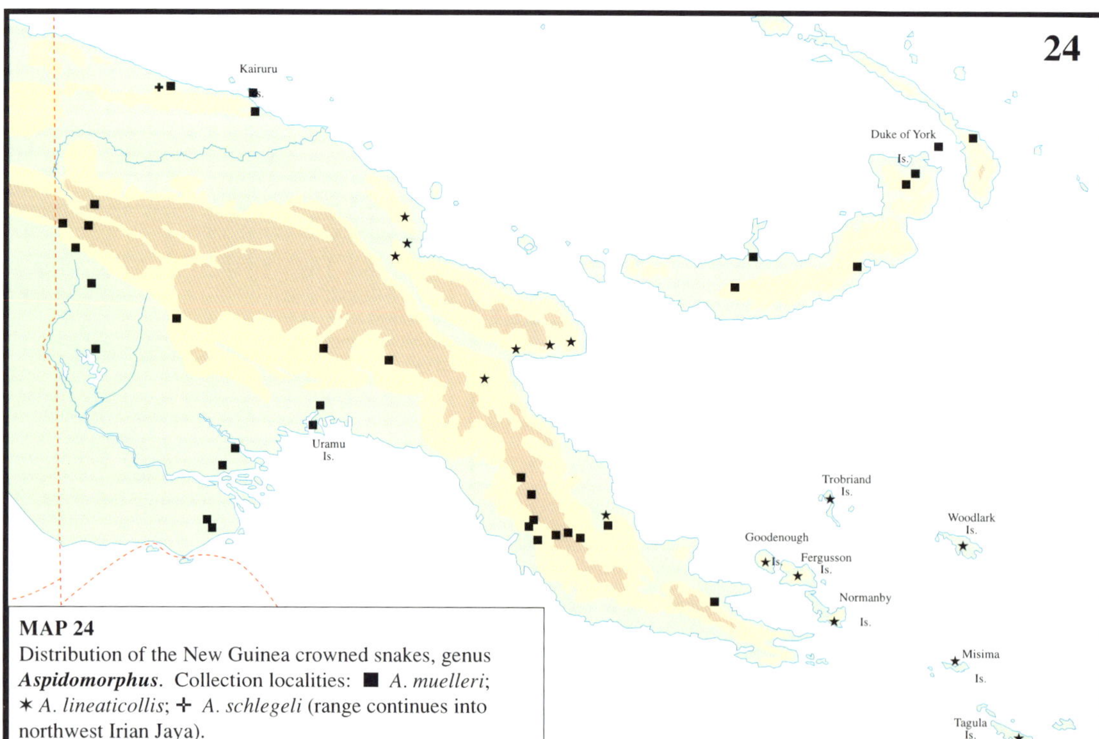

MAP 24
Distribution of the New Guinea crowned snakes, genus ***Aspidomorphus***. Collection localities: ■ *A. muelleri*; ✶ *A. lineaticollis*; ✢ *A. schlegeli* (range continues into northwest Irian Jaya).

MAP 25

Distribution of the New Guinea death adders, ***Acanthophis*** spp. ● collection localities; (!) provincial reports with imprecise locality data; (?) doubtful records; ❍ PMGH reported snakebite localities by ELISA. The extremely wide distribution of this species, or species group, is not fully illustrated by this map which is based on a limited number of collection localities, snakebite records and literature sources.

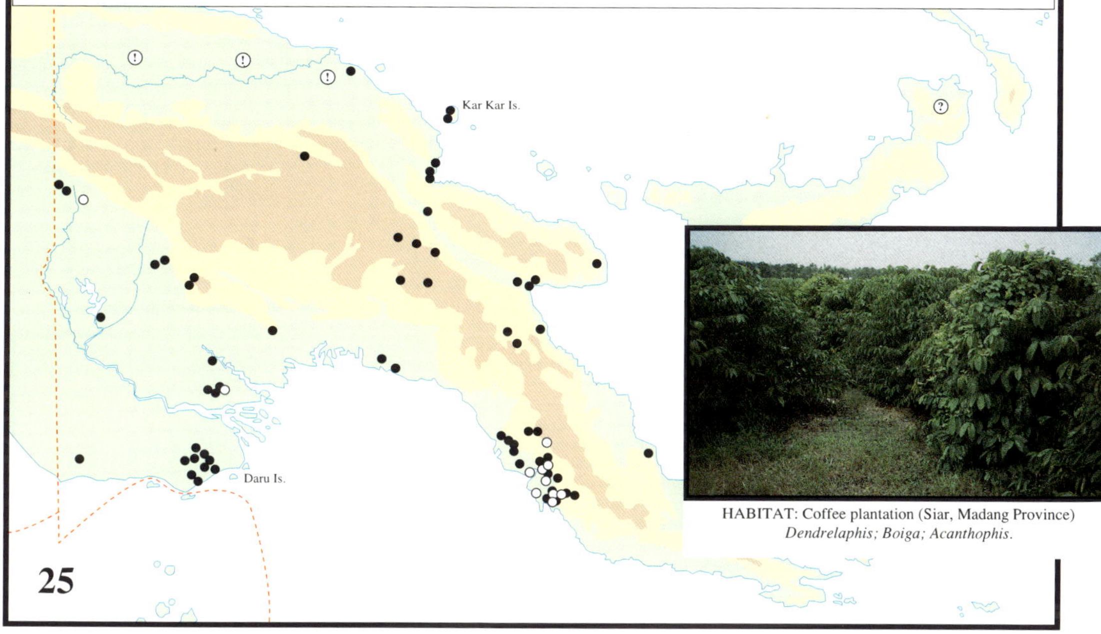

HABITAT: Coffee plantation (Siar, Madang Province)
Dendrelaphis; Boiga; Acanthophis.

MAP 26

Distribution of the New Guinea small-eyed snake, ***Micropechis ikaheka***.
● collection localities;
❍ reported snakebite localities.

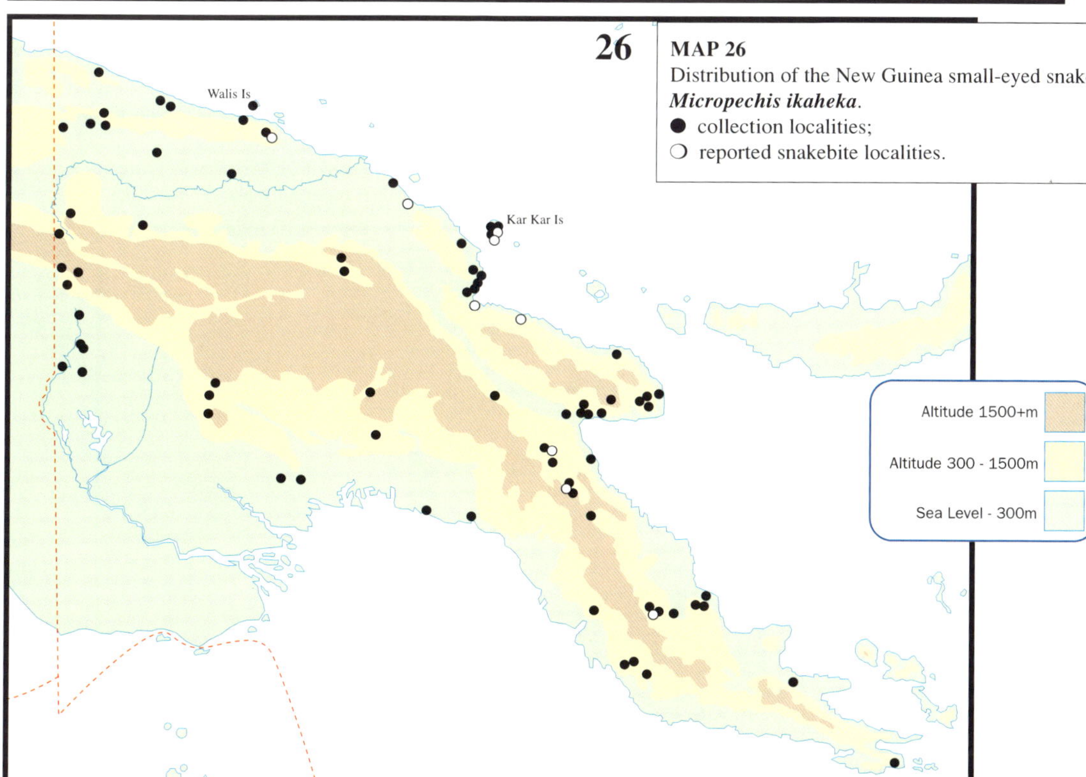

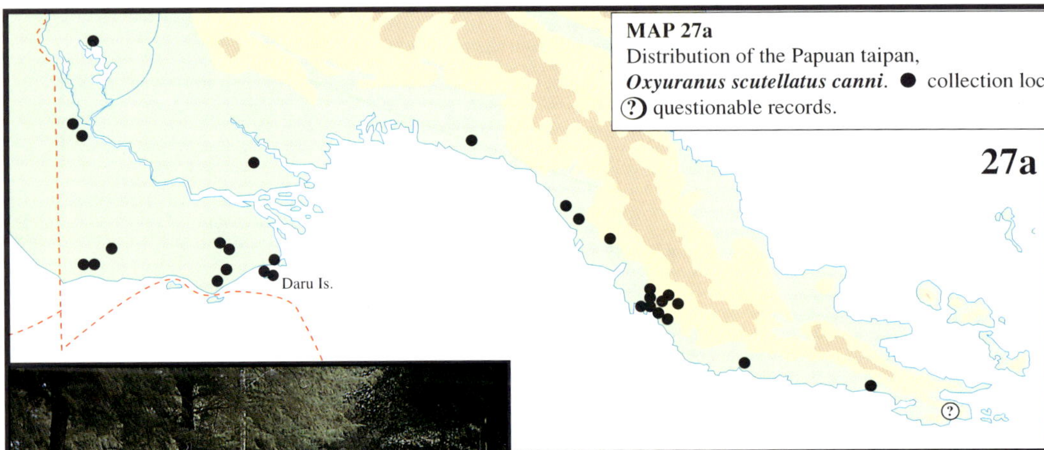

MAP 27a
Distribution of the Papuan taipan, ***Oxyuranus scutellatus canni***. ● collection localities; ? questionable records.

HABITAT: Urban situations (Ela Beach, Port Moresby, NCD) *Oxyuranus* (a dead 2m taipan was found on the Ela Beach road).

MAP 27b
Distribution of the Papuan taipan, ***Oxyuranus scutellatus canni***.
❍ PMGH reported snakebite localities by ELISA.

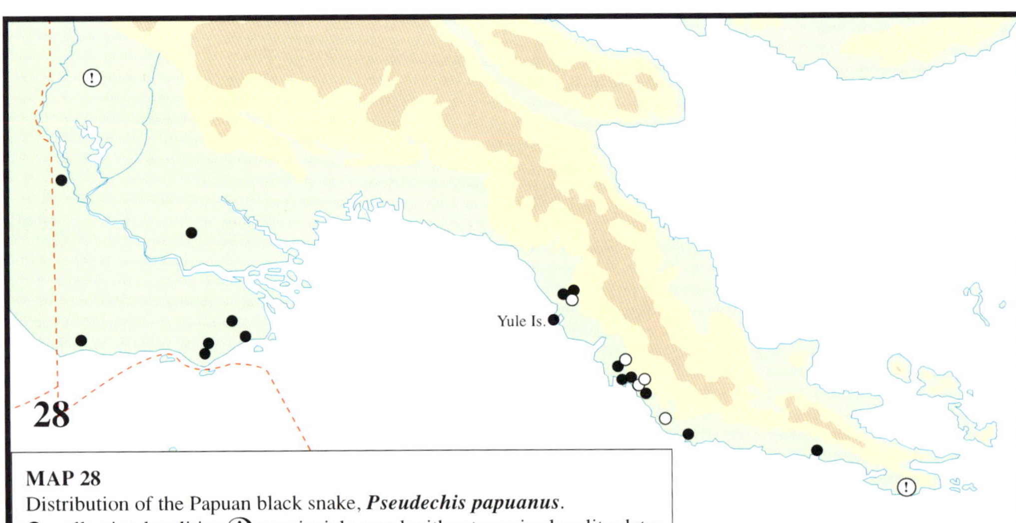

MAP 28
Distribution of the Papuan black snake, ***Pseudechis papuanus***.
● collection localities; ! provincial record without precise locality data;
❍ PMGH reported snakebite localities by ELISA.

MAP 29
Distribution of the king brown snake, ***Pseudechis australis***.
▲ Irian Jaya collection localities;
(?) possible Western Province sighting (Parker 1982).

MAP 30
Distribution of the eastern brown snake ***Pseudonaja* cf. *textilis***.
● collection localities;
❍ PMGH suspected snakebite localities by ELISA.

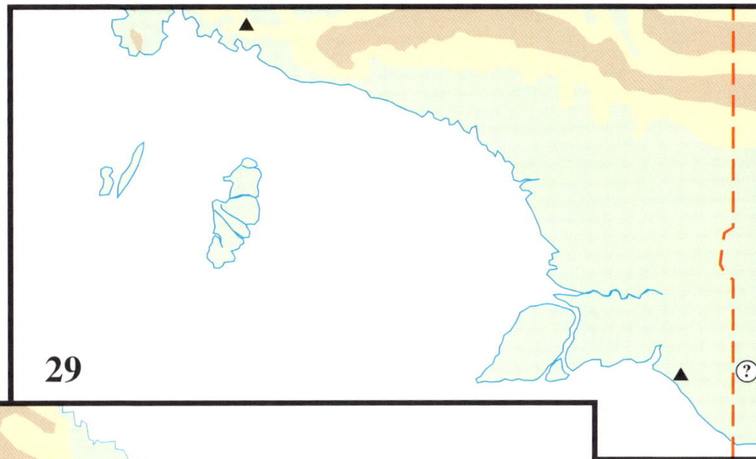

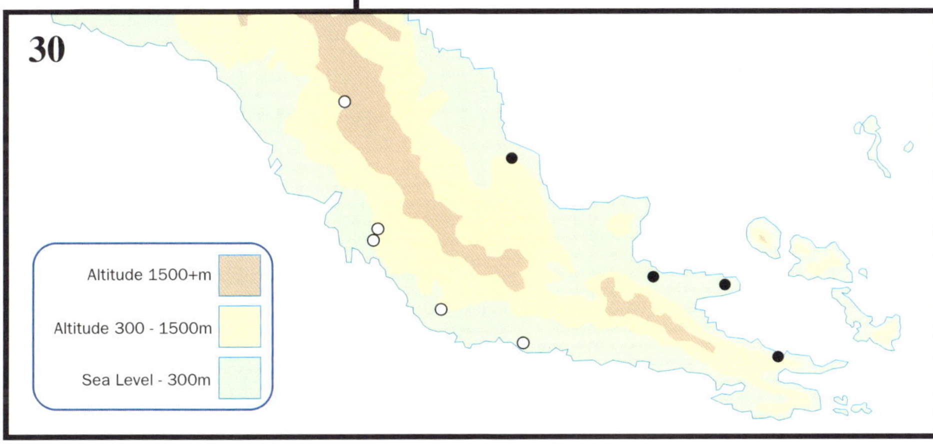

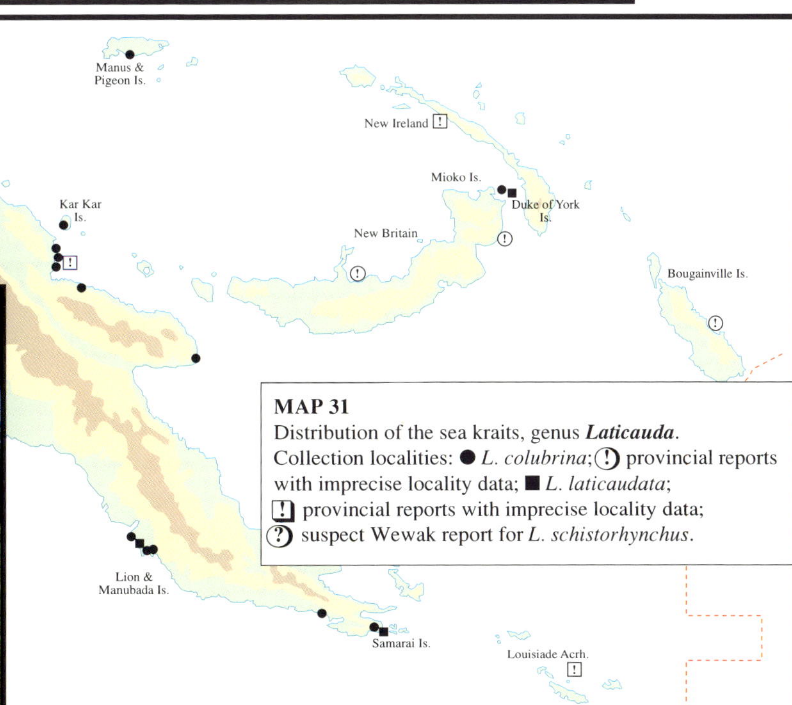

HABITAT: Rocky coastline (Nagada Harbour, Madang Province). *Acrochordus granulatus; Laticauda colubrina*.

MAP 31
Distribution of the sea kraits, genus ***Laticauda***.
Collection localities: ● *L. colubrina*; (!) provincial reports with imprecise locality data; ■ *L. laticaudata*;
[!] provincial reports with imprecise locality data;
(?) suspect Wewak report for *L. schistorhynchus*.

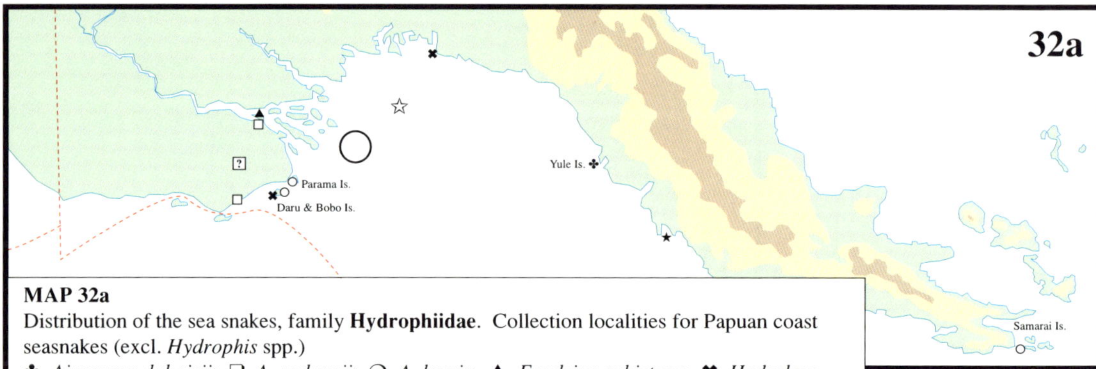

MAP 32a
Distribution of the sea snakes, family **Hydrophiidae**. Collection localities for Papuan coast seasnakes (excl. *Hydrophis* spp.)
✤ *Aipysurus duboisii*; ❑ *A. eydouxii*; ❍ *A. laevis*; ▲ *Enydrina schistosa*; ✖ *Hydrelaps darwinsensis*; ★ *Lapemis hardwickii*; [?] probable *Enhydrina schistosa* (Oriomo River at Old Zim); ☆ imprecise 'Gulf of Papua' records for *A. laevis, E. schistosa* and *L. hardwickii*; ◯ Fly River mouth ENE of Daru trawling locality for *Acalyptophis peronii; Aipysurus eydouxii*; *A. laevis; Astrotia stokesii; Disteria kingii* and *D. major*.

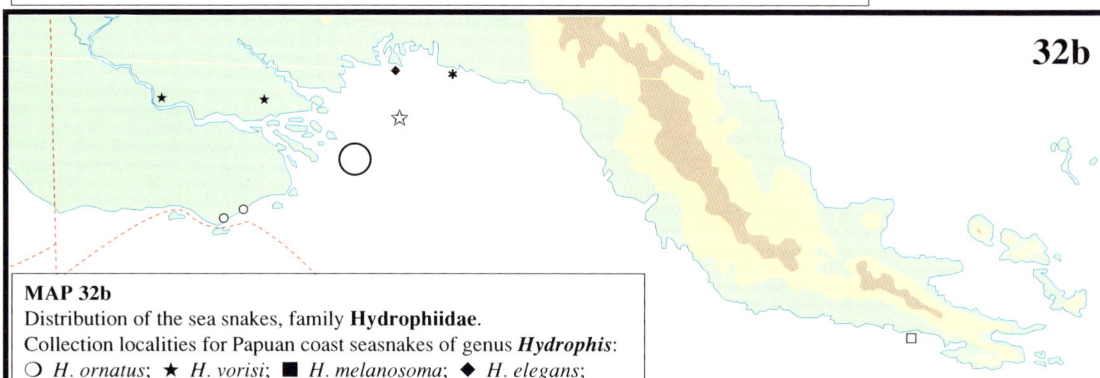

MAP 32b
Distribution of the sea snakes, family **Hydrophiidae**.
Collection localities for Papuan coast seasnakes of genus ***Hydrophis***:
❍ *H. ornatus*; ★ *H. vorisi*; ■ *H. melanosoma*; ◆ *H. elegans*; ✱ *H. pacificus*; ☆ imprecise 'Gulf of Papua' records for *H. atriceps* and *H. gracilis*; ◯ 'Fly River mouth' 60-80km ENE of Daru trawling locality for *H. elegans; H. mcdowelli* and *H. ornatus*.

MAP 32c
Distribution of the sea snakes, family **Hydrophiidae**. Collection localities for Momase and Bismarck coast seasnakes:
▼ *Enhydrina zweifeli*; (▼) *Enhydrina* sp. Ramu River snakebite fatality, possibly *E. zweifeli*; ✶ *Lapemis hardwickii*; ☆ *Lapemis hardwickii*, without precise locality data; ◆ *Hydrophis elegans*; [!] *H. ornatus* and *H. pacificus* without precise locality data; [P] *Pelamis platurus* without precise locality data.

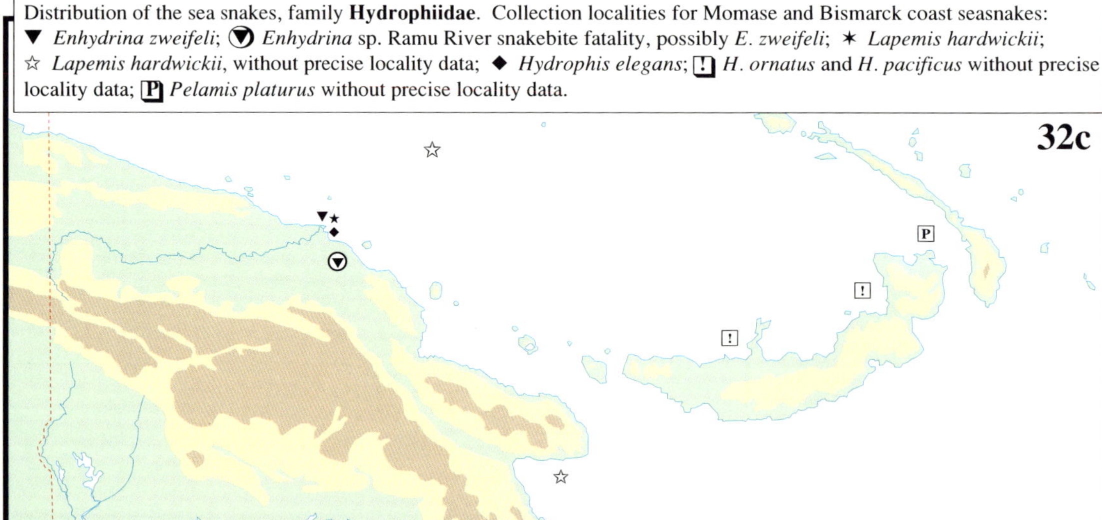

CHECKLIST OF DANGEROUS SNAKES BY PROVINCE

Central & NCD:	TAIPAN widespread coastal, excl. Yule Is. PAPUAN BLACKSNAKE coastal, esp. Mekeo, Yule Is. & s.e. Kupiano EASTERN BROWNSNAKE widespread, scattered, Tapini to Kupiano DEATH ADDERS widespread incl. mountains to Woitape SMALL-EYED SNAKE coastal, rare
Enga:	DEATH ADDERS widespread
Eastern Highlands:	DEATH ADDERS widespread, dwarf race at Henganofi SMALL-EYED SNAKE rare
East New Britain:	NO HIGHLY DANGEROUS SPECIES
East Sepik:	DEATH ADDERS widespread SMALL-EYED SNAKE common
Gulf:	TAIPAN east of Kivori, otherwise scattered PAPUAN BLACKSNAKE probable DEATH ADDERS probably widespread SMALL-EYED SNAKE rare
Madang:	DEATH ADDERS widespread SMALL-EYED SNAKE common, esp. Kar Kar Is.
Manus:	NO DANGEROUS SPECIES.
Milne Bay:	TAIPAN south coast west of Samarai PAPUAN BLACKSNAKE south coast west of Samarai EASTERN BROWNSNAKE Dogura, Cape Vogel & Moi Biri Bay SMALL-EYED SNAKE not common
Morobe:	DEATH ADDERS widespread, giant race in Markham Valley SMALL-EYED SNAKE not common
New Ireland:	NO HIGHLY DANGEROUS SPECIES
Northern Solomons:	NO HIGHLY DANGEROUS SPECIES
Oro:	EASTERN BROWNSNAKE coastal, Embogo & Popondetta DEATH ADDERS widespread SMALL-EYED SNAKE common
Simbu:	DEATH ADDERS widespread up to 1800m SMALL-EYED SNAKE localised

Southern Highlands:	DEATH ADDERS widespread SMALL-EYED SNAKE localised
Western:	TAIPAN south of Lake Murray & Balimo incl. Daru Is. PAPUAN BLACKSNAKE south of Lake Murray & Balimo DEATH ADDERS widespread SMALL-EYED SNAKE localised
Western Highlands:	DEATH ADDERS widespread
West New Britain:	NO HIGHLY DANGEROUS SPECIES
West Sepik:	DEATH ADDERS widespread, giant race in Sepik Valley SMALL-EYED SNAKE common

GAZETTEER

Gazetteer to localities referred to in the distribution sections. It has not been possible to locate some place names (*see* list at end of Gazetteer). Abbreviations: NCD (National Capital District around Port Moresby); Arch. (Archipelago); Is. (Island); Ids. (Islands); Mt. (Mount); Mts. (Mountain range); R. (River). The map references should be accurate to within 5 minutes.

	Lat.	Long.
9 Mile (NCD)	9.30	147.15
14 Mile	9.30	147.15
Abam	8.55	143.10
Abavi Is.	10.10	148.45
Adalbert Mts.	4.50	145.30
Admiralty Ids.	2.05	147.00
Afore	9.10	148.20
Agaun	9.55	149.20
Aitape	3.10	142.20
Aiyura	6.20	145.55
Albert Edward, Mt.	8.25	147.25
Alexishafen	5.05	145.45
Alotau	10.20	150.25
Ama	4.15	141.35
Amanab	3.35	141.10
Amazon Bay	10.20	149.20
Arafura Sea	10.00	141.00
Aramia R.	7.20	142.15
Arawa	6.15	155.30
Arawe Is.	6.10	148.40
Astrolabe Bay	5.25	145.50
Astrolabe Mts.	9.30	147.25
Awar	4.10	144.50
Awelkon	5.40	147.50
Babase Is.	4.00	153.40
Bagabag Is.	4.50	146.15
Baiawa	9.35	149.25
Baimuru	7.30	144.50
Baiyer R.	5.30	144.10
Balamuk	9.05	141.15
Balimo	8.00	142.55
Baluan Is.	2.30	147.15
Bamu R.	7.50	143.20
Bamustu R.	7.55	142.30
Banz	5.50	144.35
Bensbach R.	9.00	141.10
Bereina	8.40	146.30
Bewani Mts.	3.15	141.10

	Lat.	Long.
Binaturi R.	9.00	143.00
Biniguni	10.25	150.10
Bismarck Arch.	3.00	148.00
Bismarck Mts.	5.45	145.10
Bismarck Sea	4.00	148.00
Bobo (Bristow) Is.	9.10	143.15
Boboa Is.	7.05	141.35
Boera	9.25	147.00
Bogadjim	5.25	145.45
Bogia	4.15	145.00
Bomana (NCD)	9.25	147.15
Bootless Bay	9.30	147.15
Bosavi, Mt.	6.35	142.50
Boset	7.20	141.10
Bougainville Is.	6.10	155.15
Brown R.	9.15	147.15
Buin	6.45	155.40
Buka Is.	5.15	154.40
Bulolo	7.10	146.40
Buna	8.40	148.25
Bundi	5.45	145.15
Buso R.	6.40	147.10
Busu R.	6.40	147.05
Cape Hoskins	5.25	150.30
Cape Rodney	10.10	148.20
Cape Vogel	9.40	150.05
Central Mts.	5.00	142.20
Chambri Lakes	4.15	143.05
Chimbu R.	6.00	145.00
Chuave	6.05	145.05
Coral Sea	11.30	148.00
Crown Is.	5.10	146.55
D'Entrecasteaux Arch.	9.35	150.45
Dakatana Lake	5.05	150.05
Danfu R.	4.10	153.00
Daru	9.05	143.10
Daru Is.	9.05	143.10
Daviumbu Lake	7.35	141.15

	Lat.	Long.		Lat.	Long.
Dayman, Mt.	9.50	149.15	Iambom	5.50	150.00
Deception Bay	7.40	144.35	Idam R.	4.00	141.15
Diamond, Mt.	9.30	147.20	Ihu	7.55	145.20
Dinawa	8.40	147.05	Ilimo	8.55	147.55
Dobu Is.	9.45	150.50	Inawabui	8.40	146.35
Dogura Peninsula	10.05	150.05	Jacquinot Bay	5.35	151.30
Duke of York Is.	4.10	152.25	Jimi R.	5.30	143.30
Dunantina	6.10	145.50	Kainantu	6.15	145.50
Dyaul (Djaul) Is.	3.00	150.50	Kairiru Is.	3.20	143.30
East Boroko (NCD)	9.30	147.15	Kairuku	8.50	146.30
Efogi	9.10	147.35	Kaironk R.	5.15	144.25
Ela Beach (NCD)	9.30	147.10	Kalolo	6.10	147.00
Emeti	7.45	143.15	Kanudi (NCD)	9.25	147.10
Empress Augusta Bay	6.20	155.10	Kar Kar Is.	4.40	146.00
Fane	8.35	147.05	Karimui	6.30	144.50
Fergusson Is.	9.35	154.40	Kaugere (NCD)	9.30	147.10
Fife Bay	10.25	150.05	Kavieng	2.35	150.45
Finisterre Mts.	5.50	146.05	Keravat	4.20	152.05
Finschhafen	6.35	147.50	Kerema	8.00	145.45
Fissoa	2.55	151.25	Kiamari	7.45	144.50
Fly R.	8.20	143.00	Kieta	6.15	155.30
Francisco R.	7.05	147.00	Kikori	7.25	144.15
Fulleborn Harbour	6.10	150.35	Kila Kila (NCD)	9.30	147.15
Gaire	9.35	147.20	Kimbe	5.35	150.10
Garaina	7.55	147.05	Kiriwina (Trobriand) Is.	8.30	151.05
Garu	5.20	150.05	Kitava Is.	8.40	151.20
Garua Is.	5.15	150.05	Kiunga	6.05	141.15
Gasmata Is.	6.20	150.20	Koira	9.25	148.20
Gazelle Peninsula	4.35	152.00	Kokoda	8.55	147.40
Gerehu (NCD)	9.25	147.15	Kokopo	4.20	152.30
Gigabip	5.05	141.05	Kratke Mts. (N)	6.25	145.40
Ginigolo	9.50	147.41	Kratke Mts. (S)	6.55	145.50
Girua R.	8.45	148.20	Kubor Mts.	6.10	144.30
Glabi	9.00	142.45	Kukipi	8.10	146.05
Gogol R.	5.20	145.45	Kundiawa	6.00	145.50
Goldie R.	9.20	147.20	Kunini	9.10	142.55
Goodenough Is.	9.20	150.15	Kunua	5.45	154.45
Goroka	6.05	145.25	Kup	5.55	144.50
Gusiko	6.25	147.50	Kupiano	10.05	148.10
Hall Sound	8.50	146.35	Kwagira R.	9.35	149.20
Hagen, Mt.	5.45	144.05	Kwikila	9.50	147.40
Haveri	9.30	147.20	Lae	6.45	147.00
Henganofi	6.15	145.35	Lake Dakatana	5.05	150.05
Hood Bay	10.05	147.45	Lake Daviumbu	7.35	141.15
Hoskins, Cape	5.25	150.30	Lake Murray	6.50	141.25
Hunstein R.	3.10	142.20	Laloki R.	9.15	147.00
Huon Gulf	9.00	147.50	Lamington, Mt.	8.55	148.10
Huon Peninsula	6.20	147.20	Lelet Plateau	3.20	151.55

	Lat.	Long.
Lemkamin	3.20	155.55
Leonard Schultze R.	4.15	147.20
Lion Is. (NCD)	9.30	147.10
Loani	10.35	150.35
Loloata Is. (NCD)	9.30	147.10
Long Is.	5.20	147.05
Lorengau	2.00	147.20
Los Negros Is.	2.00	147.25
Lou Is.	2.25	147.20
Louisiade Arch.	10.40	153.30
Mabaduan	6.35	140.55
Maclay Coast	5.25	147.45
Madang	5.15	145.45
Madina	2.55	155.20
Madiri	8.25	143.00
Mai Kussa R.	8.55	142.05
Manam Is.	4.05	145.05
Manubada Is. (NCD)	9.30	147.10
Manus Is.	2.05	146.50
Markham R.	6.40	146.35
Marshall Lagoon	10.05	148.15
Matalau	4.10	152.10
Mawatta	9.10	142.55
Mekeo District	8.45	146.45
Melelup	5.40	154.55
Mendi	6.10	143.35
Mioko Is.	4.10	152.25
Misima Is.	10.40	152.45
Moi Biri Bay	9.35	149.30
Moka	7.25	143.30
Morehead	8.40	141.35
Mosa	5.40	150.15
Mt. Albert Edward	8.25	147.25
Mt. Bosavi	6.35	142.50
Mt. Dayman	9.50	149.15
Mt. Diamond	9.30	147.20
Mt. Hagen (mount)	5.45	144.05
Mt. Hagen (town)	5.50	144.15
Mt. Lamington	8.55	148.10
Mt. Rawlinson	6.30	147.15
Mt. Rossel	11.20	154.10
Mt. Sisa	6.10	142.45
Mt. Victoria	8.55	147.30
Mt. Wilhelm	5.45	145.00
Mt. Yule	8.15	146.45
Muliama	4.05	153.00
Murray, Lake	7.00	141.30
Musgrave R.	9.30	147.35

	Lat.	Long.
Mussau Is.	1.30	149.40
Mutuhai	5.40	154.55
Nagada Harbour	5.10	145.50
Nakanai Mts.	5.30	151.10
Namatanai	3.40	152.25
Nanu	8.50	142.45
Napa Napa	9.25	147.05
Nasioi	6.15	155.30
National Capital District	9.30	147.10
Nemadaso	6.15	142.47
New Britain	6.00	150.00
New Hanover	2.35	150.10
New Ireland	3.20	152.00
Ningerum	5.40	141.10
Nissan (Green) Atoll	4.25	145.10
Nomad R.	6.10	142.30
Nondugl	5.50	144.45
Normanby Is.	10.00	151.10
Obo	7.35	141.20
Okapa	6.30	145.35
Old Zim	8.45	143.00
Olsobip	5.25	141.30
Omati	7.30	143.55
Orangerie Bay	10.25	149.45
Oriomo R.	8.45	143.00
Orokolo Bay	7.55	145.20
Owen Stanley Mts.	9.05	147.35
Owen Stanley Mts.	10.00	149.30
Pagwi	4.05	143.00
Pahoturi R.	9.05	142.40
Palmer Junction	5.55	141.35
Palmer R.	5.45	141.35
Papua, Gulf of	8.30	145.00
Parama Is.	9.00	143.25
Pigeon Is.	2.15	146.55
Pindiu	6.30	147.30
Pio R.	6.45	144.50
Popondetta	8.45	148.15
Porebada (NCD)	9.25	147.05
Port Moresby (NCD)	9.30	147.10
Prince Alexander Mts.	3.30	142.55
Purari R.	7.35	145.10
Rabaul	4.10	152.10
Rai Coast	5.20	145.45
Rambutyo Is.	2.20	147.50
Ramu R.	4.30	144.40
Rawlinson, Mt.	6.30	147.15
Rigo	9.50	147.30

	Lat.	Long.
Rodney, Cape	10.10	148.20
Rooke (Umboi) Is.	5.40	147.55
Rossel Is.	11.20	154.10
Rossel, Mt.	11.20	154.10
Sabusa	9.25	147.15
Saidor	5.35	146.30
Samarai Is.	10.35	150.35
Sangara	8.50	148.10
Sattelberg	6.30	147.45
Schrader Mts.	5.00	144.10
Seleo Is.	3.10	142.30
Sepik R.	4.00	144.15
Sigabaduru	9.15	142.35
Simbai	5.15	144.30
Sinaeada	9.20	149.05
Sisa, Mt.	6.10	142.45
Sogeri Plateau	9.25	147.30
Solomon Sea	7.25	152.00
St. Matthias Ids.	1.35	149.50
Star Mts.	5.00	141.00
Stettin Bay	5.30	150.15
Strickland R.	7.00	142.00
Sturt Is.	8.10	142.15
Sudest (Tagula) Is.	11.30	153.40
Tabar Ids.	2.55	152.00
Tabubil	5.15	141.15
Tagula (Sudest) Is.	11.30	153.40
Talasea	5.20	150.00
Tapini	8.20	146.55
Tari	5.50	142.55
Tatau Is.	2.45	152.00
Telefomin	5.10	141.35
Tench Is.	1.40	150.40
Teptep	6.00	146.20
Tigi Plantation	5.40	144.20
Tikam	6.05	141.15
Tinputz	5.35	155.00
Toma	4.25	152.15
Tomba	5.50	144.15
Topanos	54.0	155.00
Torokina	6.10	155.00
Torricelli Mts.	3.25	142.15
Trobriand (Kiriwina) Is.	8.30	151.05
Turama R.	7.15	143.20
Umboi (Rooke) Is.	5.40	147.55
Uramu Is.	7.35	144.35
Vaifa'a	8.40	146.35
Vailala R.	7.35	145.25
Vanapa R.	9.10	147.00
Vanimo	2.40	141.15
Veimauri R.	9.05	147.00
Victoria Mt.	8.55	147.30
Vogel, Cape	9.40	150.05
Wabag	5.30	143.40
Waghi R.	6.15	145.00
Wangbin	5.40	141.10
Wapenamanda	5.40	143.50
Waro	6.25	143.05
Wassi Kussa R.	9.00	152.05
Watut R.	6.55	146.25
Wau	7.20	146.40
Wewak	3.35	143.35
Whiteman Mts.	5.50	150.00
Wilhelm, Mt.	5.45	145.00
Willaumez Peninsula	5.05	150.05
Wipim	8.40	142.55
Woitape	8.30	147.10
Woodlark Is.	9.05	152.45
Wutung	5.35	151.30
Yalom	4.25	151.45
Yule Is.	8.50	146.30
Yule, Mt.	8.15	146.45

The following PNG localities were not found and are not plotted on the maps: WEST: Alamire; Kamagato; Kavorabip; Makapa; Wawao R.; Dagagota Reef. CENT: Bavo Is.; Kabadi; Karapi; Kaugere; Koroko/Nagore camp; Subitana; Imairu; Moroka. M.BAY: Menapi. ORO: Embogo; Lego. MOR: Ivau. MAD: Bostrom Bay. E.SPK: Narimok. SHP: Waefe; Wasu; Hurria R. SIM: Dumun; Tive Plateau. EHP: Orumba, Yaiya R. WNB: Moramora; Mt. Vale. MAN: Derimat; Lengendrowa.

Irian Jayan localities cited:

	Lat.	Long.
Andai	0.55	134.00
Arfak Mts.	1.05	136.15
Aru Ids.	6.30	134.30
Batanta Is.	0.50	130.40
Biak Is.	1.00	136.00
Digul R.	5.35	140.15
Djajapura (Jayapura)	2.35	140.45

	Lat.	Long.
Doberei Pen. (Vogelkop)	1.25	133.00
Doré	0.40	136.10
Etna Bay	3.55	134.45
Fak Fak (Onin Peninsula)	2.55	132.15
Geelvink Bay	2.30	135.25
Helwig Mts.	4.30	138.40
Idenburg R.	3.10	139.00
Jamur Lake	4.00	135.10
Jayapura (Djajapura)	2.35	140.45
Jobi (Japen) Is.	1.45	135.50
Joes Is.	1.00	131.10
Kaap Valsch	8.30	136.50
Kapoar	0.55	132.40
Kei Ids.	6.00	132.50
Kepi	6.40	139.30
Kolepom (P.F.H.Is.)	8.00	138.30
Lake Jamur	4.00	135.10
Lake Sentani	2.35	140.30
Lorentz R.	5.05	138.40
Mafoor (Numfoor) Is.	1.05	134.50
Mamberamo R.	1.35	137.55
Mansinam Is.	0.45	136.10
Manokwari	0.50	134.05
McCluer Gulf	2.40	133.40
Merauke	8.30	140.20
Mimika R.	4.20	136.40
Mios Num Is.	1.25	135.15
Misool Is.	1.50	130.20
Numfoor Is.	1.00	134.50
Onin Peninsula	0.55	132.40
Otakwa (Utakwa) R.	4.40	137.20
Poiniersbivak	2.20	138.20
Pr. Frederik Hendrick Is.	8.00	138.30
Salawatti Is.	1.05	130.55
Sentani Lake	2.35	140.30
Setekwa R.	4.40	137.20
Sorong	1.00	131.25
Triton Bay	4.00	134.10
Utakwa (Otakwa) R.	4.40	137.20
Vogelkop (Doberei Pen.)	1.25	133.00
Waigeo Is.	0.10	130.35
Wildoman R.	6.00	139.40
Wissel (Paniai) Lakes	4.00	136.20

Solomon Islands localities cited:

Fauro Is.	6.55	156.05
Shortlands Is.	7.10	155.45

GLOSSARY

Anal or **cloacal plate** - A large scale located between the *ventral plates* and the *subcaudal scales*, which covers the anal or *cloacal* opening. Since some species have a single anal plate whilst others possess a divided plate this may be an important diagnostic characteristic to aid identification of species. (See p.37 fig.17).

Anticoagulant - A component of certain venoms which prevents clotting of the blood and promotes continued bleeding (see also *coagulant*). Leakage of blood into the urine, a condition known as haematuria, causes the urine to turn red and is a sign of anticoagulant. Papuan blacksnake and king brownsnake venoms are anticoagulant.

Aquatic - Living in water, usually taken to mean freshwater as opposed to marine.

Arboreal - Able to move and live in trees and bushes.

Azygous shield - An elongate scale situated between the prefrontal scales on the anterior of the upper head in some sea kraits. This scale can be used to differentiate between *Laticauda colubrina* and *L. laticaudata* (see p.171 figs.1-3).

Bradyarrhythmias - Abnormally slow heart rate.

Bronchospasm - Constriction of the smooth muscles lining the wall of the bronchi causing wheezing as in asthma.

Cardiotoxin - A poison attacking the function of the heart. Taipan venom is cardiotoxic.

Chin shields - Two pairs of elongate scales on the underside of the chin and throat posterior to the mental scale and lst infralabial scales (see p.35 fig.3).

Cloaca - The common genital and excretory opening located on the underside at the base of the tail (*see anal* or *cloacal plate*).

Coagulant or **procoagulant** - A component of certain venoms which promotes clotting of the blood within the blood vessels. This rapidly uses up all of the clotting factor in the blood and leads to prolonged bleeding similar to that caused by an *anticoagulant*. Early signs include continual bleeding from the site of the bite and from any razor-cuts to the bitten limb which may lead to excessive blood loss and are potentially dangerous. Taipan and eastern brownsnake venoms are strongly coagulant.

Coagulopathy - Disturbance in blood clotting, resulting from depletion of clotting factors such as fibrinogen. In severe coagulopathy the blood becomes incoagulable.

Concave - Curved inwards

Crepuscular - Active at dusk and dawn (see also *diurnal* and *nocturnal*).

Cytotoxin or **proteolytic** - A venom or component of a venom which causes the destruction of tissue resulting in necrosis or tissue death. A number of dangerous PNG snakes are thought to posses cytotoxic qualitites in their venoms.

Diurnal - Active by day (see also *nocturnal* and *crepuscular*).

Dorsal and **dorsum** - The upper surface; the back.

Dorsal scales - The relatively small scales across the back, usually arranged in an odd number of longitudinal rows including the central or vertebral row (see p.37 fig.16). Dorsal scales may be smooth or *keeled*.

Dorsolateral - The upper flank or upper side (see *dorsal* and *lateral*).

ELISA - Enzyme Linked Immunosorbent Assay, a method of determining which snake species was responsible for a particular snakebite by comparison of the victim's blood serum with venoms from known snake specimens.

Endemic species - A species (or genus etc.), the distribution of which is confined to a definable geographical or political location ie. New Guinea, PNG, Misima Island, Rossel Island, etc.

Epiphytic - A plant which grows above the ground, often on the trunk of a tree from which it gains support but not nourishment; usually an orchid or a fern.

Fossorial - Burrowing; living below the ground or in leaf-litter.
Frontal scale - The single scale situated in the centre of the top of the head between the eyes (see p.35 fig.1).
Glottis - The muscular airway in the inside of the mouth which permits continued breathing whilst large prey is swallowed. The glottis may also be closed to prevent drowning whilst a snake is submerged.
Gular fold - An area of loose skin between the *chin sheilds* which allows expansion of the lower jaws to swallow large prey (see p.35 fig.3). Also termed *mental groove.*
Haematuria - Passage of blood in the urine.
Haemoglobinaemia - Passage of the blood pigment (haemoglobin) in the urine giving a pinkish, reddish or blackish colour depending on the amount of haemoglobin present.
Haemolytic - A component of a venom (a haemotoxin) which destroys red blood corpuscles, the result of which is haemoglobin (the oxygen carrying component of the blood) being lost in the urine which turns red, a condition known as haemoglobinuria. Most of the dangerous snakes of PNG possess haemolytic components in their venoms.
Haemorrhagic - A component of a venom which causes internal bleeding into the tissues and subsequent blood loss due to damage to blood vessels. Papuan blacksnake venom is believed to be partially haemorrhagic.
Haemostatic disturbances - Excessive bleeding and defective clotting of the blood.
Imbricate - Overlapping, as in scales of most snakes (see p.179 fig.12) (see also *juxtaposed*).
Infralabial scales - The scales of the lower lip not counting the very front scale or mental scale (see p.35 fig.3).
Interspecifically - Between different species (see *intraspecifically*).
Internasal scales - A pair of scales on the top of the snout, situated between the nasal scales and anterior to the prefrontal scales (see p.35 fig.1).
Interstitial skin - The skin between the scales which allows expansion of that part of the body during swallowing or display. For threat display interstitial skin is often brightly coloured and in contrast to the scale colouration.
Intertidal zone - The area between the mean high and low tide levels.
Intraspecifically - Within a single species (see *interspecifically*).
Juxtaposed - Non-overlapping as in the scales of seasnakes which cannot move on land (see p.179 fig.11) (see also *imbricate*).
Keeled scales - Scale with a ridge down the centre resembling the keel on the underside of a boat and giving the snake a rough texture. Especially evident in snakes such as the keelbacks and ground boas (see p.37 fig.16).
Lateral - The flank or sides.
Lingual fossa - The small opening in the rostral scale which permits the tongue to be extended without opening the mouth (see p.35 fig.4).
Longitudinal - running along the length of the body.
Loreal scale - The scale between the *preocular scale* and the *nasal scale* (see p.35 fig.2). This scale is the singular most important scale for determining whether a snake is venomous or non-venomous in PNG. All elapids (venomous *terrestrial* snakes) lack a loreal scale while all other species (except the white-bellied mangrove snake) possess one or several loreal scales.
Lymphadenopathy - Painful enlargement of the lymph glands.
Mental groove - See *gular fold.*
Mental scale - The small scale at the front of the lower jaw and in contact with the *infralabial scales* (see p.35 fig.3).
Myogobinuria - The passage of muscle pigment (myoglobin) in the urine causing a dark brown or 'coca-cola' colour.

Myolytic - Causes destruction of muscle (see *rhabdomyolytic*).
Myotoxin - A poison which attacks the muscle, particularly the pulmonary (lungs) and cardiac (heart) muscles (see *myolytic* and *rhabdomyolytic*).
Nasal scale - The scale surrounding the nostril (see p.35 fig.2).
Nephrotoxic - Causes damage to the kidneys.
Neurotoxin - A poison which causes paralysis of the voluntary muscles including those controlling breathing so death may be caused through asphyxsia (an inability to breathe). All dangerous snakes in PNG possess primarily neurotoxic venoms and the effects of some of these neurotoxins are more easily reversed than others (see also *pre-* and *post-synaptic*).
Nocturnal - Active by night (see also *diurnal* and *crepuscular*).
Oblique scale rows - Scales arranged in rows running at an acute angle at approx. 30° to the horizontal as in the treesnakes *Boiga* and *Dendrelaphis* (see p.91 fig.5). Most other species have their scales arranged in 'normal rows' at approx. 45° to the horizontal.
Ontogenetic changes - Any changes in patterning or colouration or morphology with increasing age or maturity i.e. between juvenile and adult. Especially evident in the green tree python.
Ophthalmoplegia - Paralysis of the muscles responsible for moving the eyes.
Oviparous - The reproductive condition when the female lays thin fertilised leathery shelled eggs which may either be incubated externally by the female in her coils (some pythons) or abandoned to incubate under environmental conditions (most oviparous species). Most of the land snakes of PNG and the sea kraits are oviparous (see also *ovoviviparous*).
Ovoviviparous - The reproductive condition, often misnamed 'live-birth', when the female retains the developing eggs within her body. They hatch in the oviduct and are deposited as fully formed babies (known as neonates). The boas, the file snake, the death adders and the true seasnakes are the only ovoviviparous snakes to be found in PNG. True live-birth or viviparity is where the female possesses a placenta which nourishes the embryo until its birth. Only a few of the more advanced reptiles are viviparous. (see also *oviparous*).
Parietal scales - A pair of large scales positioned on top of the head, posterior to the single *frontal scale* (see p.35 fig.1).
Parthenogenetic - Existing as female-only populations capable of producing fertile eggs without the presence of a male. Several lizards are known but only a single snake, the Brahminy blindsnake, is parthenogenetic.
Pelvic spurs - A pair of sharp projections found on either side of the cloaca or anal opening in boas and pythons only (see p.38 fig.21). They are the only external evidence of the presence of vestigal hindlimbs and with the internal remnants of the pelvic girdle they demonstrate the lizard ancestry of these more primitive snake groups. Usually larger in males, the claws may even be absent in females of some populations eg. the boas. Males stroke the females with these claws prior to and during mating, often leaving considerable scars.
Plumbeus - The colour of lead, blue-grey.
Post-synaptic neurotoxin - A *neurotoxin* which works on the muscle side of the synaptic gap (*see* diagram) and causes a form of paralysis which is more easily reversed using antivenom than that caused by a *pre-synaptic neurotoxin*. The venom of the death adder is post-synaptic.
Postocular scales -The scale or scales immediately behind the eye (see p.35 fig. 2) .
Prefontal scales - A pair of scales on the anterior top of the head, anterior to the *frontal scale* (see p.35 fig.1).
Pre-synaptic neurotoxin- A *neurotoxin* which works on the nerve side of the synaptic gap and causes a form of paralysis which is more difficult to reverse using antivenom than that caused by a *post-synaptic neurotoxin*. The venom of the taipan, the blacksnakes and the brownsnakes are pre-synaptic.
Preanal pores - Small pores located in a semi-circle just anterior to the cloaca in certain lizards, notably the pygopodid or snake-lizards (see p.48 figs. 3-4).

Prehensile tail - A tail which can be used to secure the animal to an object and which serves as a limb.
Preocular scales - The scale or scales immediately in front of the eye (see p.35 fig.2).
Ptosis - Drooping of the eyelids resulting from paralysis of the muscles which normally elevate them. This is one of the first signs of neurotoxic poisoning.
Rhabdomyolysis - Generalised destruction of skeletal muscle with release of muscle pigment (myoglobin), muscle enzymes and other constituents.
Rhabdomyolytic or **myolytic** - The effect of a venom which attacks skeletal muscle resulting in the loss of myoglobin (muscle protein) into the urine and possible kidney failure. The venoms of seasnakes and possibly the small-eyed snake are particularly rhabdomyolytic and other snake venoms may also cause rhabdomyolysis to a lesser degree.
Rostral scale - The single scale situated at the very front of the upper jaw, in front and in contact with the *supralabial* and *nasal scales* (see p.35 fig.4).
Scutes - The large, well defined scales of the head (see p.35 figs.1-4).
Septal ischaemia - Inadequate blood supply to the septum dividing the two ventricles of the heart.
Subcaudal scales - The scales under the tail running from the anal plate to the tail tip. These scales may be paired, single or a combination of the two and are important diagnostic characteristics to aid in the identification of species (see p.37 fig.17).
Subocular scales - Scales which separate the eye from the supralabial scales (see p.124 fig.4). Present in death adders *Acanthophis* spp.
Supralabial scales - The scales of the upper lip not counting the very front scale or *rostral scale* (see p.35 fig.2). The number of supralabials on each side, counting from front to back, noting which, if any, are in contact with the eye, is an important diagnostic characteristic aiding identification of species.
Supraocular scales - Enlarged scales directly above the eye (see p.35 fig.1), particularly important in the recognition of the taipan which has protruding 'shelf-like' supraoculars (see p.124 fig.5).
Syncope - Fainting, collapse and unconsiousness, usually resulting from a fall in blood pressure.
Synonymy - A scientific name of a former species which is no longer considered valid as a distinct species and which is therefore placed within another, earlier named, species.
Tachycardia - Abnormally rapid heart rate.
Temporal scales - A series of scales posterior to the *postocular scales* and separating the *parietal scales* from the *supralabial scales* (see p.35 fig.2). Usually arranged as two rows, antrior and posterior temporals (see *temporolabial scale*).
Temporolabial scale - A large scale which projects downwards between the 5th and 6th *supralabials* (see p.125 fig.7). Present in most PNG elapids including the dangerous species but excluding the eastern brownsnake.
Terrestrial - Living on land.
Thrombocytopenia - Decrease in the number of circulating blood platelets or thrombocytes.
Thrombosis - A condition where blood thickens and clots within a blood vessel; a potentially fatal result of certain venoms.
Transverse - Running diagonally across the body.
Tuberculate - Possessing small raised tubercules or fleshy projections.
Urticaria - A raised, itchy rash, usually with white lumps against a red background, caused by leakage of plasma into the skin as a result of an allergic reaction.
Ventral and **venter** - The lower surface, the underbelly.
Ventral plates - The broad belly scales running from the underside of the chin to the *anal plate* (see p.37 fig.17).
Ventrolateral - The lower flank or lower side (see *ventral* and *lateral*).
Vertically elliptical - An eye like a cat.
Vestigal - Rudimentary remnants of organs which are fully developed in other organisms. The remnants of the hind limbs and pelvic bones are found in boas, pythons and whales (see p.38 fig.21).

REFERENCES AND FURTHER READING

General: Loveridge 1948; Slater 1954; 1959; 1960; Parker 1982; Cogger, Cameron & Cogger 1983; McDowell 1984; Wilson & Knowles 1988; Shine 1991b; Cogger 1992; Ehmann 1993a

Pygopodidae: Shine 1986a; 1986d; Patchell & Shine 1986b

Lialis: Rooji N. de 1915; Kinghorn 1924; Brongersma 1953a; Neill 1957; Kluge 1974; Weber & Werner 1977; Patchell & Shine 1986a

Typhlopidae: McDowell 1974; Ehmann & Bamford 1993

Ramphotyphlops: McDowell 1974; Shine & Webb 1990; Webb & Shine 1993; Wallach 1995 *in press.*

Typhlops: McDowell 1974; Wallach & Wynn *in prep.*

Boidae: Stull 1932a

Boinae: McDowell 1979

Candoia: Stull 1932b; Stickel & Stickel 1946; Bustard 1969; McDowell 1979; Wynn & Zug 1985; Harlow & Shine 1992; O'Shea 1995

Pythoninae: McDowell 1975; Underwood & Stimson 1990; Kluge 1993; Ehmann 1993b; Shine & Slip 1990; Barker & Barker 1994; *in prep.*

Apodora: Brongersma 1953b; 1956; McDowell 1975; Barker & Barker *in prep.*

Bothrochilus: McDowell 1975; Barker & Barker *in prep.*

Leiopython: Brongersma 1953b; 1956; McDowell 1975; Barker & Barker 1994; *in prep.*

Liasis: McDowell 1975; Barker & Barker 1994; *in prep.*

Morelia: Brongersma 1953b; 1956; 1969; McDowell 1975; Barker & Barker 1994; *in prep.*

Acrochordidae: McDowell 1979; Shine & Houston 1993

Acrochordus: McDowell 1979; Shine 1985; 1986b; 1986c; 1986e; Houston & Shine 1993; 1994a; 1994b; 1994c

Colubridae: Shine 1991a; Ehmann 1993c

Boiga: Fritts 1988; Fritts *et al* 1990; Shine 1991a

Cantoria: Gyi 1970

Cerberus: Gyi 1970; Shine 1991a

Dendrelaphis: McDowell 1984; Shine 1991a

Enhydris: Gyi 1970; Shine 1991a

Heurnia: Gyi 1970

Fordonia: Gyi 1970; Shine 1991a

Myron: Gyi 1970; Shine 1991a

Stegonotus: McDowell 1972a; 1984; Shine 1991a

Tropidonophis: Malnate 1960; Malnate & Underwood 1988; Shea 1990; Shine 1991a

Elapidae: Worrell 1961; Slater 1968; Cogger 1971; Mengden 1983; Hutchinson 1990; O'Shea 1990; Golay 1993; Shea *et al* 1993; David & Ineich *in prep.*

Acanthophis: Campbell 1966; Shine 1980a; Storr 1981b; McDowell 1984

Aspidomorphus: Brongersma 1934; de Rooji 1917; McDowell 1967

Demansia: Shine 1980b; Storr 1978

Furina: de Rooji 1917; Shine 1981

Laticauda: Smith 1926; Brongersma 1956; Cogger 1975; McDowell 1967; 1972; Burger & Natsumo 1974; Kharin 1984a; Heatwole & Guinea 1993; McCarthy 1993

Loveridgelaps: McCoy 1980; McDowell 1970

Micropechis: Lesson 1830; Blasco & Hornabrook 1972; McDowell 1970; Hudson 1988; Hudson & Pomat 1988; O'Shea 1995

Oxyuranus: Slater 1956; 1959; Campbell 1967a; Shine & Covacevich 1983

Parapistocalamus: McCoy 1980; McDowell 1967; 1969; Williams & Parker 1964
Pseudechis: Loveridge 1927; Campbell 1967b; McDowell 1970; Smith 1982; Mengden, Shine & Moritz 1986; Shine 1987
Pseudonaja: McDowell 1967; Shine 1989
Rhinoplocephalus: Brongersma & Knaap-van Meeuwen 1964; McDowell 1970; Parker 1972; Shine 1988; Storr 1981a
Salomonelaps: McCoy 1980; McDowell 1970
Toxicocalamus: Bogert & Matalas 1945; de Rooji 1917; McDowell 1969a

Hydrophiidae: Smith 1926; McDowell 1969b; 1972b; Burger & Natsumo 1974; Cogger 1975a; 1975b; 1992; Dunson 1975; Culotta & Pickwell 1993; Heatwole & Cogger 1993; McCarthy 1993; David & Ineich *in prep.*
Acalyptophis: Brongersma 1956
Aipysurus: Kharin 1981
Astrotia: McDowell 1972b
Disteria: McDowell 1972b
Enhydrina: McDowell 1972b; Kharin 1985
Hydelaps: Brongersma 1956
Hydrophis: McDowell 1972b; Kharin 1983; 1984b
Lapemis: Voris 1969
Pelamis: Hecht, Kropach & Hecht 1974

Snakebite:
Colubrine: Fritts 1990; Fritts *et al* 1990
Elapine: Campbell 1966; 1967a; 1967b; 1969; 1976; Campbell & Young 1961; Brian & Vince 1987; Currie *et al* 1988a; 1988b; 1990; Hudson & Pomat 1988; Malasit *et al* 1986; O'Shea 1990; Sutherland 1983; Sutherland *et al* 1979; Warrell 1987; 1990; Watt *et al* 1986; Kamiguti *et al* 1994; Lalloo *et al* 1994
Hydrophine: Sutherland 1983; Hudson & Frohm 1986: Hudson & Pomat 1988; Reid 1979
Antivenoms: Theakston & Warrell 1991

Sources of information:

Barker D.G. & T.M. Barker 1994 *Pythons of the World, Volume I: Australia.* The Herpetocultural Library, Advanced Vivarium Systems, Lakeside, California. pp. 171.

" " *in prep. Pythons of the World, Volume II: Indonesia and New Guinea.* The Herpetocultural Library, Advanced Vivarium Systems, Lakeside, California.

Blasco P. & R.W. Hornabrook 1972 A neglected but potentially dangerous New Guinea snake - the small-eyed snake (*Micropechis ikaheka*). *PNG Med. J.* **15**: 155-156.

Bogert C.M. & B.L. Matalas 1945 Results of the Archbold Expditions No. 53: A review of the elapid genus *Ultrocalamus* of New Guinea. *Amer. Mus. Novit.* **1284**: 1-8.

Brian M.J. & J.D. Vince 1987 Treatment and outcome of venomous snake bite in children at Port Moresby General Hospital, Papua New Guinea. *Trans. Roy. Soc. Trop. Med. Hyg.* 1987 **81**: 850-852.

Brongersma L.D. 1934 Contributions to Indo-Australian herpetology. *Zool. Meded. Leiden* **17** (3-4): 161-251.

" 1953a Notes on New Guinean Reptiles and Amphibians I. *Proc. Ned. Acad. Wetn.* **56** (2): 137-142.

" 1953b Notes on New Guinean Reptiles and Amphibians II. *Proc. Ned. Acad. Wetn.* **56** (3): 317-325.

" 1956 Notes on New Guinean Reptiles and Amphibians IV. *Proc. Ned. Acad. Wetn.* **59** (4):

447-453.

" 1956 Notes on New Guinean Reptiles and Amphibians V. *Proc. Ned. Acad. Wetn.* **59** (5): 599-610.

" 1969 Notes on *Liasis boeleni* Brongersma. *Proc. Ned. Acad. Wetn.* **72** (2): 124-128.

" & M.S. Knaap-van Meeuwen 1964 On a new species of *Denisonia* (Reptilia, Serpentes) from New Guinea. *Zool. Meded. Leiden* **39**: 550-554.

Burger W.L. & T. Natsuno 1974 A new genus for the Arafura smooth seasnake and redefinitions of other seasnake genera. *The Snake* **6**: 61-75.

Bustard H.R. 1969 Defensive behaviour and locomotion of the Pacific boa, *Candoia aspera*, with a brief review of head concealment in snakes. *Herpetologica* **25** (3): 164-170.

Campbell C.H. 1966 The death adder (*Acanthophis antarcticus*): The effects of the bite and its treatment. *Med. J. Aust.* **ii**: 922-925.

" 1967a The taipan (*Oxyuranus scutellatus*) and the effect of its bite. *Med. J. Aust.* **i**: 735-738.

" 1967b The Papuan blacksnake (*Pseudechis papuanus*) and the effects of its bite. *Papua and New Guinea Med. J.* **10** (4): 117-121.

" 1969 *A Clinical Study of Venomous Snake Bite in Papua.* PhD Thesis Univ. Sydney.

" 1976 *Snake Bite, Snake Venoms and Venomous Snakes of Australia and New Guinea, An annotated bibliography.* Univ. Sydney Serv. Publ. 13.

" & L.N.Young 1961 The symptomatology, clinical course and successful treatment of Papuan elapine snake envenomation. *Med. J. Aust.* **i**: 489.

Cogger H.G. 1971 The venomous snakes of Australia and Melanesia. *in* Bucherl & Buckley (eds.) *Venomous Animals and Their Venoms.* Academic Press. p. 35-77.

" 1975a Seasnakes of Australia and New Guinea. *in* Dunson W.A. (ed.) *The Biology of Seasnakes.* Baltimore Univ. Press. p.59-139.

" 1975b *Reptiles and Amphibians of Australia* (1st Edition.). Reed.

" 1992 *Reptiles and Amphibians of Australia* (5th Edition.). Reed.

" , E.E.Cameron & H.M.Cogger 1983 *Zoological Catalogue of Australia. 1. Amphibia and Reptilia.* Aust. Govt. Pub. Serv.

Culotta W.A. & G.V. Pickwell 1993 *The Venomous Seasnakes: A comprehensive bibliography.* Kreiger.

Currie B., J. Vince & S. Naraqi 1988 Snakebite in Papua New Guinea. *PNG Med. J.* **31**: 195-198.

" , M. Fitzmaurice & J. Oakley 1988 Resolution of neurotoxicity with anticholinesterase therapy in death adder envenomation. *Med. J. Aust.* **148**: 522-525.

" , J. Richens, A. Korinihona & J. Worthington 1990 Anticholinesterase therapy for death adder envenomation. *Aust. NZ J. Med.* **20**: 190.

David P. & I. Ineich *in prep.* Les serpents venimeux du monde: systématique et répartition. *Dumerilia* **3**.

Dunson W. 1975 *The Biology of the Seasnakes.* Baltimore : University Park Press.

Ehmann H. 1993a *Encyclopedia of Australian Animals: Reptiles.* Angus & Robertson.

" 1993b Family Boidae *in* Glasby, Ross & Bleesley (eds.) *Fauna of Australia Vol. 2A: Amphibia and Reptilia.* Aust. Govt. Pub. Serv. pp. 284-289.

" 1993c Family Colubridae *in* Glasby, Ross & Bleesley (eds.) *Fauna of Australia Vol. 2A: Amphibia and Reptilia.* Aust. Govt. Pub. Serv. pp. 290-294.

" & M.J. Bamford 1993 Family Typhlopidae *in* Glasby, Ross & Bleesley (eds.) *Fauna of Australia Vol. 2A: Amphibia and Reptilia.* Aust. Govt. Pub. Serv. pp. 280-283.

Faliu B. 1987 Les morsures de serpents chez les Mekeo de Papouasie-Nouvelle-Guinée. *Bull. Société des Océanistes* pp. 19-43.

Fritts T.H. 1988 The brown treesnake, *Boiga irregularis*, A Threat to Pacific Islands. US. Fish &

Wildlife Biol. Report 88(31).

" , M.J. McCoid & R.L. Haddock 1990 Risks to infants on Guam from bites of the brown treesnake (*Boiga irregularis*). *Am. J. Trop. Med. Hyg.* **42** (6): 607–611.

Gajdusek D.C. 1977 *In* Symposium on Health and Diseases in Tribal Societies. London, Ciba Found. Symp. New Series No. 49, New York, Elsevier.

Golay P. 1993 Australasian Elapid accounts *in* Golay *et al* (eds.) *Endoglyphs and other Major Venomous Snakes of the World, A Checklist.* Elapsoidea, Switz.

Gow G.F. 1976 *Snakes of Australia.* Angus and Robertson.

Groombridge B. 1988 *World Checklist of Threatened Amphibians and Reptiles.* IUCN / NNC, UK.

Gyi K.K. 1970 A revision of colubrid snakes of the subfamily Homalopsinae. *Univ. Kans. Mus. nat. Hist.* **20** (2): 47-223.

Harlow P. & R. Shine 1992 Food habits and reproductive biology of the Pacific island boas (*Candoia*). *J. Herp.* **26** (1): 60-66.

Heatwole H. & H.G. Cogger 1993 Family Hydrophiidae. *in* Glasby, Ross and Beesley (eds.) *Fauna of Australia Vol. 2A: Amphibia and Reptilia.* Aust. Govt. Pub. Serv. pp.310-318.

" & M.L. Guinea 1993 Family Laticaudidae. *in* Glasby, Ross and Beesley (eds.) *Fauna of Australia Vol. 2A: Amphibia and Reptilia.* Aust. Govt. Pub. Serv. p.319-324.

Hecht M.K., C. Kropach & B.M. Hecht 1974 Distribution of the yellow-bellied seasnake, *Pelamis platurus*, and its significance in relation to the fossil record. *Herpetologica* **30** (4): 387-396.

Houston D. & R. Shine 1993 Sexual dimorphism and niche divergence: feeding habits of the Arafura filesnake. *J. Anim. Ecol.* **62** :737-748.

" & " 1994a Low growth rates and delayed maturation in Arafura filesnakes (Serpentes: Acrochordidae) in tropical Australia. *Copeia* **1994** (3): 726-731.

" & " 1994b Movements and activity patterns of Arafura filesnakes (Serpentes: Acrochordidae) in tropical Australia. *Herpetologica* **50** (3): 349-357.

" & " 1992c Population demography of Arafura filesnakes (Serpentes: Acrochordidae) in tropical Australia. *J. Herp.* **28** (3): 273-280.

Hudson B.J. 1988 The small-eyed snake (*Micropechis ikaheka*): A review of current knowledge. *PNG Med. J.* **31**: 173-178.

" & T. Frohm 1986 Bites due to seasnakes in the Ramu River system. *Proc. 22nd Ann. Symp. PNG Med. Soc.* (abstract)

" & K. Pomat 1988 Ten years of snakebite in Madang Province, Papua New Guinea. *Trans. Roy. Soc. Trop. Med. Hyg.* 1988 **82**: 506-508.

Hutchinson M.N. 1990 The generic classification of the Australian terrestrial elapid snakes. *Mem. Q'land Mus.* **29** (2): 397-405.

Kamaguti A.S., C.D.Laing, G.M.Lowe, M.Zuzel, D.A.Warrell & R.D.G.Theakston 1994 Biological properties of the venom of the Papuan blacksnake (*Pseudechis papuanus*): Presence of a phospholipase A_2 platelet inhibitor. *Toxicon* **32** (8): 915-925.

Kharin V.E. 1981 A review of the seasnakes of the genus *Aipysurus* (Serpentes, Hydrophiidae). (in Russian). *Zool. Zhur.* **60** (2): 257-264.

" 1983 A new species of the genus *Hydrophis* sensu lato (Serpentes, Hydrophiidae) from the north Australian shelf. (in Russian) *Zool. Zhur.* **62** (11): 1751-1753.

" 1984a Revision of seasnakes of subfamily Laticaudinae Cope 1879 sensu lato (Serpentes, Hydrophiidae). (in Russian) *Proc. Zool. Inst. Leningrad.* **1984**: 128-139.

" 1984b Seasnakes of the genus *Hydrophis* sensu lato (Serpentes, Hydrophiidae). On taxonomic status of the New Guinea *H. obscurus*. (in Russian) *Zool. Zhur.* **63** (4): 630-632.

" 1985 A new species of seasnakes of the genus *Enhydrina* (Serpentes, Hydrophiidae) from waters of New Guinea. (in Russian) *Zool. Zhur.* **64** (5): 785-787.

Kinghorn J.R. 1924 A short review of the lizards belonging to the genus *Lialis* Gray. *Rec. Aust. Mus.* **14**: 184-188.

Klemmer K. 1963 Liste der rezenten Giftschlangen. In. *Die Giftschlangen der Erde*. :255-464. Marburg an der Lahn. N.G.Elwert Univ. und Verlag Buch.

Kluge A.G. 1974 A taxonomic revision of the lizard family Pygopodidae. *Misc. Publ. Mus. Zool. Univ. Mich.* **147**: 1-221.

" 1993 *Aspidites* and the phlyogeny of pythoninae snakes. *Rec. Aust. Mus.* **1993** (19): 1-77.

Lalloo D., A.Trevett, J.Black, J.Mapao, S.Naraqi, D.Owens, R.Hotton, R.D.G.Theakston & D.A.Warrell 1994 Neurotoxicity and haemostatic disturbances in patients envenomed by the Papuan blacksnake (*Pseudechis papuanus*). *Toxicon* **32** (8): 927-936.

Lesson M. 1830 *Coluber ikaheka* and *Acanthophis tortor* in Voyage Autour du Monde, Exécuté par Ordre du Roi, Sur La Corvette de la Majesté, La Coquille Pendant les Années 1822, 1823, 1824 and 1825 by M.L.I. Duperrey, Zoologie Vol. 2, Arthus Bertrand, Paris pp. 54-58.

Leviton A.E., R.H. Gibbs Jr., E. Heal & C.E. Dawson 1985 Standards in herpetology and ichthology: Part 1. Standard symbolic codes for institutional resource collections in herpetology and ichthology. Copeia **1985**:802-832.

Loveridge A. 1927 On *Pseudechis australis* (Gray). *Bull. Antivenin Inst. America* **1**: 58.

" 1948 New Guinean reptiles and amphibians in the Museum of Comparative Zoölogy and United States National Museum. *Bull. Mus. Comp. Zoöl Harvard.* **101** (2): 307-430.

Malasit P, D.A. Warrell, P. Chanthavanich *et al* 1986 Prediction, prevention and mechanism of early (anaphylactic) antivenom reactions in victims of snake bites. *Brit. Med. J.* **292**: 17–20.

Malnate E.V. 1960 Systematic division and evolution of the colubrid snake genus *Natrix*, with comments on the subfamily Natricinae. *Proc. Acad. Nat. Sci. Phil.* **112** (3): 41-71.

" & G.A. Underwood 1988 Australasian natricine snakes of the genus *Tropidonophis*. *Proc. Acad. Nat. Sci. Phil.* **140** (1): 59-201.

McCarthy C. 1993 Hydrophiid and Laticaudine accounts *in* Golay *et al* (eds.) *Endoglyphs and other Major Venomous Snakes of the World, A Checklist.* Elapsoidea, Switz.

McCoy M. 1980 *Reptiles of the Solomon Islands.* Wau Ecol. Inst. H.book 7 pp.80.

McDowell S.B. 1967a *Aspidomorphus*, a genus of New Guinea snakes of the family Elapidae, with notes on related genera. *J. Zool. Lond.* **151**: 497-543.

" 1967b Notes on the Australian seasnake *Ephalophis greyi* M.Smith (Serpentes: Elapidae, Hydrophiinae) and the origins and classification of seasnakes. *Zool. J. Linn. Soc.* **48**: 333-349.

" 1969 *Toxicocalamus*, a New Guinea genus of snakes of the family Elapidae. *J. Zool. Lond.* **159**: 443-511.

" 1970 On the status of the Solomon Island elapid snakes. *J. Zool. Lond.* **161**: 145-190.

" 1972a The species of *Stegonotus* (Serpentes, Colubridae) in Papua New Guinea. *Zool. Meded.* **47**: 6-26.

" 1972b The genera of seasnakes of the *Hydrophis* group (Serpentes: Elapidae). *Trans. Zool. Soc. Lond.* **32** :189 - 247.

" 1974 A catalogue of the snakes of New Guinea and the Solomon Islands, with special reference to those in the Bernice B. Bishop Museum Part 1. Scolecophidia. *J. Herp.* **8** (1): 1-57.

" 1975 A catalogue of the snakes of New Guinea and the Solomon Islands, with special reference to those in the Bernice B. Bishop Museum Part 2. Anilioidea and Pythoninae. *J. Herp.* **9** (1): 1-79.

" 1979 A catalogue of the snakes of New Guinea and the Solomon Islands, with special reference to those in the Bernice B. Bishop Museum Part 3. Boinae and Acrochordoidea. *J. Herp.* **13** (1): 1-92.

" 1984 Results of the Archbold Expeditions No. 112. The Snakes of the Huon Peninsula, Papua New Guinea. *Am. Mus. Nov.* **2775**: 1-28.

Mengden G.A. 1983 The taxonomy of Australian elapid snakes: A review. *Rec. Aust. Mus.* **35**: 135-222.

" R. Shine & C. Moritz 1986 Phylogenetic relationships within the Australasian venomous snakes of the genus *Pseudechis*. *Herpetologica* **42** (3):215-229.

Neill W.T. 1957 Notes on the pygopodid lizards *Lialis burtoni* and *L. jicari*. *Copeia* **1957** (3): 230-232.

O'Shea M.T. 1987a The pythons of New Guinea *in* Coote J. (ed.) *Reptiles: Proceedings of the 1986 UK Herpetological Societies Symposium on Captive Breeding*. Brit. Herp. Soc. pp.61-75.

" 1987b Snakes of the Homalopsinae (Serpentes: Colubridae) from the Southern Trans-Fly , Papua New Guinea. *Herptile* **11** (4):155-162.

" 1988 *Liasis papuanus* (Papuan olive python) Feeding. *Herp. Review*. **19**(2): 36.

" 1990 The highly and potentially dangerous elapids of Papua New Guinea. in Gopalakrishnakone & Chou (eds.) *Snakes of Medical Importance (Asia-Pacific)*. Univ. Singapore Press. pp.585-640.

" 1991 The Reptiles of Papua New Guinea. *Brit. Herp. Soc. Bull.* 37: 15-32.

" 1994a *Micropechis ikaheka* (Small-eyed or Ikaheka snake) Ophiophagy. *Herp. Review.* **25**(2): 70.

" 1994b *Micropechis ikaheka* (Small-eyed or Ikaheka snake) Cannibalism. *Herp. Review.* **25**(2): 70-71.

" 1995 The herpetofauna of coconut husk piles on Kar Kar Island, Madang Province, Papua New Guinea, *in* Reason & Reason (eds.) *ASRA Journal 1994* (updated proceedings of ASRA Symposium 1993) pp. 51-72.

" & I.Bigilale 1991 *Bothrochilus papuanus* (Papuan olive python) Keraphophagy. *Herp. Review*. **22**(2): 60.

Parker F. 1972 Snakes of the elapid genus *Suta* in New Guinea. *Papua New Guinea Sci. Soc. Proc. 1971* **23**: 13-17.

" 1982 *The Snakes of Western Province*. PNG Dept. of Lands & Environment 82/1. pp 78.

Parker H.W. & A.G.C. Grandison 1977 *Snakes - A Natural History*. BM(NH), Cornell Univ. Press.

Patchell F.C. & R. Shine 1986a Feeding mechanisms in pygopodid lizards: How can *Lialis* swallow such large prey ? *J. Herp.* **20** (2): 59-64.

" & " 1986b Hinged teeth for hard-bodied prey: A case of convergent evolution between snakes and legless lizards. *J. Zool.* **208** (2): 269-275.

Reid H.A. 1979 Symptomatology, pathology and treatment of the bites of seasnakes. *in* Lee C–Y (ed) *Venoms. Handbook of Experimental Pharmacology* **52**: 922–955. Berlin: Springer Verlag.

Rooji N. de 1915 *The Reptiles of the Indo-Australian Archipelago*. Leiden. vol. 1, Lacertilia, Chelonia and Emydosauria. pp.384.

" 1917 *The Reptiles of the Indo-Australian Archipelago*. Leiden. vol. 2, Ophidia. pp.344.

Shea G. 1990 On the status of *Katophis plumbea* Macleay (Serpentes: Colubridae). *J. Herp.* **24** (3): 313-314.

" , R. Shine & J.C. Covacevich 1993 Family Elapidae *in* Glasby, Ross & Bleesley (eds.) *Fauna of Australia Vol.2A: Amphibia and Reptilia*. Aust. Govt. Pub. Serv. pp. 295-309.

Shine R. 1980a Ecology of the Australian death adder *Acanthophis antarcticus* (Elapidae): Evidence for convergent evolution with the Viperidae. *Herpetologica* **36** (4): 281-289.

" 1980b Ecology of Eastern Australian whip snakes of the genus *Demansia*. *J. Herp.* **14** (4): 381-389.

" 1981 Ecology of Australian elapid snakes of the genera *Furina* and *Glyphodon*. *J. Herp.* **15** (2): 219-224.

" 1985 A radiotelemetric study of movements, thermoregulation and habitat utilization of Arafura filesnakes (Serpentes: Acrochordidae). *Herpetologica* **41** (3): 351-361.

" 1986a Food habits and reproductive biology of the Australian legless lizards (Pygopodidae). *Copeia* **1986** (1): 30-39.

" 1986b Predation upon filesnakes by aboriginal hunters: Selectivity with respect to body size, sex and reproductive condition. *Copeia* **1986** (1): 238-239.

" 1986c Ecology of a low-energy specialist: Food habitats and reproductive biology of the Arafura filesnake. (Acrochordidae). *Copeia* **1986** (2): 424-437.

" 1986d Evolutionary advantages of limblessness: Evidence from the pygopodid lizards. *Copeia* **1986** (2): 525-529.

" 1986e Sexual differences in morphology and niche utilization in an aquatic snake, *Acrochordus arafurae*. *Oecologia* **69**: 260-267.

" 1987 The evolution of viviparity: Ecological correlates of reproductive mode within a genus of Australian snakes (*Pseudechis*: Elapidae). *Copeia* **1987** (3): 551-563.

" 1988 Food habits and reproductive biology of small Australian snakes of the genera *Unechis* and *Suta* (Elapidae). *J. Herp.* **22** (3): 307-315.

" 1989 Constraints, allometry, and adaptation: Food habits and reproductive biology of Australian brownsnakes (*Pseudonaja*: Elapidae). *Herpetologica* **45** (2): 195-207.

" 1991a Strangers in a strange land. *Copeia* **1991**(1): 120-131.

" 1991b *Australian Snakes: A Natural History*. Cornell Univ.

" & J. Covacevich 1983 Ecology of highly venomous snakes: The Australian genus *Oxyuranus* (Elapidae). *J. Herp.* **17** (1): 60-69.

" & D. Houston 1993 Family Acrochordidae *in* Glasby, Ross & Bleesley (eds.) *Fauna of Australia Vol.2A: Amphibia and Reptilia*. Aust. Govt. Pub. Serv. pp. 322-324.

" & J.K. Webb 1990 Natural history of Australian typhlopid snakes. *J. Herp.* **24** (4): 357-363.

Slater K.R. 1954 A brief account of the snakes of the Port Moresby area. *Papua New Guinea Scient. Soc. a Rep. Proc.* pp.51-56.

" 1955 Reptiles of Western Province. *Papua New Guinea Scient. Soc. a Rep. Rec.* pp. 47-51.

" 1956 On the New Guinea taipan. *Mem. Natn. Mus. Vict.* **20**: 201-205.

" 1959 Incubation of New Guinea taipan eggs. *Aust. Zool.* **12**: 306-307.

" 1961 Reptiles in New Guinea. *Aust. Territories* **1**: 27-35.

" 1968 *A Guide to the Dangerous Snakes of Papua*. Bloink, Port Moresby.

Smith L.A. 1982 Variation in *Pseudechis australis* (Serpentes: Elapidae) in Western Australia and description of a new species of *Pseudechis*. *Rec. West. Aust. Mus.* **10** (1): 35-45.

Smith M.A. 1926 Monograph of the Seasnakes. Brit. Mus. (Nat. Hist.).

" 1943 *Fauna of British India: Reptilia and Amphibia Vol. III Serpentes*. Taylor & Francis.

Slip D.J. & R. Shine 1990 Biological aspects of the adaptive radiation of Australasian pythons (Serpentes: Boidae). *Herpetologica* **46** (2): 283-290.

Stickel W.H. & L.F. Stickel 1946 Sexual dimorphism in the pelvic spurs of *Enygrus*. *Copeia* **1946** (1): 10-12.

Storr G.M. 1978 Whip snakes (*Demansia*, Elapidae) of Western Australia. *Rec. West. Aust. Mus.* **6** (3): 287-301.

" 1981a The *Denisonia gouldii* species-group (Serpentes, Elapidae) in Western Australia. *Rec. West Aus. Mus.* **8** (4): 501-515.

" 1981b The genus *Acanthophis* (Serpentes, Elapidae) in Western Australia. *Rec. West Aus. Mus.* **9** (2): 203-210.

Stull O.G. 1932a A checklist of the family Boidae. *Proc. Bost. Soc. Nat. Hist.* **40**: 387-408.

" 1932b Five new subspecies of the family Boidae. *Occ. Pap. Boston Soc. Nat. Hist.* **8**: 25-30.

Sutherland S.K. 1983 *Australian animal toxins. The creatures, their toxins and care of the poisoned patient.* Melbourne: Oxford University Press.

" , A.R. Coulter, R.D. Harris 1979 Rationalisation of first–aid measures for elapid snake bite. *Lancet* **i**: 183–186.

Theakston R.D.G. & D.A.Warrell 1991 Antivenoms: A list of hyperimmune sera currently available for the treatment of envenoming by bites and stings. *Toxicon* **29** (12): 1419-1470.

Underwood G. and A.F. Stimson 1990 A classification of the pythons (Serpentes, Pythoninae). *J. Zool Lond.* **221**: 565-603.

Voris H.K. 1969 *The evolution of the Hydrophiidae with a critique on methods of approach* (thesis). Univ. Chicago 429pp.

Wallach V. 1995 A new genus for the *Ramphotyphlops subocularis* species group (Serpentes: Typhlopidae), with a description of a new species. *Asiatic Herpetological Research* **6**:132-150.

" *in press* The systematic status of the *Ramphotyphlops flaviventer* (Peters) species group (Serpentes: Typhlopidae). *Amphibia-Reptilia.*

" *in prep. Typhlops fredparkeri,* a new species of blind snake from Papua New Guinea (Serpentes: Typhlopidae).

Warrell D.A. 1987 Animal Poisons. in Manson–Bahr PEC, Bell DR (eds.) *Manson's Tropical Diseases. 19th Edition.* London: Balliere Tindall, Chapter 48: pp 855–989.

" 1990 Treatment of snake bite in the Asia–Pacific region: a personal view. *in* Gopalakrishnakone P. and Chou L.M. (Eds) *Snakes of Medical Importance (Asia–Pacific region).* Singapore University Press, pp 641–670.

Watt G., R.D.G. Theakston, C.G. Hayes *et al* 1986 Positive response to edrophonium in patients with neurotoxic envenoming by cobras (*Naja naja philippinensis*). A placebo–controlled study. *N. Engl. J. Med.* **315**: 1444–8.

Webb J.K. & R. Shine 1993 Dietary habits of Australian blindsnakes (Typhlopidae). *Copeia* **1993** (3): 762-770.

Weber E. & Y.L.Werner 1977 Vocalizations of two snake-lizards (Reptilia: Sauria: Pygopodidae). *Herpetologica* **33** (3): 353-363.

Wells R.W. & C.R.Wellington 1984 A synopsis of the class Reptilia in Australia. *Aust. J. Herp.* **1** (3-4): 73-129.

" & " 1985 A classification of the Amphibia and Reptilia of Australia. *Aust. J. Herp.* Supp. series 1: 1-61.

Werner F. 1936 Nova species boidarum. *Ann. Nist. Nat. Mus. Natl. Hungarici* **30**: 105.

Williams E.E. & F. Parker 1964 The snake genus *Parapistocalamus* on Bougainville, Solomon Islands (Serpentes, Elapidae). *Senck. Biol.* **45** (3/5): 543-552.

Wilson S.K. & D.G. Knowles 1988 *Australia's Reptiles.* Cornstalk Pub.

Worrell E. 1961 Dangerous Snakes of Australia and New Guinea. Angus and Robertson.

" 1963 *Reptiles of Australia.* Angus and Robertson.

Wynn A.H. & G.R. Zug 1985 Observations on the reproductive biology of *Candoia carinata* (Serpentes, Boidae). *The Snake* **17**: 15-24.

Zweifel R.G. 1976 *Herpetological Expedition to New Guinea.* National Geographic Research Report 1968 Reports :503-510.

HOW YOU CAN CONTRIBUTE TO OUR KNOWLEDGE OF THE SNAKES OF PAPUA NEW GUINEA

If you find a snake which does not fit the descriptions of the snakes in this Guide you may have found an Irian Jayan, Solomons or Australian species not previously known to occur in PNG, or even a species new to science! We would very much like to know about your find.

You may also find species from this book in areas not marked on the distribution maps and we would also like to hear about these snakes since there are large areas of PNG which have been poorly sampled or studied by herpetologists in the past.

The opposite page is designed to be photocopied and forwarded to the National Museum of Papua New Guinea in Port Moresby. Please provide as much information as possible and if photographs of the snake are available please also send these. In some cases if snakes are found dead on the road or killed in villages it may be possible to preserve the specimen by soaking it and injecting it with 10% formulin or 70% industrial alcohol for 12 hours before transporting it, wrapped in old cloth or rags soaked in the same fluid, inside a strong sealed plastic bag and a secure leak-proof box. Alternatively the specimen may be deep frozen and kept safe until it can be delivered or collected. Preserved or frozen specimens should be sent or delivered to the National Museum in Port Moresby.

We do not encourage the unnecessary killing of identifiable specimens and although the National Museum does maintain a collection of live snakes, contact should be made with the Curator of Natural History before specimens are sent to Port Moresby. <u>The collection of live snakes by inexperienced persons, especially where potentially dangerous or protected species are concerned, is strongly discouraged.</u>

<u>PROTECTED SPECIES SHOULD NOT BE HARMED.</u>

Enquiries and any preserved specimens (do not send unpreserved dead specimens) should be addressed to Mr Ilaiah H. Bigilale, Acting Curator of Natural History, at the following address:

Papua New Guinea National Museum and Art Gallery
PO Box 5560
Boroko
National Capital District Phones: (675) 325 2458; 325 2405; 325 2422 Fax: (675) 325 1779

Correspondence may be sent to myself at the address given below:

Mark O'Shea
46 Buckingham Road
Penn, Wolverhampton
WV4 5TJ England Phone: (05 44) 1902 338916 Fax: (05 44) 1902 338916
Email: oshea@snakemos.demon.co.uk

<u>(Do not send any specimens to the U.K. address as this may contravene CITES, Customs and airline regulations)</u>

or

c/o The Independent Group
PO Box 168
Port Moresby
National Capital District Fax: (675) 325 2506

PHOTOCOPY THIS PAGE
(DO NOT REMOVE FROM THE BOOK)

Please provide as much information as possible. Please write clearly. Photographs, drawings or sketches are useful.

ABOUT YOU

YOUR NAME: ______________________ DATE: __________

YOUR POSTAL ADDRESS: ______________________

______________________ PHONE: __________

YOUR OCCUPATION: ______________________

ABOUT YOUR SNAKE

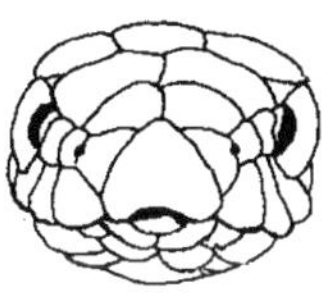

WHERE DID YOU FIND THE SNAKE?

Village Name: ______________________

Map Ref. (if known) ______________________

Province: ______________________

Altitude (if known) ______________________ Time of Day: __________

Habitat (ie. garden, savanna, forest): ______________________

What was the snake doing? ______________________
(ie. climbing, swimming, eating)

____

DESCRIPTION OF SNAKE:

Length of snake: ______________ Size of eye and pupil type: ______________

General body and head shape: ______________________

____

Colouration and patterning: ______________________

USEFUL SCALE DATA (see scalation data for species accounts in main body of the book)

No. of DORSALS at midbody: __________ Dorsals smooth or keeled: __________

No. of VENTRALS: __________ ANAL PLATE entire or divided: __________

SUBCAUDALS paired or single: __________ No. of SUBCAUDALS: __________

No. of SUPRALABIALS: __________ SUPRALABIALS touching eye: __________

LOREAL present or absent: __________